Action and Reaction

"Sir Isaac Newton." Engraving by William Sharp, after a painting by Sir Godfrey Kneller, 6⅝ × 8⅜ inches. From *The Copperplate Magazine*, London, 1778; included as part of the exhibit "Isaac Newton and the *Principia*: 300 Years," mounted at the Smithsonian's National Museum of American History in Washington and at the IBM Gallery of Science in New York City in 1987 and 1988.

CENTER FOR RENAISSANCE AND BAROQUE STUDIES

Director: S. Schoenbaum
Executive Director: Adele F. Seeff

Advisory Board

The Center for Renaissance and Baroque Studies at the University of Maryland, College Park, sponsors programs in all disciplines of the arts and humanities as well as in such allied fields as the history and philosophy of science. Designed primarily for faculty and graduate students, the Center's scholarly programs include an annual Scholar-in-Residence program, conferences and colloquia, lectures, an annual interdisciplinary symposium, and a series of summer institutes for secondary school teachers. Programs in the arts include concerts, lecture-demonstrations, exhibitions, and the annual Maryland Handel Festival. The Center is administered by its executive director and its director in conjunction with an advisory board of outside consultants and a faculty advisory committee.

Action and Reaction

Proceedings of a Symposium
to Commemorate the Tercentenary
of Newton's *Principia*

EDITED BY
Paul Theerman
and
Adele F. Seeff

DELAWARE

Newark: University of Delaware Press
London and Toronto: Associated University Presses

Associated University Presses
440 Forsgate Drive
Cranbury, NJ 08512

Associated University Presses
25 Sicilian Avenue
London WC1A 2QH, England

Associated University Presses
P.O. Box 338, Port Credit
Mississauga, Ontario
Canada L5G 4L8

The paper used in this publication meets the requirements
of the American National Standard for Permanence of Paper
for Printed Library Materials Z39.48-1984.

Library of Congress Cataloging-in-Publication Data

Action and reaction : proceedings of a symposium to commemorate the
tercentenary of Newton's Principia / edited by Paul Theerman and
Adele F. Seeff.
     p.   cm.
   Includes bibliographical references and index.
   ISBN 0-87413-446-3 (alk. paper)
   1. Physics—History—Congresses. 2. Science—History—Congresses.
3. Newton, Isaac, Sir, 1642–1727.   Principia—Congresses.
I. Theerman, Paul Harold, 1952–   . II. Seeff, Adele F.
QC6.9.A38   1993
530′.09—dc20   91-50589
                                                              CIP

PRINTED IN THE UNITED STATES OF AMERICA

Contents

Director's Preface

Established in 1981 through the vision of Dr. Shirley S. Kenny, provost of arts and humanities at the University of Maryland, and the beneficence of the Maryland legislature during an all-too-familiar period of retrenchment in high education, the Center for Renaissance and Baroque Studies held its Inaugural Conference on 11 and 12 March 1982. From the outset the university has envisaged the center as multidisciplinary. Music and the visual arts, literature in several modern European languages, philosophy, and history—indeed all the appropriate disciplines in the humanistic pantheon—come within our sphere. Each autumn we have a Handel festival—performed music, panels, and papers; in the spring, a symposium. Throughout the academic year the center reminds staff and students of its presence with a continuing program of scholars-in-residence, public lectures, and musical recitals.

It was perhaps inevitable that a multidisciplinary operation like the Center should one day contemplate the universe. History—one of our disciplines—provided us with this opportunity on the three hundredth anniversary of Isaac Newton's *Principia*. The Center celebrated the anniversary at its April 1987 conference in College Park, at the National Museum of American History, and at the Albert Einstein Planetarium of the Air and Space Museum (the most appropriate of venues). I. Bernard Cohen initiated the event with a paper on the *Principia* and the Newtonian Revolution in Science, followed by Richard S. Westfall's discourse on Newton and the Scientific revolution. We were launched. There followed talks on the Newton corpus, on Newton's cosmology and political theology, on algebraic vs. geometric techniques in Newton's determination of planetary orbits, and on Newton's unification of heaven and earth; on the *Principia* and the rhetoric of science, and on eighteenth-century theories of abstraction in art and science. All this and much more. We ended up by meditating on scientific creativity and on Newton's place in history. Thus, for those all too brief and crowded days, did we celebrate the three hundredth anniversary of the publication of Isaac Newton's *Principia mathematica philosophiae naturalis,* and thus did the Center extend its overview—in the Albert Einstein Planetarium that Saturday—to the very cosmos.

Introduction

STEPHEN G. BRUSH, ADELE F. SEEFF,
AND PAUL THEERMAN

This book is the outcome of a symposium commemorating the three-hundredth anniversary of the publication of Isaac Newton's *Philosophiae naturalis Principia mathematica,* held at the University of Maryland and the Smithsonian Institution in 1987. During the planning of this symposium we learned that several other *Principia* meetings and exhibits were being organized in America and Europe. Scientists and historians of science obviously considered this an important anniversary, comparable to those of Copernicus, Darwin, Einstein, Galileo, and Kepler (see table 1). Yet in our century Newton's theories have been superseded as fundamental laws of nature, if not as useful methods of calculation, by those of Einstein, Bohr, and Schrödinger. Nevertheless the *Principia* is widely acknowledged to be the most important scientific book ever written. Was its publication similarly honored in earlier centuries before relativity and quantum mechanics had stolen some of its authority? And what were the achievements for which it was celebrated?

There were no celebrations of the fiftieth and hundredth anniversaries of the *Principia,* at least none that resulted in publications listed in the comprehensive bibliography of Peter and Ruth Wallis.[1] During the period from 1687 to 1787 the *Principia* was a guide and a challenge to research; the scientific world was busy digesting, applying, and extending this work, but was not yet ready to celebrate it as a historical landmark.[2] In the 1730s the French were organizing expeditions to the arctic and equator to test Newton's theory of the shape of the earth (*Principia,* Book 3, Propositions 18–20), while Voltaire was writing his popular account of Newton's scientific philosophy; in Russia, the Swiss mathematician Daniel Bernoulli was setting forth the foundations of hydrodynamics and proposing a kinetic theory of gases.[3] In the 1780s the Italian mathematician Lagrange was working in Berlin and Paris to reformulate Newtonian mechanics in terms of algebra and calculus, while the French mathematician Pierre Simon de Laplace was demonstrating that Newtonian mechanics could be used to solve the outstanding problems in the theory of lunar and planetary motion, thereby confirming to his satisfaction the thesis of the "clockwork universe."[4]

The sesquicentenary passed without much notice in 1837, but a year later there appeared one of the first major scholarly studies of the *Principia*, perhaps the first to treat it as a document in the history of science rather than as a scientific treatise: the Oxford astronomer Stephen Peter Rigaud's *Historical Essay on the First Publication of Sir Isaac Newton's Principia*.[5] According to I. Bernard Cohen, Rigaud was "the principal instaurator of serious modern Newtonian scholarship" and his *Essay* "set a new standard of Newtonian scholarship for the simple reason that it was based on a careful examination of manuscript sources."[6] Rigaud gave a detailed account of the path leading to the publication of the first edition of the *Principia*, beginning with the famous story of the falling apple, reviewing the controversies with Robert Hooke, praising the role of Edmond Halley in facilitating the publication, and concluding with remarks on the preparation of the second and third editions. Rigaud also published with his *Essay* several relevant documents such as Newton's early treatise, *De Motu*, and his correspondence with Halley. But he provided little in the way of background information to relate Newton's achievements to seventeenth-century science and society, did not discuss the contents of the *Principia* itself, and made no attempt to assess its impact on later developments.

Soon afterward, Henry Brougham produced a different kind of essay on the *Principia*. A lord chancellor of Great Britain, Brougham is known to the history of science because he severely criticized Thomas Young's wave theory of light, a sharp departure from Newtonian particle theory. Brougham published his essay on the *Principia* in 1839 as part of his *Dissertations on Subjects of Science Connected with Natural Theology*. He later expanded it, with the help of E. J. Routh, a tutor at Cambridge, into the *Analytical View of Sir Isaac Newton's Principia*, published in 1855. The Brougham-Routh book was intended primarily as an aid to university students who needed to understand the demonstrations in order to pass their examinations, although it was also directed to readers "who only desire to become acquainted with the discoveries of Newton, and the history of the science, but without examining the reasoning."[7] Thus the authors recast Newton's geometrical deductions into the language of differential and integral calculus.

Brougham and Routh characterized the *Principia* as containing "the exposition of the laws of motion in all its varieties, whether in free space or resisting media, and of the action exerted by the masses or the particles of matter upon each other" and especially "the most magnificent discovery that was ever made by man—the Principle of Universal Gravitation, by which the system of the universe is governed under the superintendence of its Divine Maker." But they note that Newton refrained from giving "any opinion respecting the nature or cause" of gravitational force and did not ascribe attraction "to mere centres or mathematical points," using the word "force" only "to express certain known and observed facts." In addition to working out the consequences of this principle, the *Principia* "laid a deep and solid foundation for subsequent discoveries in the science of physical astronomy," it "gave a complete system of dynamics applicable to all subjects connected with motion and force and statics," and it "propounded and showed the application of a new calculus, or method of mathematical investigation."[8]

Rigaud's monograph and Brougham's essay appeared in the midst of an early nineteenth-century wave of interest in Newton, which focused on Newton's character and used his achievements to glorify British science. Erasmus Darwin and William Blake each used Newton to symbolize the cultural pattern they saw for England's future, but for Darwin it was a positive vision, and quite the opposite for Blake.[9] In 1794 appeared the French physicist Jean Baptiste Biot's unflattering essay, suggesting that Newton lost his scientific competence and turned to theology after losing his mental acuity following his 1693 breakdown. From its initial publication in the *Biographie universelle,* it was widely reprinted and translated in the following decades and prompted a British response. Francis Baily's *Life of Flamsteed* (1835) resurrected the Newton-Flamsteed epistolary debate of the late seventeenth century. The physicist David Brewster, defender of Newton's particle theory of light against the new wave theory of Thomas Young and Augustin Fresnel, and of Newton's heroic character against the aspirations of Biot and other skeptics, published the first of several editions of his *Life of Newton* in 1831. In turn, the mathematician Augustus De Morgan challenged Brewster's interpretations and agonized over the implications of the affair between Newton's niece, Catherine Barton, and his patron, Charles Montagu (Lord Halifax) for Newton's own moral purity.[10]

Yet most of these early nineteenth-century commentators agreed on what ground Newton's greatness was to be assessed: *discovery.* According to Brewster, "The great discovery which characterizes the *Principia,* is that of the principle of universal gravitation, *that every particle of matter in the universe is attracted by, or gravitates to every other particle of matter, with a force inversely proportional to the squares of their distances.*" From this principle Newton could estimate "the quantity of matter in the sun and in all the planets that had satellites" and "the density or specific gravity of the matter of which they were composed." To a modern reader, Newton's deduction of Kepler's laws from his law of gravity is conspicuous by its absence from Brewster's account, although he mentions Newton's inference that comets move in parabolic or hyperbolic orbits. Brewster ascribed more importance to Newton's determination of the shape of the Earth and other planets, his explanation of the tides produced by lunar and solar gravity, his analysis of the moon's motion, and his calculation of the rate of precession of the equinoxes. Brewster did not discuss Newton's conceptions of space, time, and mass or his rules of reasoning as presented in the *Principia*—perhaps these were thought self-evident and hence not of the same importance as the principle of universal gravitation. He did not hesitate to ascribe to Newton the idea that gravitational force is a real action at a distance—an idea that Newton's contemporaries resisted, Brewster maintained, merely because of their own prejudice or ignorance.[11]

The first explicit recognition of a *Principia* anniversary apparently took place in 1887. George Gabriel Stokes, in his Presidential Address to the Royal Society of London, briefly mentioned the bicentenary but considered the Jubilee of Queen Victoria's accession to the throne, the invention of the electric telegraph, and W. H. Perkin's research leading to the foundation of the dye industry to be events more worthy of notice.[12] But the *Times* (London) devoted a two-column leader to the *Principia:* "It may be asserted without fear of contradiction that of all the anniversaries celebrated this year none is more noteworthy than the bicentenary of the publication of Newton's 'Principia.' No single work has asserted a more signal influence on science and on the progress of civilization." After summarizing the history of the *Principia*'s publication, the writer called for an exhibition of all available Newton manuscripts and books, especially including the yet-unpublished papers in the possession of the Earl of Portsmouth.[13]

Despite this encouragement, the British scientific community failed to mount any formal celebration of the bicentenary in 1887. The anniversary was not noticed in the published proceedings of the British Association for the Advancement of Science, the Cambridge Philosophical Society, or the Physical Society of London.

James Glaisher failed to mention it in his Presidential Address to the Royal Astronomical Society, even though he discussed the connection between recent advances in lunar theory and Newton's work,[14] but he did schedule a commemorative lecture at Trinity College, Cambridge, in December 1887. But this had to be postponed until the following April because of the death of Coutts Trotter, vice-master of the College. In the meantime the bicentenary had been celebrated at a meeting of scientists and mathematicians in Moscow on 1 January 1888, translated into 20 December 1887 "old style" for the occasion.[15]

According to Florian Cajori, Glaisher and John Couch Adams used the occasion of the *Principia* bicentenary to propose that the reason for Newton's twenty-year delay in publishing his discoveries on gravity was not his use of an incorrect value for the size of the earth, as often stated, but rather the need for a proof that the gravitational force exerted by a sphere on an external point is the same as if the entire mass of the sphere were concentrated at its center. This proposal does not, however, appear to have been actually published in 1887, but is found only in a local newspaper account of Glaisher's 1888 lecture at Cambridge.[16]

Glaisher's 1888 address dealt with Newton's lunar theory (a topic then of interest to astronomers because of George William Hill's recent work) and his table of refractions, and referred to papers in the Portsmouth Collection that throw light on the second corollary to Proposition 45 in Book 1 (on the motion of the apsides in nearly circular orbits). Glaisher, in agreement with Brewster, asserted that Newton always believed that attraction is an inherent property of matter.[17]

Physicists in the 1880s acknowledged the power of Newton's mathematical physics in dealing with problems in planetary and lunar motion, but were no longer so sure that his concept of force should be used as a part of fundamental expressions of the laws of nature. Peter Guthrie Tait had attacked the lax usage of "force" in 1876, to the applause of James Clerk Maxwell expressed in humorous verse.[18] The widespread acceptance of the principle of energy conservation had made it more convenient to use energy rather than force in doing calculations, but one could still insist on the existence of "potential energy" derived from forces. In a book published in 1888, J. J. Thomson denied even this statement and argued that all potential energy could be reduced to the kinetic energy of some kind of matter in motion.[19] On the continent, the Newtonian concept of force was being vigorously attacked by Ernst Mach, Heinrich Hertz, and others.[20] Mach in particular had recently published a major critique of Newtonian mechanics in which not only force but also absolute space, time, motion, and mass were denounced as unscientific residues of medieval philosophy.[21] The Michelson-Morley experiment conducted in the bicentennial year 1887 was later regarded as a disproof of Newton's assumption of the reality of motion through absolute space.

During the half-century between 1887 and 1937, the quantum-relativity revolution dethroned Newtonian principles as the fundamental basis for physics. But by demonstrating explicitly the limits of validity of Newtonian physics, the new theories reconfirmed its virtual correctness within those limits. Scientists, far from throwing out Newton's laws, could, even more than before, be confident of their truth in the realm of the not-too-large, the not-too-small, and the not-too-fast.[22] Commenting on the *Principia* at a meeting commemorating the bicentenary of Newton's death, the Harvard mathematician George David Birkhoff called attention to Newton's statements about absolute space, time, and motion (statements that had been ignored—as self-evident?—by the nineteenth-century commentators mentioned above). Even though Newton's views are contradicted by relativity, Birkhoff asserted that his dynamics and his gravitational theory, as first approximations that happen to be extremely accurate in most situations, are likely to stand permanently.[23]

By the time of the two-hundred-and-fiftieth anniversary of the *Principia* in 1937, the development of the history of science as a scholarly discipline had helped focus attention on the seventeenth-century context of Newton's work. The most famous (or notorious) example was Boris Hessen's Marxist interpretation of the *Principia,* presented to the Second International Congress of the History of Science in 1931.[24] Another influential work from this period was Robert Merton's Harvard thesis on seventeenth-century English science, which included several references to Newton and the possible influences of religious and practical considerations on his research.[25]

In 1934 two major works of Newtonian scholarship appeared in the United States: Florian Cajori's revision of Andrew Motte's English translation of the *Principia,* and Louis Trenchard More's biography. In his Appendix, Cajori discussed the relationship of Newton's theories to those of Descartes, and pointed out that (in spite of the impression created by the first edition) there is good documentary evidence to show that "Newton was no more a believer in gravity as an innate property of bodies than was Descartes." But Cajori also discussed the bearing of developments in the nineteenth and twentieth centuries on Newton's ideas about gravity, space, time, light, and causality. Since Cajori died before he could compose a general preface, and his Appendix is in the form of a series of notes keyed to specific pages, there is no summary statement of his views as to the major achievements of the *Principia.*[26]

A professor of physics at the University of Cincinnati, More was inspired to write his biography because the 1927 bicentenary of Newton's death "called attention to the fact that we are without any satisfactory critical biography of the man who is still regarded as the greatest of scientific geniuses."[27] Though willing to admit Newton's personal faults and the existence of technical errors in the *Principia,* More asserted that the basic principles laid down in this book remained valid in spite of the discoveries and theories of the twentieth century. The *Principia* "cannot be superseded, and it must remain the fixed corner-stone of all mechanistic science"—a point he wanted to emphasize all the more because "of late there has been a renewed outburst of unrestrained speculation by mathematical physicists, who have forsaken the sober paths of scientific achievement in order to indulge in pure symbolism, and have invented a universe whose phenomena and laws have no correspondence with our sense perceptions." In particular, More attacked Einstein's general theory of relativity as an attempt toward "a philosophy of pure idealism" that, if followed, "will cause the decadence of science as surely as the medieval scholasticism preceded the decadence of religion."[28]

Although there were no major conferences to celebrate this anniversary of the *Principia,* a few articles on Newton and his masterpiece did appear in 1937.[29] The most substantial of these was an address read by Charles S. Slichter to the Mathematical Association of America. In line with the 1930s' emphasis on the social context of scientific discoveries, Slichter began by asserting that "To understand the *Principia,* one must first recall something of the times in which it was written and the temper of the group of young English scientists which gave character to the age in which Newton was born." He noted the association between scientists and men of the world such as Samuel Pepys, "an example of the animalism of the seventeenth century," whose name appears on the title page of the *Principia* because of his position as president of the Royal Society. Slichter pointed out that the equivalent of the Michelson-Morley experiment in Newton's day was the problem of the moon's motion, which challenged the best scientists and mathematicians of the seventeenth and eighteenth centuries. It was hoped that the

solution of this problem would provide a means for accurate determination of longitude at sea: "to Newton the determination of
longitude at sea meant safe journeys, and the stimulation of inter-
course, and the spread of civilization among the races of men."
Thus the success of the *Principia* was not only the intellectual
triumph of showing "the sufficiency of the laws of mechanics and
the principle of universal gravitation to explain the motions and all
the irregularities of those motions in the phenomena of the heavens" but also a step toward a practical benefit to humanity.[30]

But in answering his own question, "What is the most profound
concept in the *Principia?*" Slichter dropped his contextualism and
revealed his twentieth-century perspective: "Perhaps it is the observation that the quantity of matter is always proportional to
weight—or, as we now say, that gravitational mass and inertial mass
are identical. Two centuries were required to bring out the full
meaning of this fact in the theory of relativity."[31] The theory that
More had seen as a dangerous threat to Newtonian science, Slichter
saw presaged in Newton's own work.

World War II prevented timely celebration of the tercentenary of
Newton's birth in 1942, but in July 1946 the Royal Society of
London hosted a major conference in delayed recognition of this
event. The week-long meeting in London and Cambridge was attended by official delegates from academies and universities of
thirty-four foreign countries (including Italy but not Germany, Aus-
tria, or Japan). The overt theme, international cooperation in science, was promoted by a schedule dominated by parties, tours, and
buffet lunches—there were only eight formal lectures during the
five days.[32] But the texts of the lectures and welcoming speeches
must be read against the background of the spectacular demonstration, less than a year before, of the ability of scientists to unlock the
terrifying destructive powers of nature.

The 1946 tercentenary is best known to historians of science for
the presentation of John Maynard Keynes's paper, "Newton, the
Man." (It was actually read by his brother Geoffrey Keynes, Lord
Keynes having died two months earlier.) Contrary to the view
inherited from the eighteenth century that depicted Newton as a
rationalist, Keynes saw him as "the last of the magicians." Having
reassembled a portion of the Portsmouth Collection that had been
sold at auction in 1936 and brought it to Cambridge, he took the
trouble to read some of the papers documenting Newton's mystical
and religious beliefs, and tried to understand how Newton's mind
worked. According to Keynes, Newton

looked on the whole universe and all that is in it *as a riddle,* as a secret which could be read by applying pure thought to certain evidence, certain mystic clues which God had laid about the world to allow a sort of philosopher's treasure hunt to the esoteric brotherhood. He believed that these clues were to be found partly in the evidence of the heavens and in the constitution of elements (and that is what gives the false suggestion of his being an experimental natural philosopher), but also partly in certain papers and traditions handed down by the brethren in an unbroken chain back to the original cryptic revelation in Babylonia. He regarded the universe as a cryptogram set by the Almighty.[33]

Keynes had apparently written this paper in 1942 for the purpose of giving tercentenary lectures; he gives Newton credit for reading the riddle of the heavens, but lists the constitution of the elements as an unsolved problem. But in 1946 the Soviet physicist S. I. Vavilov, well aware that "atoms have graduated from the category of hypotheses into the class of the most potent realities," raised the question: did Newton anticipate nuclear physics?[34] On the basis of his own retranslation of a passage in Newton's *De natura acidorum,* Vavilov argued that he understood the distinction between the atom and the nucleus and guessed that the alchemists had failed to achieve transmutation only because they lacked "the kind of un-usual agents (mesotrons, neutrons, etc.) which are needed" to penetrate the inner region of the atom.[35] Thus Newton, "as if divining the future," was able to conjecture the outlines of twen-tieth-century knowledge about the ultimate structure of matter and the way in which one element could be changed to another.

The image of Newton as mystic or magician, trying (sometimes successfully) to grasp the mysteries of nature by some faculty that transcends logic or empirical observation, appears several times in the 1946 tercentenary volume.[36] But scientists (especially nuclear physicists) promoting this image of themselves would soon see it twisted into a far more sinister picture: the amoral genius who puts humanity at risk by selling his soul to the devil in return for forbidden knowledge.[37]

While Keynes's paper encouraged historians to explore Newton's theological and alchemical manuscripts,[38] the proceedings of the 1946 conference contain no advances in scholarly understanding of the *Principia.* But during the next twenty years, a remarkable flowering of scholarly research transformed the situation (table 1). In 1966 a conference was held at the University of Texas (Austin) to commemorate a second Newton tercentenary: the "Annus Mirab-

	Page citations			Item citations				
	1953–57	1958–62	1963–67	1968–72	1973–77	1978–82	1983–87	1988–89
Aristotle (384–22 BC)	26	34	58	196	204	173	169	77
Copernicus (1473–1543)	9	27	21	54	174*#	63	51	16
Darwin (1809–82)	10	46*#	43	112	121	119	190*	66
Descartes (1596–1650)	20	32	35	107	124	137	145	82
Einstein (1879–1955)	20	27	36	99	91	154*	105	46
Franklin (1706–90)	26*	25	10	10	15	11	3	2
Freud (1856–1939)	16*	23	19	142	170	156	178	79
Galileo (1564–1642)	20	41	73*	132	134	107	155	65
Kepler (1571–1630)	9	18	34	62*	103	41*	49	18
Leibniz (1646–1716)	9	25	32*	114	104	88	82	53
Newton (1642–1727)	27	49	60	172	148	132*	114	81#

Table 1. Relation between publications about selected scientists and their anniversaries, based on the name index to the *Isis Critical Bibliography (CB)*. Before 1968, the name index cited only the page on which publications were described, and there could be several items on a page. Starting in 1968, when John Neu took over the editorship of the *CB,* the name index cited individual items, and coverage was expanded to include publications on the history of philosophy and medicine; this presumably accounts for part of the increase in the number of items about Descartes, Leibniz, and Freud. Before 1952 the name index generally cited only authors of publications, not subjects. Most of the items in the *Critical Bibliography* for a given year were published in the preceding year. No bibliography was published in 1954 but we assume those publications were included in 1955.

The right-hand column includes only two *CB*s rather than five. We thank John Neu for providing the counts for 1989 in advance of publication.

The number of articles associated with anniversaries is underestimated here because a commemorative volume containing many articles is generally counted as only one item in the name index entry for the scientist who is the subject of the volume.

*indicates that an anniversary of birth or death falls within the time period (e.g., 1957–61 for the 1958–62 Bibliography); # indicates an anniversary of a major publication (Darwin's *Origin of Species,* Newton's *Principia;* Copernicus's *De Revolutionibus*).

ilis" of 1666 in which Newton claimed to have made his first great discoveries in calculus, optics, and gravitational theory. As the proceedings editor noted, the recent historical studies that provided much of the justification for this conference had, among other things, "effectively questioned the accuracy of Newton's memory" about what he had done in that year. But it could still "be taken to symbolize one of the decisive turning-points in the history of modern science."[39] This was perhaps the first major anniversary celebration for Newton or any other scientist to be dominated by historians rather than scientists, and it offered many valuable results and provocative insights on the origin, structure, and influence of the *Principia*.

Scientists writing about Newton's concept of force at the time of earlier anniversaries had failed to give a satisfactory account, partly because they had not examined the relevant manuscripts and partly because they did not try to disentangle Newton's views from their own. By 1966, historians of science, perhaps less encumbered by strong opinions about what force "really" is, had begun to make progress in understanding Newton's ideas. In his discussion of the second law and the concept of force in the *Principia*, I. Bernard Cohen argued that Newton "was able to produce a physics of central forces upon the more generally acceptable base on the contact 'forces' of impact or percussion."[40] Cohen argued that the distinction between impact and continuous forces is essential to understanding how Newton's dynamics evolved from but is different from Descartes's dynamics. As John Herivel pointed out in his paper at the 1966 conference, in order to understand Newton's achievements in the *Principia* we need to see how he arrived at them in the preceding years. In particular, the derivation of a mathematical connection between Kepler's laws and an attractive force directed toward the sun, varying with distance, can be traced back to Newton's treatment of the problem of centrifugal force in the mid-1660s. Herivel showed that the original argument, traces of which survive in the *Principia*, depends on replacing a continuous force *pulling* the moving object toward a center by a set of equally spaced impulses exerted by an imaginary external boundary *pushing* it toward the center.[41]

In 1966, as before and since, most discussions of Newton's achievements were limited to celestial mechanics, optics, the atomic structure of matter, and the analysis of fundamental concepts like space, time, motion, and force. But Clifford Truesdell broached a subject rarely touched by historians of science: the influence of Book 2 of the *Principia* on eighteenth-century research in the mechanics of solids and fluids. He argued that while Newton had successfully applied his axioms and mathematical methods to the motion of point masses, his treatment of extended bodies was largely incomplete and partly deficient: a "bewildering alternation of mathematical proof, brilliant hypothesis, pure guessing, bluff, and plain error."[42] Those who tried to construct rational mechanical theories of solids and fluids—Euler, the Bernoullis, and d'Alembert—were not mere "disciples of Newton" applying his theories, but brilliant mathematicians who responded to the challenge of Newton's work by creating new theories that eventually became part of what we call Newtonian physics.

The two philosophers who discussed the *Principia* at the 1966 conference exemplified a major trend in postwar philosophy of science: the use of primary historical sources to formulate arguments about the foundations and methods of science. Howard Stein analyzed Newton's statements about space and time, not simply to compare them with modern theories, but to illuminate Newton's own views in relation to those of his contemporaries.[43] Dudley Shapere surveyed Newton's ideas about scientific explanation and the subject-matter of science in order to address the question (raised by Thomas Kuhn in 1962): is there a single "background framework" that scientists must adopt?[44]

In this volume we present a sampling of the best current scholarship on the *Principia,* its context, and its influence. We have not attempted a balance between different topics and approaches. The papers presented at the symposium, most of which appear here, are the products of the entire last generation of work on Newton. They reflect the depth of inquiry and diversity of research of that generation.

Several distinctions characterize present-day Newtonian scholarship. First of all, it is marked by a close reading of texts, a trait that appears throughout the papers below. Close reading of texts is hardly a novelty. Indeed it marks the great continuity of scholarship within our discipline. But the categories of texts have certainly broadened, including not only the published volumes, but also the private correspondence and the unpublished manuscripts; not only Newton, but also self-professed Newtonians. Beyond this extension of the "primary matter" of study, the interpretive strategies have widened as well. There is increased appreciation that scientific texts and scientists' correspondence are complexly constructed documents. They always repay scrutiny for the content of their ideas. But they are also studied to understand their rhetoric, the structures of their arguments, and their creative reworking of older patterns into new ones. Close reading of Newton and Newtonian thinkers has led to a deeper understanding of physics and the ways that physics was persuasive in its own time.

The second great trait of Newtonian scholarship in recent times has been, in an overworked but apt phrase, science in context. The influence of Newton's ideas is now judged on the basis of the books it begot. Nor do his ideas float freely, to interact of their own accord with other "great ideas" or to drop down willy-nilly into the *tabulae rasae* of succeeding generations. Instead biographical details as well as intellectual influence determine the course of the Newtonian doctrine. The universe that Newton and his followers inhabited is now seen to be larger than considered before. The concerns that scholars have brought to bear in their interpretations have always included the world of mathematical physics and—thanks to Koyré and others—seventeenth-century philosophy. Now the mentality of the scientist—the fruitful source of creative analogy and the basis of the historian's interpretation—includes chemistry, alchemy, biblical interpretation, and political philosophy among others—metaphors of conservation as well as dissipation. The effect of the ideas reaches into the medical and social sciences, as well as to the natural sciences more narrowly conceived. Present-day Newtonian scholarship appreciates his achievement in physics while locating his work in a larger cultural context.

The conclusions that this volume reaches are perhaps not exceptional in the light of the past twenty years' work. But after all, our contributors pretty much created that body of work! The present papers extend and deepen those judgments, however, and the results bear repeating. Newton was a man very much of the seventeenth century, yet extraordinarily sophisticated and subtle in his manipulation of the concepts and techniques that that century bequeathed him. His researches extended seamlessly into areas quite far removed from present-day definitions of mathematics, astronomy, and physics. His ideas, though not possible of infinitely various interpretation, could yet be put to quite different uses in an assortment of disciplines. And finally, and three centuries of interpretive work on Newton would not gainsay this, Newton's science and its methodological style were decisively new, became emblematic for science, and continue to influence practice to the present day.

In the pages that follow, I. Bernard Cohen illustrates the importance of the close examination of a text and its structure rather than the superficial commentary characteristic of so much writing on the classic works of scientists, and shows the difference between "mathematical principles" and "natural philosophy." Two complementary papers give new insights into the Newtonian foundations of celestial mechanics: William L. Harper analyzes Newton's argument for universal gravitation from the perspective of a philosopher of science; Michael S. Mahoney discusses the mathematical aspects of Newton's use of the force law to determine planetary orbits.

The significance of the central achievements of the *Principia* can be generalized to the corpus of Newton's work and to science in the seventeenth century: Betty Jo Teeter Dobbs uses her research on alchemy to develop an integrated view of Newton's work, while Richard S. Westfall presents a new justification for the commonplace remark that Newton was the "culmination of the scientific revolution." Newton's interest in alchemy is further pursued by Peter Spargo in his paper on chemical experiments. Studies of comets are linked to the seventeenth-century context in a novel way by Simon Schaffer. Newton's concepts of the structure of matter and aether inspired speculations about the nature of insanity, as shown by Anita Guerrini. His formulation of laws of nature provided a model for Adam Smith's formulation of economic laws, according to Norriss S. Hetherington.

But science moved on, not always in the directions or with the methods indicated by Newton. Arthur Donovan argues that Lavoisier's formulation of chemistry was not carried out in imitation of Newtonian natural philosophy but initiated a new tradition of "positive science." For much of the nineteenth century it was expedient to ignore Newton's grander visions such as the unity of light and matter; these could then be revived, on a more solid empirical basis, in the twentieth century. Thus Frank Wilczek can look back from the perspective of contemporary physics and see the seeds of modern ideas about transformations in Newton's admittedly speculative Queries. Dudley Shapere has provided a concluding comment on the different views of Newton and Newtonian science that this volume encompasses.

Notes

1. Peter Wallis and Ruth Wallis, *Newton and Newtoniana 1672–1975: A Bibliography* (Folkestone, Kent, England: Wm. Dawson & Sons Ltd., 1977).

2. C. Truesdell, "Reactions of Late Baroque Mechanics to Success, Conjecture, Error, and Failure in Newton's *Principia*," *Texas Quarterly* 10, no. 3 (1967): 238–58; reprinted with other articles on eighteenth-century science in his *Essays in the History of Mechanics* (New York: Springer-Verlag, 1968).

3. Tom B. Jones, *The Figure of the Earth* (Lawrence, Kansas: Coronado

Press, 1967). Voltaire, *Eléments de la philosophie de Newton mis à la portée de tout le monde* (1738); trans. John Hanna *The Elements of Sir Isaac Newton's Philosophy* (1738); reprint ed. (London: Frank Cass, 1967). Daniel Bernoulli, *Hydrodynamica, sive de Viribus et Motibus Fluidorum Commentarii* (1738), trans. Thomas Carmody and Helmut Kobus, *Hydrodynamics* (New York: Dover Publications, 1968).

4. Lagrange, *Mécanique analytique* (Paris, 1788). Charles Coulston Gillispie, "Laplace in His Prime, 1778–1789," *Dictionary of Scientific Biography,* ed. C. C. Gillispie (New York: Charles Scribner's Sons, 1978), 15:301–33.

5. Stephen Peter Rigaud, *Historical Essay on the First Publication of Sir Isaac Newton's Principia* (1837); facsimile reprint with an introduction by I. Bernard Cohen (New York: Johnson Reprint Corporation, 1972).

6. I. B. Cohen, "Introduction to the Reprint Edition" of Rigaud's *Essay,* pp. v, vi, 6. The other major scholarly Newtonian work in the nineteenth century, also reprinted with an introduction by Cohen, was Rouse Ball's *Essay* (1893), cited below in n. 16.

7. Henry Lord Brougham and E. J. Routh, *Analytical View of Sir Isaac Newton's Principia* (London: Longman, Brown, Green, and Longmans, 1855); reprinted with an Introduction by I. Bernard Cohen (New York: Johnson Reprint Corp., 1972), pp. xx–xxi.

8. Ibid., pp. 1, 10, 14.

9. M. McNeil, "Newton as National Hero," in *Let Newton Be!,* ed. John Fauvel et al. (New York: Oxford University Press, 1988), pp. 223–39. McNeil argues that in the twentieth century, by contrast, Newton came to stand for England's past, which is to be preserved rather than changed.

10. These works are cited and discussed by Paul Theerman, "Unaccustomed Role: The Scientist as Historical Biographer—Two Nineteenth-Century Portrayals of Newton," *Biography* 8 (1985): 145–62. See also Richard S. Westfall's "Introduction" to the reprint edition of David Brewster's *Memoirs of the Life, Writings, and Discoveries of Sir Isaac Newton,* from the Edinburgh edition of 1855 (New York: Johnson Reprint Corp., 1965).

11. David Brewster, *Memoirs,* ch. 12.

12. G. G. Stokes, "Address of the President," *Proceedings of the Royal Society of London* 43 (1887): 185–95.

13. "Newton's 'Principia'—Bicentenary of Publication," *Times* 30 July 1887, p. 4. The Portsmouth Collection had been presented to Cambridge University in 1872 but a catalogue of it was not published until 1888 and the materials were not fully exploited by scholars until after World War II; see I. B. Cohen, *Introduction to Newton's 'Principia'* (Cambridge: Harvard University Press, 1971), pp. 10–11.

14. J. W. L. Glaisher, "Presidential Address," *Monthly Notices of the Royal Astronomical Society* 47 (1887): 203–20.

15. A. Stoletow, "Newton's 'Principia,'" *Nature* 37 (1888): 273. The publication resulting from this celebration is cited by Wallis & Wallis, *Newton and Newtoniana,* item 400.5.

16. F. Cajori, "Newton's Twenty Years' Delay in Announcing the Law of Gravitation," in *Sir Isaac Newton 1727–1927, A Bicentenary Evaluation of His Work* (Baltimore: Williams & Wilkins Co., 1928), pp. 127–88. The full text of Glaisher's address was published as "The Bicentenary of Newton's Principia" in the *Cambridge Chronicle and University Journal,* 20 April 1888, pp. 7–8 (reference given in Wallis and Wallis, *Newton and Newtoniana.* See also W. W. Rouse Ball, *An Essay on Newton's Principia* (1893), reprinted with an Introduction by I. Bernard Cohen, (New York: Johnson Reprint Corp., 1972), pp. 14–17.

17. "Bicentenary of Newton's 'Principia,'" *Times* (London), 21 April 1888, p. 15. For the proposition on lunar motion see I. B. Cohen, *Introduction to Newton's 'Principia,'* pp. 116–119.

18. P. G. Tait, "Force," *Nature* 14 (1876): 459–63. Tait blamed Leibniz and his followers for confusing the meaning of the word. Max Jammer, *Concepts of Force* (Cambridge: Harvard University Press, 1957), pp. 232–40 (Jammer's citation of Tait in his n. 63 is incorrect). On Tait's 1876 lecture and Maxwell's poetic response see C. G. Knott, *Life and Scientific Work of Peter Guthrie Tait* (Cambridge: Cambridge University Press, 1911), pp. 252–55; and Lewis Campbell and William Garnett, *The Life of James Clerk Maxwell* (London: Macmillan and Co., 1882), pp. 646–48.

19. J. J. Thomson, *Applications of Dynamics to Physics and Chemistry* (1888; reprint ed., London: Dawsons of Pall Mall, 1968). David R. Topper, "Commitment to Mechanism: J. J. Thomson, the Early Years," *Archive for History of Exact Sciences* 7 (1971): 393–410.

20. Jammer, *Concepts*, pp. 215–31.

21. Ernst Mach, *Die Mechanik in ihrer Entwicklung historisch-kritisch dargestellt* (first edition, 1883), trans. from the Sixth Edition by Thomas J. McCormack as *The Science of Mechanics: A Critical and Historical Account of Its Development* (reprint ed., LaSalle, Ill.: Open Court Publishing Company, 1960); the remark about medieval philosophy is on p. 272.

22. The view that a new theory does not discredit an old one but reinforces and builds on it was articulated by Ludwig Boltzmann and is widely accepted by physicists; see Erhard Scheibe, "The Physicists' Conception of Progress," *Studies in History and Philosophy of Science* 19 (1988): 141–59.

23. G. D. Birkhoff, "Newton's Philosophy of Gravitation with Special Reference to Modern Relativity Ideas," in *Sir Isaac Newton 1717–1927: A Bicentenary Evaluation of His Work* (Baltimore: Williams & Wilkins Co., 1928), pp. 51–64. For the British celebration of this bicentenary see citation in Wallis & Wallis, *Newton and Newtoniana*, item 400.502.

24. B. Hessen, "The Social and Economic Roots of Newton's *Principia*," in *Science at the Crossroads* (London: Kniga, 1931), separately paginated. Hessen's interpretation was criticized by G. N. Clark, "Social and Economic Aspects of Science in the Age of Newton," *Economic History* 3 (1937): 362–79. Loren Graham has recently discussed the conditions under which this paper was written pointing out that it was intended to help protect Einstein's relativity theory, currently under attack in the USSR; see "The Socio-Political Roots of Boris Hessen: Soviet Marxism and the History of Science," *Social Studies of Science* 15 (1985): 705–22.

25. Robert K. Merton, *Science, Technology and Society in Seventeenth-Century England* (New York: Harper & Row, 1970), originally published in *Osiris* (1938), reprinted with a new introduction by the author.

26. *Sir Isaac Newton's Mathematical Principles of Natural Philosophy and His System of the World*, trans. into English by Andrew Motte in 1729. The translations revised, and supplied with an historical and explanatory appendix, by Florian Cajori, late Professor of the History of Mathematics Emeritus in the University of California (Berkeley: University of California Press, 1934); see Appendix, pp. 627–80 (quotation from p. 633). Cajori died in 1930 and the book was seen through the press by R. T. Crawford.

27. L. T. More, *Isaac Newton, A Biography* (London: Constable, 1934); reprint ed., New York: Dover Publications, Inc., 1962,) p. v.

28. Ibid., pp. 287, 305.

29. See for example the publications by Clark, Fueter, Hiscock, Hobhouse, Lippmann, and L'Vovsky cited by Wallis and Wallis, *Newton and Newtoniana,* items 76.8, 109.3, 378.4, 378.404, 393.641, 393.65, 398.7 and the papers by Ohlsson and Sonnenstrahl, cited in *Astronomische Jahresbericht* 38 (1936): 20; 39 (1937): 12, 19. Of these, only L'Vovsky mentions the two-hundred-and-fiftieth anniversary in the title.

30. C. S. Slichter, "The Principia and the Modern Age," *American Mathematical Monthly* 44 (1937): 433–44; reprinted in his *Science in a Tavern: Essays and Diversions on Science in the Making* (Madison: University of Wisconsin Press, 1938; second edition, second printing, 1958), pp. 74–95; quotations from pp. 75, 78, 83–84.

31. Slichter, *Science in a Tavern,* p. 89.

32. The Royal Society, *Newton Tercentenary Celebrations, 15–19 July 1946* (Cambridge: Cambridge University Press, 1947). The lectures were by E. N. da C. Andrade (biographical), Keynes (see below), J. Hadamard (calculus), S. I. Vavilov (see below), N. Bohr (modern atomic mechanics), H. W. Turnbull (algebra and geometry), W. Adams (observational astronomy), and J. C. Hunsaker (fluid mechanics).

33. J. M. Keynes, "Newton, the Man," in *Newton Tercentenary Celebrations,* pp. 27–34; quotation from p. 29.

34. S. I. Vavilov, "Newton and the Atomic Theory," in *Newton Tercentenary Celebrations,* pp. 43–55 (quotation from p. 44).

35. Ibid., p. 51.

36. *Newton Tercentenary Celebrations,* pp. 1, 18–21, 27–29, 51–54, 60, 65.

37. Spencer R. Weart, *Nuclear Fear: A History of Images* (Cambridge: Harvard University Press, 1988).

38. Frank Manuel, *Isaac Newton, Historian* (Cambridge: Harvard University Press, 1963); *The Religion of Isaac Newton* (New York: Oxford University Press, 1974). Betty Jo Teeter Dobbs, *The Foundations of Newton's Alchemy* (New York: Cambridge University Press, 1975). For an elaboration of Keynes's thesis about Newton's self-perception, see J. E. McGuire and P. M. Rattansi, "Newton and the 'Pipes of Pan,' " *Notes and Records of the Royal Society of London* 21 (1966): 108–43.

39. Robert Palter, "Introduction" to "The *Annus Mirabilis* of Sir Isaac Newton Tricentennial Celebration," *Texas Quarterly* 10, no. 3 (Autumn 1967): 7. The proceedings are reprinted as *The Annus Mirabilis of Sir Isaac Newton,* ed. Robert Palter (Cambridge: MIT Press, 1970).

40. I. B. Cohen, "Newton's Second Law and the Concept of Force in the *Principia,* " *Texas Quarterly* 10, no. 3 (1967): 127–59; quotation from p. 128.

41. J. W. Herivel, "Newton's Achievements in Dynamics," *Texas Quarterly* 10, no. 3 (1967): 103–18. The relevant documents had been published in Herivel's *The Background to Newton's Principia* (Oxford: Clarendon Press, 1965).

42. Truesdell, "Reactions," n. 2.

43. H. Stein, "Newtonian Space-Time," *Texas Quarterly* 10, no. 3 (1967): 174–200.

44. D. Shapere, "The Philosophical Significance of Newton's Science," *Texas Quarterly* 10, no. 3 (1967): 201–18. T. S. Kuhn, *The Structure of Scientific Revolutions* (Chicago: University of Chicago Press, 1962). As Shapere indicates in n. 26 to this paper, he is presenting a somewhat more favorable view of Kuhn's theory than in his review, "The Structure of Scientific Revolutions," *Philosophical Review* 73 (1964): 383–94.

Action and Reaction

The Culmination of the Scientific Revolution: Isaac Newton

Richard S. Westfall

Three hundred years ago, early in March 1687, Edmond Halley received the manuscript of Book 2 of Isaac Newton's *Philosophiae naturalis principia mathematica.* Nearly a year earlier, as clerk to the Royal Society, he had received the manuscript of Book 1, and when he had been perhaps overly forward for a clerk in promoting the publication of a work he recognized to be monumental, the Council of the Society had simply divorced itself from the treatise and handed it over to Halley. Finding himself unexpectedly a publisher, but determined that such a book must appear, Halley engaged a printer, and through the summer of 1686 work progressed slowly but then came to a halt in October. There were good reasons why it did not proceed apace. Halley's position at the Royal Society was in doubt; moreover, he was by no means clear as to what further copy he might expect from Newton. The heading, Book 1, implied at least a Book 2, but in an outburst against Hooke's charge of plagiarism, Newton had threatened in June 1686 to suppress a projected third book. When he said nothing further to remove the threat, and indeed said very little at all, Halley was left wholly in the dark. Finally, in February, the crisis at the Royal Society resolved itself in Halley's favor, and when he set the press in motion once more, Newton responded by sending the manuscript of Book 2. In it Halley found a reference to a Book 3, *De systemate mundi,* and he chanced an oblique question to a man he now recognized to be prickly as to whether Newton intended to send it as well. Thus on 4 April Halley received the copy of the concluding book, and learned at last what it was he was publishing. Three months later, on 5 July, he was able to announce the *Principia*'s completion to its author.

Late in April, precisely three centuries ago, Halley was running himself ragged as he supervised two presses that were printing the work. He was willing. As each successive installment of the manuscript had revealed new wonders, he had come to think of it as more

than merely monumental. It was now Newton's "divine Treatise."[1]
It may justly be said, he did not hesitate to state in the review he
composed for the *Philosophical Transactions,* "that so many and
so Valuable *Philosophical Truths,* as are herein discovered and put
past Dispute, were never yet owing to the Capacity and Industry of
any one Man."[2]

If Halley was absorbed in the *Principia* exactly three centuries
ago, Newton had moved on to other affairs. As it happens, he was in
London also, but not for anything that concerned his book. Because of his leading role in the University's resistance to King
James's effort to intrude Father Alban Francis into the body of
teaching masters, he had been elected as one of eight men to
represent Cambridge in a hearing before the Court of Ecclesiastical
Commission. The first hearing was on 21 April, and there would be
three more during the following three weeks. It is interesting to note
that there is no shred of evidence to indicate that Newton made any
effort to meet with Halley while he was in London. Do not, however, conclude that Newton was unconcerned with his treatise, or
that he valued it less than Halley did. The two men, who at that time
three centuries ago were the only ones who knew anything about it,
both understood that the *Principia* was an epochal book which was
destined to transform scientific thought.

It did not take long after its publication for the British scientific
community to reach the same conclusion. Already at the beginning
of September, David Gregory was writing from Scotland to thank
Newton "for having been at the pains to teach the world that which
I never expected any man should have knowne."[3] John Locke, then
a political refugee in the Netherlands, could not cope with the
mathematics, but accepting Huygens's assurance that it was reliable, he applied himself to the *Principia*'s prose. When he returned
to England in 1689, he made it one of his first items of business to
meet Newton. If continental scientists were slower to accept the
philosophic innovation the work embodied, they did not fail to
recognize the book's power, and for them, too, Newton installed
himself at once as more than merely a leader of scientific thought.

It is safe then to say that through 300 years, from 1687 until 1987,
there has never been a time when Newton was not received as one
of the giants of the human race. Although it does not appear
explicitly in their statements, it is also safe to say that even his
contemporaries, who lacked the advantage of perspective, understood that Newton's greatness was connected to the fact that he did
not stand alone, that he came after Copernicus, Kepler, Galileo,
Huygens, and others, in a word, that he was part and parcel,

perhaps the culmination, of a movement we have learned to call the scientific revolution. As a historian of the scientific revolution who does have the advantage of pespective, I remain convinced that this is the only adequate light in which to view Newton, and the task I set myself in this paper is to explicate precisely what I mean when I speak of him in these terms. Newton was a man of manifold accomplishment, and the development of modern science was a complex affair. The concept of him as the culmination offers us the opportunity to clarify our understanding both of the scientific revolution and of Newton's participation in it.

First and foremost, Newton was the culmination of a radical restructuring of Western mankind's understanding of nature. There are various categories in terms of which one can discuss this intellectual movement, but in my opinion the fundamental one is the development of a quantitative physics that stood at the very heart of the scientific revolution. Half a century ago, Alexandre Koyré summed up this feature of the scientific revolution as the geometrization of nature.[4] During the seventeenth century, one cannot seriously speak of a quantitative science beyond the realms of physics and astronomy. In the eighteenth century, however, quantitative modes of thought were central to the reform of chemistry, and in our own age they are reshaping biology. To be a scientist today is to know and to do mathematics. Such is perhaps our most distinctive legacy from the scientific revolution.

For science had not always been so. Consider, for example, a manuscript commentary on Aristotle's *De caelo* composed by Galileo, when he was a young man, from the notes on lectures given by professors at the Jesuits' Collegio Romano. We can date the manuscript to about 1589, very close to one hundred years before Newton's *Principia.*[5] One looks in vain for any mathematics in it. To be sure, it had some vague quantitative features that adumbrated a different natural philosophy. Thus at one point the commentary stated that the highest degree of intensity of a quality is eight.[6] In the commentary, the number did not lead anywhere, however. It did not correspond to any empirical measurement, and it remained simply a way of talking about qualities, another dimension of the verbal, nonmathematical nature of the manuscript. In this respect especially the commentary, as a product of Galileo, is striking to the modern student. It was nevertheless entirely typical of the tradition that had dominated natural philosophy for 2,000 years. One must not myopically deride that tradition. On the contrary, the commentary contained much penetrating insight, such as the effective distinction between temperature and quantity of heat, another adumbration of the quantitative science that a later age would construct.[7] I want only to insist that the commentary represented an approach to natural philosophy that differed from modern science, and that differed primarily in its nonmathematical character.

At the time when Galileo compiled the commentary, another tradition was already beginning to take shape, primarily within astronomy. Near the dawn of the seventeenth century, the work of Kepler, who believed that the mind of God thinks geometrically and that the world, as the creation of God, embodies geometrical relations as its fundamental structure, would carry the new tradition a significant step forward. And it was Galileo himself, in his mature work, who brought geometry down from the heavens to the earth and in his mathematical science of local motion effectively invented the concept of a quantitative physics. The very writing in which he expounded it, the treatise *De motu locali* inserted in the third and fourth days of the *Discourses on Two New Sciences,* modeled itself on Euclid, beginning with Definitions and Axioms and proceeding via Propositions, Theorems, Problems, and Scholia to ever more complex conclusions.[8] What Galileo began others would take up— his Italian disciples, such as Castelli, Torricelli, and Borelli, and to the north Descartes, Pascal, Mariotte, Marci, Wallis, Wren, Beeckman, Snel, and Huygens, to name no more. Nevertheless, had anyone queried an aspiring historian of science, say a Sprat or a Fontenelle, in April 1687, about the central characteristics of the reformulation of natural philosophy they observed around them, it is extremely unlikely that they would have even considered the geometrization of nature. The book published less than three months later would change that. With the *Principia* to guide us, we can look back at earlier scientists and perceive the precedents, but the *Principia* itself was the book that established the mathematical paradigm. In this sense, surely, Newton was a culmination of the scientific revolution.

One hardly needs to justify the assertion that the *Principia* was an exercise in mathematical physics. Open the book at random, and what do you meet? Mathematical demonstrations, a book modeled, as Galileo's treatise on local motion had been, on Euclid, with Axioms, Propositions, Scholia, and the like, so that a reader like John Locke, who did not want to waste his time and effort, had first to seek assurance about the mathematics before he ventured to grapple with the ideas. Even before we open the book, the title proclaims the contents—not the *Principles of Philosophy,* such as Descartes had offered a generation earlier, but the *Mathematical Principles of Natural Philosophy.*

Let us explore the *Principia* in more detail from this perspective. Its opening passage, preceding Book 1, consists of Definitions and Axioms, which together constitute a quantitative science of dynamics, the foundation of everything in the three books that follow. If one is to appreciate the *Principia,* he cannot remember too clearly that before Newton modern science had not developed a viable science of dynamics. In the work of Galileo and of Kepler it had achieved a kinematics of terrestrial and celestial motions, but it still lacked a dynamics to undergird the kinematics. Kepler, of course, had not looked upon his own celestial mechanics as kinematics, but the Aristotelian dynamics by which he explained it did not survive, while the kinematic laws did. Students of mechanics had attempted to extend the law of the lever into dynamics, but its ambiguities had confused them about the measure of force, and the venture was doomed in any case by its very nature. Others had attempted to develop a dynamics based on impact, but the complexities concealed in that model had thwarted their efforts. At least two scientists, Torricelli and Huygens, at one time or another, had explored the possibilities that the alternative model of free fall might offer, and each had derived relations indistinguishable from Newton's second law for the case of rectilinear acceleration. However, neither one had pursued the idea far enough that one can even begin to speak of it as a science of dynamics.[9] When Newton began to compose the *Principia,* there was no established dynamics on which he could draw.

Perhaps nothing illustrates this fact better than Newton's first drafts of what grew to be his masterpiece, that is, the tract *De motu* and its various revisions. The tract was Newton's response to the visit he received in August 1684 from Halley, who asked him what the shape of an orbit would be if an orbiting body were attracted by a central one with a force that varies inversely as the square of the distance. In *De motu,* Newton demonstrated that an elliptical orbit around an attracting center located at one focus (that is, Kepler's first law) entails an inverse square attraction, that Kepler's third law is a consequence of an inverse square attraction, and that Kepler's second law follows from any central attraction. Nevertheless the principles of dynamics stated at the beginning of the tract—for the most part received concepts not seriously examined—were sufficiently crude that one cannot imagine either that they could have withstood close scrutiny to see if the following demonstrations truly built upon them or that they could have supported the greatly elaborated structure of the later *Principia.*

Newton did start with one important concept that he never

altered, centripetal force, an idea he had picked up from Robert Hooke some five years earlier.[10] Before Hooke's insight, analyses of circular motion had universally spoken of the force that a body moving in a circle exerts as it endeavors to recede from the center, what Huygens had labeled centrifugal force. Without that name, Newton had also initially approached circular motion in those terms. Hooke had realized that if bodies tend to move in straight lines, as the new conception of motion held, some force toward the center is necessary to hold them in closed orbits, and this concept Newton now baptized with the term he coined, centripetal force. Beyond that concept, however, the overt dynamics of *De motu* contained little that we would recognize as Newtonian dynamics. *De motu* did not include any one of the three laws of motion we associate with Newton. It did not state any general force law whatever or anything that corresponded to the third law. Superficially, one of its two "Hypotheses" appears similar to Law 1, the principle of inertia, but in fact it stated that a body moving uniformly in a straight line is carried by its internal force. The tract's implicit dynamics rested on the unstated concept of the parallelogram of forces, which Newton used to calculate the resultant of the joint exertion of the internal force of a body that carries it with a uniform motion and a centripetal force that operates to change its direction.[11]

The initial stage in the composition of the *Principia* witnessed the elaboration of this inchoate beginning into the science of dynamics that continues to bear Newton's name. The manuscripts of his effort survive, and we can follow his progress in detail, a progress we can best define as the geometrization of nature in microcosm. Partly it was a work of definition. In its original form, *De motu* contained two definitions, centripetal force and the internal force of a body. In subsequent papers, as he pondered the foundations of the science of dynamics, Newton tentatively set down definitions of no less than nineteen other quantities—absolute space, absolute time, rest, motion, velocity, quantity of motion, the force of a body resulting from its motion, impressed force, the force exerted by a body, and the like.[12] No definition was more critical than that for quantity of a body, which at a later stage became quantity of matter or mass. "The quantity of a body is measured by the bulk of corporeal matter, which is usually proportional to its weight. The oscillations of two equal pendulums with bodies of equal weight are counted, and the quantity of matter in each will be inversely as the number of oscillations made in the same time." Then Newton paused. He crossed out the final sentence, and in a blank space opposite he wrote down the following: "When experiments were carefully made with gold, silver, lead, glass, sand, common salt, water, wood, and wheat, however, they resulted always in the same number of oscillations."[13] The precise proportionality of quantity of matter to weight—beyond its theoretical significance the proportionality supplied a way to measure mass. At a still later stage in the revision of the definitions, Newton emphasized their role in a quantitative dynamics by adding the notion of measure to the definitions; thus the *Principia* stated that quantity of matter is the "measure" of matter.[14] Meanwhile, the concept of mass, which established the proportion between an impressed force and the change of motion it produces, became one of the keystones in the structure of Newton's dynamics.

The revisions of *De motu* also supplied a general law of force. "The change in the state of motion or of rest [of a body] is proportional to the force impressed and is made in the direction of the right line in which that force is impressed."[15] Newton was not satisfied with his first statement, and he worked it over at length before he arrived at the form, more similar to the first statement than to some of the intervening revisions, that appeared in the published book. Neither in its first nor in its final form did it say that we expect to see, for it spoke of the change of motion and not of the rate of change of motion. That is, the literal statement of the second law in the *Principia* dealt with the quantity that today we call impulse.[16] In the context of the work, this modest conceptual unclarity, a legacy of the century's effort to define what it meant by "force," caused no serious difficulties. In a number of propositions, such as the analysis of pendulums, Newton did set force proportional to acceleration.[17] When he used force in the exact sense of Law 2, he understood the application of successive impulses at uniform intervals of time, so that the dimension of time in the context supplied the dimension absent from the concept of force.[18]

And one other change occurred. The concept of the internal force of a body transformed itself from a force that carries a body in uniform motion into a force of reaction, a force that a body "exerts only in changes of its state produced by another force impressed upon it," and with this change Newton clarified once and for all his understanding of the concept of inertia as we find it in Law 1.[19] Newton was now in a position to perceive the full implications of the notion implicitly present in his concept of centripetal force from the beginning, that uniform circular motion is dynamically equivalent to uniformly accelerated motion in a straight line. This may well be the central insight on which the whole of Newtonian dynamics stands.

The quantitative science of dynamics that Newton elaborated

during the final months of 1684 and the early months of 1685 not only supplied a causal foundation both to Galileo's terrestrial kinematics and to Kepler's celestial kinematics and was in this sense, by itself, a culmination of the scientific revolution; it also supplied Newton with the tool necessary to construct the *Principia.* One could, of course, follow it through every detail of its application in the book. Within the confines of this paper, let me call your attention only to two passages. Toward the end of Book 2, Newton turned to consider the rival system of the universe that the concept of universal gravitation intended to supplant, the Cartesian system of vortices. He assumes, as the premise of his analysis, that a vortex must have a dynamic structure that produces it, because vortical motion is not natural to bodies. He accepted the proposition that friction within the fluid, whereby motion is transmitted from one concentric layer to the next, drives a vortex. With the additional assumption that the friction is proportional to the relative velocity of two layers, he then proceeded to examine the quantiative elements of vortical motion.

Two conclusions emerged from the analysis. First, the distance-velocity relation in a vortex is incompatible with the demands of Kepler's third law. One can argue that the conclusion is arbitrary since it follows directly from the assumption that friction is proportional to the relative velocity of layers, and with a different assumption, a highly improbable one to be sure, one can derive precisely the distance-velocity relation of Kepler's third law. Since no assumption about friction can derive the two incompatible distance-velocity relations of Kepler's first law and his third, however, Newton's critique was not without point. And second, his analysis showed that a vortex cannot be a self-sustaining entity. The necessary condition of its existence is the continuous input of energy—in Newton's terminology, the input of motion—at the center. Without the constant input, motion will diffuse itself evenly through the whole vortex, or more precisely, since a vortex has no boundary, through the infinity of space.[20] "The hypothesis of vortices," Newton concluded, "is pressed with many difficulties."[21] The difficulties flowed directly from his quantitative science of dynamics.

Let me also call your attention to Proposition 66, Book 1, one of the most important propositions in the entire *Principia.* In the form in which it appears, with twenty-two corollaries, it was one of the latest additions Newton made to his work as he continuously expanded its scope and intent during 1685 and the early months of 1686. In a much briefer form, with only a single corollary, it had appeared in the earlier draft misleadingly known as the *Lectiones de motu,* where it functioned as a necessary factor in the argument for universal gravitation.[22] Newton's strategy in the *Principia* was to start with a single body, in effect a planet, moving in an abstract force field around an attracting center, what Professor Cohen has aptly named the one-body system.[23] Newton demonstrated that for any force law the motion of the single planet will conform to Kepler's second law, and that Kepler's first entails an inverse-square force. Multiple planets in an inverse-square force field, when we ignore any mutual influence on each other, will conform to Kepler's third law. Newton accepted Kepler's three laws as established facts, and in the proceeding investigation they functioned as empirical constraints to which valid conclusions had to conform.

If the law of universal gravitation was true, however, the simple demonstrations of the one-body system could not suffice, for planets would not merely be attracted by the body at the center of a force field, but they would attract it in return and attract each other. Would not these mutual attractions upset conformity to Kepler's laws? Newton dealt with this issue in two stages. First he attacked the two-body problem, in effect, the sun and a single planet that attracted each other, and he demonstrated that although, for a given intensity of attraction, the period and the size of the orbit will differ from those for the single-body problem, Kepler's three laws are not affected.

However, the solar system, which for all practical purposes was equivalent to the universe in Newton's work, consists of more than the sun and one planet. In Proposition 66 Newton examined the question in terms of the three-body problem. Although he failed to achieve a solution to the problem—indeed we can now demonstrate that an analytic solution is impossible—he did develop an analysis that allowed him to justify the concept of universal gravitation. Suppose that **S** is the sun and **P** and **Q** two planets circling it (fig. 1).

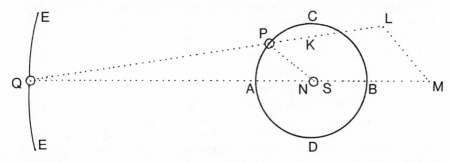

Figure 1

The problem is to determine the effect of **Q** on the motion of **P**.[24] Let **LQ** represent the attraction of **Q** for **P**. Since we are dealing with an inverse-square force, it varies in intensity according to the position of **P**, being equal to **KQ** when **P** is at its mean distance from **Q**. Newton analyzed **LQ** into two components, **LM** (parallel to **PS**) and **MQ**. Because **LM** is radial, it does not upset Kepler's second law, but when it is added to the attraction of **S** for **P** (which is measured by a different scale, of course, and is hundreds of times larger than **PS**), so that the centripetal force on **P** no longer varies precisely inversely as the square of the distance from **S**, it does upset Kepler's first. The effect of **MQ**, which is neither radial nor inversely as the square of the distance **PS**, is still more disturbing. However, its effect is largely offset by the fact that **Q** also attracts **S** with a force proportional to **SQ** in intensity, so that only the segment **MS** operates to perturb the motion of **P**. Newton asked under what condition the perturbations caused by the presence of **Q** will be minimal, in effect, below the threshold of observation as it then stood. Under conditions like those that in fact prevail, that is, when distances between planets are large, and above all when attraction is universal so that **Q** attracts not only **P** but also **S**. When several smaller bodies revolve around a great one, he concluded, the orbits will approach nearer to ellipses and the descriptions of areas will be more nearly uniform, "if all the bodies mutually attract and agitate each other in proportion to their weights and distances, and if the focus of each orbit is placed in the common center of gravity of all the interior bodies . . . than if the innermost body were at rest, and were made the common focus of all the orbits."[25]

Before he was done, however, Newton saw that he could invert the analysis, turn it around a full 180 degrees, and make a proposition designed originally to argue that most perturbations arising from the presence of a third are too small to be perceived into one that subjected some perturbations to exact quantitative calculation. The diagram changed into the form that appears in the published work (fig. 2). The central body became **T** (Terra). The satellite circling the earth remained **P** because it represented a number of different bodies in the varied applications of the analysis. Most importantly it was the moon. Thirteen of the twenty-one new corollaries to Proposition 66 concerned lunar theory; for them Newton relabeled the third body **S** (Sol). **P** could also represent a ring of water contiguous to the earth or the bulge of matter around the equator that results from the centrifugal effect of the earth's rotation; in these cases the third, perturbing body was primarily the moon, with the sun adding an additional increment of disturbance.

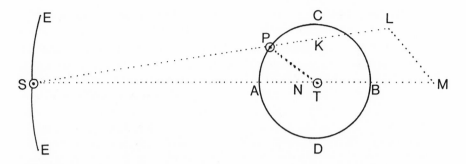

Figure 2

In its new form, Proposition 66 became the foundation of the first explanation of inequalities of the moon's motion that had been established empirically by astronomers before Newton's time in terms of the dynamic factors Western science has continued to accept, and it offered the first satisfactory accounts both of the tides and of the conical motion of the earth's axis, which produces the phenomenon known as the precession of the equinoxes. More than this, it inaugurated a new epoch in science. Here indeed we can see the *Principia* as the culmination of the geometrized conception of nature, for through Proposition 66 quantitative science attained a new level that it had not hitherto dared to dream possible. From this time mathematical science ceased to be confined to ideal cases and accepted the proposition that if the world is mathematically structured, science must give a precise quantiative account, not only of ideal cases, but of the extent to which concrete reality fails to conform to the ideal cases. Perhaps more through Proposition 66 than through anything else, the *Principia* became for following generations the paradigm of a scientific work, a paradigm that embodied in its fully realized form the new view of nature that the entire scientific revolution had struggled to bring forth.

Beyond a geometrized conception of nature, the scientific revolution also elaborated a new method of studying nature, and again Newton played a prominent role. I need to be careful in speaking of Newton as the culmination of a new experimental procedure. On the one hand, I do not mean to suggest that no one experimented before the scientific revolution; I know very well that such was by no means the case. Nevertheless, it was during the seventeenth century that experimental procedure became, in a way that it had never been before, the distinctive method by which science pursues the knowledge it seeks, and in this sense I am willing to speak of a new experimental method. I must be careful on the other hand not to advance the claim that Newton perfected or completed the method, analogous to the claim I have just advanced in relation to the quantitative conception of nature. If Newton was a culmination in method, he was a culmination in a different sense. I do think it is correct to assert that Newton was one of the leading exemplars of experimentation at a time when it was still a novel procedure, and that by lending it the weight of his authority, he helped to insure the perception of experimentation as the distinctive method of science.

Looking back no further than the dawn of his own century, Newton would readily have found examples of how one pursued an experimental investigation. Thus in 1611 Galileo, who had already used experiments in exploring his new science of motion, plunged into a dispute with Aristotelian philosophers about why bodies float in water.[26] In a discussion which asserts that experimentation became the distinctive method of science during the seventeenth century, it is significant to note that Galileo's opponents were the ones who made experiments central to this dispute by showing that bodies heavier than water, chips of ebony, can be made to float on the surface. For his part, Galileo made a more sophisticated appeal to experiment in order to counter their claim that ice, which forms in sheets on the surface of water, floats because, like the chips of ebony, it cannot overcome the water's resistance to division; ice does not float, they argued, because it is lighter than water. Very well, Galileo responded, let us thrust a piece of ice to the bottom and see if the water's resistance to division will prevent it from rising.[27] There, beyond doubt, we see an experimental method at work, not a passive observation of the phenomena that nature presents, but an active interrogation of nature in which the experimenter sets a specific question and defines conditions under which it can be answered. We see an experimental method at work in Pascal's argument that in the barometer, and in similar phenomena, a mechanical balance between two equal weights is in play.[28] We find it in Harvey's use of ligatures as he investigated the circulation of the blood.[29] We find it in many other places as well. By the time Newton discovered the world of science while he was an undergraduate in Trinity, there was a goodly supply of examples that could instruct him in the new procedure.

No small part of Newton's genius lay in his superb command over techniques that differed widely. On the one hand, as I have argued, the ideal of a quantitative science coincided with his natural bent for mathematics; on the other hand, he responded such as readily to the radically different concept of experimental investigation. Turn to any part of his scientific activity, and you encounter experiments. The *Principia,* the masterpiece of theoretical physics, is filled with them. Nevertheless, his work in optics has always appeared as the supreme manifestation of Newton the experimental scientist.[30] "My design in this Book," he stated in the first sentence of the published *Opticks,* "is not to explain the Properties of Light by Hypotheses, but to propose and prove them by Reason and Experiments."[31]

Forty years before those words appeared in print, Newton had forecast their intent at the very beginning of his investigation of colors. In his early reading in the new natural philosophy, he had encountered, among other questions, the problem of colors, and as he pondered the various explanations of colors he had found, he began to entertain his first suggestion that phenomena of colors may appear, not through some modification of light, but through the isolation of individual rays that undergo different refractions and reflections. Almost his first step, recorded on the same page with the initial suggestion of the central idea, was to expose it to experiment. If the idea were correct, then if he observed a thread that was red on one end and blue on the other through a prism, he ought to see the two ends disjoined. He tried it, and he did see them so.[32]

Apparently some time passed before the full import of the experiment sank in, but in his later writings on the theory of colors, that is, in all of his works on optics, he always cited this experiment near the beginning as a principal piece of evidence.[33] Once it did sink in, Newton began to elaborate what would become his mature theory. Central to the theory was the concept that light is heterogeneous. Sunlight is a mixture of rays that differ in their degree of refrangibility, in their tendency to reflect readily from various surfaces, and in the sensations of color they provoke. White is the sensation caused by the mixture. Individual colors are the sensations caused by the individual rays, and phenomena of colors arise from the analysis of sunlight into its component rays either by refraction or by reflection. There were quantitative consequences of the theory, and Newton the mathematician did not fail to pursue them. From the beginning, however, he saw that the central propositions of his theory were empirical assertions, and it never occurred to him that they could be substantiated by anything except experiments.

The investigation of colors was my first serious encounter with Newton the scientist. What impressed me most was his command of experimental procedure—his immediate command, as though by instinct, for he was a young man just crossing the threshold of his career. Meeting the investigation again more than a quarter of a century later, I am perhaps more impressed. Everyone knows the basic experiment, a modification of one Newton had met in his reading, carefully redesigned so his idea that phenomena of colors arise from the analysis of sunlight into its components would be put to the test: a prism in a dark room disperses a narrow beam of light into an elongated spectrum displayed on the wall opposite the window. And everyone knows the elaboration of the basic experiment into the *experimentum crucis* by the addition of a second prism, to test whether dispersion itself may be a modification of light caused by the prism instead of a consequence of light's analysis. Since people are less familiar with other experiments he developed to probe his theory, let me concentrate rather on them.

Newton had fully assimilated the concept of experimentation. If his theory were correct, predictable results had to follow under defined conditions. In defining conditions, that is, in defining experiments to test the theory, the only constraint was the limit of his imagination, and apparently its boundary was infinite. One of the known properties of light was the existence of a critical angle of refraction from one medium into another of lesser optical density. Thus if the incidence on a surface of a ray of light within ordinary

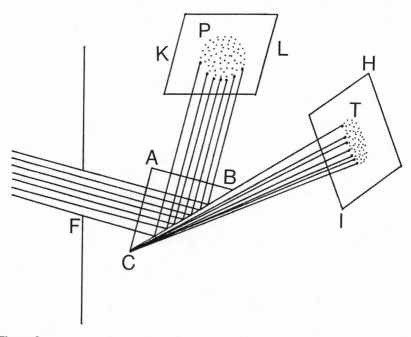

Figure 3

glass is greater than about 42°, it cannot pass out of the glass into air but will be reflected internally. If Newton's theory that sunlight is composed of rays that differ in refrangibility were correct, however, they ought to differ as well in their critical angles, and one ought to be able to use the difference to effect an analysis of sunlight. Once the idea presented itself, the means to pursue it followed quickly. Let a beam of light enter one face of a prism and fall upon the base **CB** (fig. 3). Most of the beam is refracted by the base, and being separated by the refraction appears on a screen at **T** as a compressed spectrum. There is always partial reflection at a refracting surface, however, and that part of the beam appears on another screen at **P** as a dim white spot. Now rotate the prism clockwise on its axis until the incidence of the beam on the base approaches the critical angle. The purple-blue end of the spectrum, being most refrangible, reaches the critical angle first. Those rays disappear from **T**, which becomes predominantly red. The spot at **P** grows much brighter and also turns bluish. As you continue the rotation, all of the colors successively disappear from **T**, which wholly vanishes in the end as the spot at **P**, now much brighter, turns white once more.[34]

Or look at the sky reflected from the base of the same prism (fig. 4). In this case there is no narrow beam. Light falls upon the base from every angle, and one must start from the eye, and consider only those rays incident on different parts of the base at angles such that they are reflected into the eye. Light reflected to the eye from that part of the base, **tB**, farthest removed from it, is incident on the base at an angle greater than the critical one. Hence all that light is reflected, and the portion of the base, **tB**, appears bright. Since the

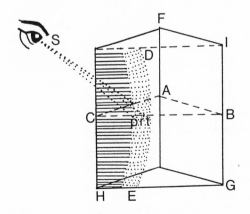

Figure 4

eye is close to the prism, light that enters it from the near side of the base, **pC**, must be inclined at a lesser angle. Most of it is refracted through the base, and that portion of the base appears much darker, "being as it were shaded wth a thin curtaine." The partition between the two sides of the base, a thick curved line, **pt**, all segments of which are equidistant from the eye, is near the critical angle. Thus the red end of the spectrum is transmitted through the base, and the partition appears bluish.[35]

Since Newton's theory intended to account for all phenomena of colors, it had necessarily to posit that reflection can also analyze sunlight into its components. The colors of solid bodies are the sensations caused by the rays they reflect most readily. In the end, the experiments with thin films that revealed "Newton's Rings" would be his principal means to investigate colors produced by reflection, but because he came upon that device, his imagination was up to the task of devising other tests. He could expose bodies that in sunlight appear to be of one color to monochromatic rays from another part of the spectrum—for example, a blue body to monochromatic red light. If the theory were correct, a body could only appear red, perhaps not brilliant red, but at least a dull red, in monochromatic red light. And so it did.[36] In a darkened room he could reflect a beam of sunlight from colored bodies onto a white screen, and note that the screen always appeared then in the color of the body.[37] What would happen, he asked himself, if he reflected sunlight from a shiny black body onto a white screen? One of the features of Newton's theory was the separation of the scale of brilliance from the scale of colors. The theory of colors that Newton replaced identified the two scales. It held that brilliant red was light modified by the smallest admixture of darkness, the color closest to pure white; dull blue was darkness with the smallest admixture of light, the final step in the dilution of light by darkness before pure black. For Newton, the scale of brilliance ran from white through the shades of grey to black and had nothing to do with color. The apparently bizarre idea to reflect sunlight from a black surface onto a white screen therefore presented itself as a crucial experiment; if Newton's theory were correct, the screen ought to appear white. He tried the experiment, and the screen did.[38]

It also followed from Newton's theory that he ought to be able to reconstitute the mixture he analyzed and to reproduce whiteness. White was merely the sensation caused by the heterogenous mixture of rays that is sunlight. This concept, so contrary to the common sense notion that white is the color of simple, pure light, "above the rest appeares *Paradoxicall*," Newton became persuaded, "& is with most difficulty admitted."[39] One way to overcome the reluctance to accept this concept was to reconstitute the mixture and demonstrate that it caused the sensation of white. The experiment required equipment, and initially all he had was the prism with which he analyzed sunlight. Never mind; used with imagination, the same prism could recombine what it separated. He covered one face of the prism with a sheet of heavy paper in which six parallel oblong slits had been cut (fig. 5). Each slit functioned as an individual prism and projected its own spectrum onto a screen. When all six were employed, the spectra merged until white appeared in their midst where rays from all of the slits fell. A further elaboration offered itself. Newton reasoned that this white differed from the white associated with a direct beam of sunlight because the different rays arrived from different angles. He pierced the screen in the middle of the white spot with a needle and held a second screen behind the hole. Colors appeared on the second screen as the rays that passed through the hole diverged (fig. 6).[40]

Newton quickly lit upon a better device for reconstituting the heterogeneous mixture of light, the familiar lens that focused the divergent beam back to a small circle. When the screen was between the lens and the focus, it revealed a reduced spectrum in the standard order. As the screen was moved back, the spectrum shrank into a white spot and then reappeared in the reverse order (fig. 7). Immediately he saw a possible extension of the experiment. Newton was aware that impressions on the retina persist for a finite time; thus a red-hot coal, whirled fast enough, appears as a circle.[41] Would it be possible to reconstitute the mixture that causes the sensation of white, solely through the persistence of images on the retina? He set up a wheel so that its spokes interrupted the converging beam and prevented it all from reaching the focus together (fig. 8). When he turned the wheel slowly, a succession of colors appeared at the focus. When he turned it fast enough that the eye could no longer distinguish the succession, the sensation of white appeared once more.[42]

Newton was able, not only to devise difficult experiments, but to carry them out. I recall repeating a number of these experiments with a group of students. Our circumstances were not ideal—a darkened room in the basement of my home—but they were certainly no worse than the circumstances of Newton's original experiments, and our prisms were undoubtedly much better. Working after three centuries of uninterrupted corroboration, which left us in no doubt the experiments must work, we nevertheless had some

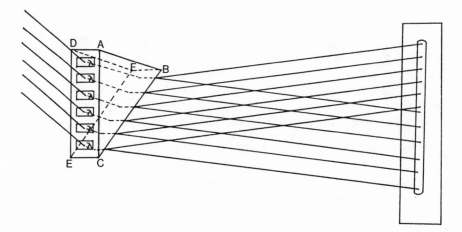

Figure 5

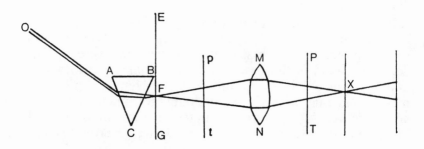

Figure 6

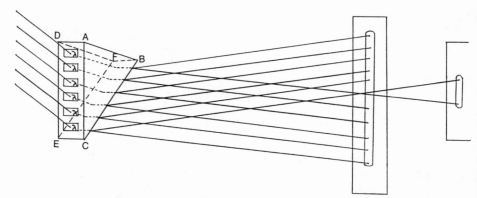

Figure 7

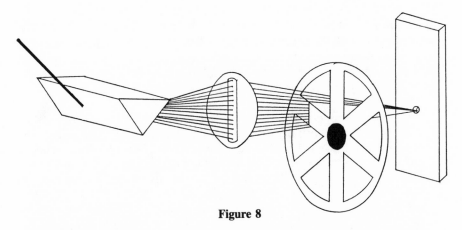

Figure 8

difficulty in achieving, for example, analysis via the critical angle of refraction. Newton, in contrast, testing for the first time a theory which challenged a tradition that had scarcely been questioned over a span of 2,000 years, was skillful enough to make the experiments work. We cannot seriously imagine that they were thought experiments. Consider, for example, his detailed discussion of the one in which a lens focuses the diverging spectrum back into a small circle. He had found that the circle of white had a green fringe at one edge and a purplish red one at the other. Think of the refracted beam, he reasoned, as a series of circles that repeat the shape of the sun (fig. 9). The circle of red rays, **ST,** at one end of the spectrum comes to a focus at **Z.** The circle of purple rays, **PQ,** which have a difficult refrangibility, cannot be focused by the same

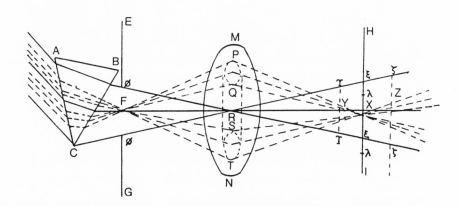

Figure 9

lens at the same point, but rather at the different point **Y**. It is impossible then to think of a point as the single focus for the entire beam that enters through the hole, φFφ. Rather one must think of the focus as a three dimensional space, YζζY, and careful analysis of the circles of different colors revealed to Newton why the experiment had to turn out the way it did.

When I had recognized these things [he reported], I then tested whether they would agree with my preconceptions. Although the test turned out poorly at first when I employed a narrow lens, yet afterward, when I consequently used a wider lens so that the angle **XY1** or εYλ and hence **X1** or ελ (that is, the width of that colored band) would become greater, it turned out as I hoped. Therefore, use a lens whose width or aperture is three inches or greater but whose focal length (as you choose) is three or four feet; then place it six or eight feet away from the hole φFφ, so that the colors **PQRST** flowing into it extend up to its edges, yet with none flowing beyond. Next shift the paper **HI** placed behind it back and forth, and at the end of the image toward **H** you will see all the successive prismatic colors continuously from purple to red; but at the other parts of the image toward **I**, between the purple visible at ζ and

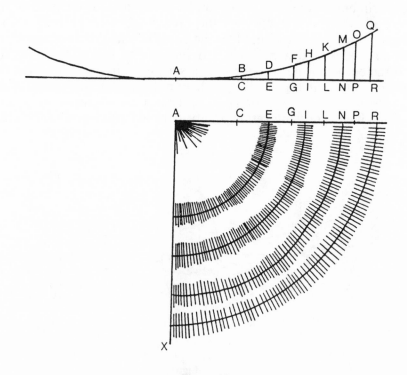

Figure 10

the red at Y, neither green nor any of the other intermediate colors will appear, except perhaps those that arise from a mixture of red and purple.[43]

Some of the experiments made common cause with the movement toward a quantified science. Quantitative precision was the very essence of the experiments with Newton's Rings. He pressed a lens of known curvature down on a flat sheet of glass, illuminated the apparatus from above, and looking down on it, observed the now familiar pattern of successive colored and dark rings (fig. 10). Newton's primary purpose in the experiment was to establish that difform rays of light reflect from films of different thicknesses and to measure the thicknesses of the films of air that correspond to the different colors. In the process he also investigated the suggestion made by Robert Hooke that the color phenomena of thin films are periodic. Precision was the necessary condition to achieve any of these goals.

The level of precision he demanded of himself was wholly in keeping with the man. Using a set of pointers, he measured the diameters of rings, not just to hundredths of an inch, but to fractions of a hundredth. Thus one experiment records the following set of diameters: 23½, 34⅓, 42⅓, 49, 54¾, 60. Was Newton deluding himself with these figures? It is difficult to check them against modern measurements of wave lengths because one must make an arbitrary decision about the wave length to compare to his dark and bright circles. The arbitrary choice is confined within limits, however, and it is impossible to extend them enough to bring Newton's results into agreement with our measurements. It would appear that his calculated thicknesses of films were about ten percent too large, and from this perspective the hundredths and fractions of hundredths were delusions. Because the error was systematic and not random, it must have flowed from a cause common to all the measurements, probably his own evaluation of the curvature of the lens. If we look instead at the issue of periodicity, in which the curvature of the lens, being common to all the measurements in a set, cancels out, the picture of accuracy changes dramatically. In Newton's Rings, the thickness of the film varies as the square of the diameter. Hence he squared the measured diameters and subtracted one square from the next to see if they formed an arithmetic series. If they did, the thicknesses of the films for successive rings likewise formed an arithmetic series; that is, the color phenomena of thin films were periodic. In the set of measurements I quoted above, the differences of squares were 626, 613, 609, 596, 603. The first ring, always a problem in these experiments, as he had to press the lens down hard on the sheet of glass, distorting the two near the point of contact, varied somewhat more. To me, the level of precision is extraordinary. It enabled Newton to track down a consistent divergence that he found between sets of measurements and to establish that one face of the lens ("wch hath ye 5 little short scratches together") had a shorter radius of curvature than the other. The difference in curvature, as he established it, corresponded to a difference of less than one hundredth of an inch in the diameter of the innermost ring and about two hundredths of an inch in the diameter of the innermost ring and about two hundredths of an inch in the diameter of the sixth. Although the concept of periodicity was not Newton's, it is only just to say that these experiments, through their precision, first established the fact of periodicity in one optical phenomenon.[44]

There are then numerous features worthy of notice in Newton's optical experiments. Nevertheless, it is his command of the procedure of experimentation that rivets my attention. Two thousand years earlier, Aristotle had asserted there is nothing in the mind that was not first in the senses. It remained for the scientists of the seventeenth century fully to realize that what is in the senses need not be confined to the phenomena that nature spontaneously presents. Given a proposed theory, one should deduce its consequences, the more varied, the more unexpected, the better. The theory is either true or false; if properly deduced consequences do not appear in carefully performed experiments, it must be false. Newton did not invent this procedure, but he seized it and made it his own. In devising experiments to put his theory of colors to the test, his imagination was fertility itself. There were undoubtedly other experimenters in the seventeenth century equally skilled, but I have yet to find an experimental investigation from the age that surpassed Newton's in its intellectual excitement. The *Opticks,* published seventeen years after the *Principia* had established his leadership in the scientific world, placed his imprimatur on the new procedure, and offered to following generations an outstanding examplar of its use. In this sense, Newton the experimental scientist can be called a culmination of the scientific revolution.

Beyond the content of science and its method, a third major issue characterized the scientific revolution—a reappraisal of the locus of authority in the intellectual world. In concrete terms this amounted to a redefinition of the relation of science to Christianity. Newton's profound absorption in religious issues is well known. It cannot be understood properly, I am convinced, unless one realizes that on this front as well he was the culmination of the scientific revolution.

Consider, if you will, two examples from an age not so much earlier than Newton's. I refer first to Galileo's commentary on Aristotle's *De caelo* that I mentioned before. "Everything eternal," Galileo stated in a typical passage, as he discussed whether the universe could have existed from eternity, "from the ninth [book of Aristotle's] Metaphysics, text 17, lacks the potency of contradiction to existence and non-existence, and this is opposed to the omnipotent power of God, on which all created things must depend for their existence."[45] Along with the entire tradition of medieval philosophy from which his commentary descended directly, and along with the nearly unanimous opinion of everyone in the Latin West for more than a millenium, Galileo accepted the truth of Christianity as the ultimate ground on which all other truth stood. Natural philosophy could only be pursued within the boundaries defined by Christian truth, and no conclusion of natural philosophy that contradicted Christian doctrine could be true. The second example also concerned Galileo. In 1616, as the Holy Office of the Church pursued a charge of heresy against him, it considered two "propositions of the Mathematician Galileo," that the sun stands still in the center of the universe, and that the earth moves with a diurnal and annual motion. The theological consultants of the Inquisition judged the propositions to be heretical or at least suspect because they were opposed to the manifest meaning of passages in scripture.[46] As a result of their judgment, the Church placed Copernicus's book on the Index. The moving force behind the decision was Cardinal Roberto Bellarmino, a man we should not scorn in any matter of learning more readily than we would scorn Galileo. Like Galileo himself in the commentary I quoted, Bellarmino articulated the almost unanimous position of the Christian West. Virtually no one was willing to accept even Galileo's limited argument in the "Letter to the Grand Duchess Christina" that scripture was not concerned with natural philosophy and that its literal word did not have higher authority in matters of science than the evidence of nature, which is also God's work.[47]

By the closing years of the century such was hardly still the case. Among English scientists, the immediate ambiance in which Newton moved, the popularity of natural theology, with its need to demonstrate fundamental assertions of Christianity, testified to the fact that propositions once taken for granted no longer appeared beyond challenge. Almost exactly a century after Galileo set down his judgment about the eternity of the universe that I just cited, Robert Boyle endowed a lecture series to prove the Christian religion against notorious infidels, although he himself, a most prolific author, had already done just that at least fifty times. In the same year John Ray concluded his *Wisdom of God Manifested in the Works of the Creation,* which had assembled the evidence of design in organic nature to demonstrate the existence of a creator, with an appeal to Pascal's wager. Even if the existence of God is not demonstrated beyond doubt, the prudent man will place his bet there, for he loses nothing if he is wrong; whereas the nonbeliever will suffer eternal punishment if he is mistaken.[48] Could Ray have displayed the collapse of certainty more prominently? Shortly after Boyle and Ray, the explosion of deism confirmed their manifest uneasiness. In the short span of less than a hundred years, European civilization had moved from one era into another.

At first glance Newton appears to have belonged more to the earlier age than to the later. To be sure, he inserted the themes of natural theology into his works, in the final Queries of the *Opticks,* for example, and in the General Scholium to the *Principia.* "When I wrote my treatise about our systeme," he began a famous letter to Richard Bentley, "I had an eye upon such Principles as might work with considering men for the beleife of a Deity & nothing can rejoyce me more than to find it usefull for that purpose."[49] On the whole, however, Newton seems less troubled in the assurance of his faith than a number of his contemporaries. Perhaps nothing appears to support this more than the fact—in our eyes, the quaint fact, the wholly pre-modern fact—that he devoted years to the interpretation of the Biblical prophecies.

In order to correct our first impression we have to go behind the printed word to Newton's manuscripts. There we find that Newton held beliefs that he was at great pains to keep out of his published works and out of the public eye, beliefs that would have served to expel him from Cambridge and to ostracize him from society. The beliefs looked toward the future rather than the past. By about 1670 he had rejected the full divinity of Christ, a position from which he never retreated.[50] Examined from inside his theological manuscripts, even the interpretation of the prophecies seems less anarchronistic to us, for its central theme was closely tied to Newton's Arian theology. The prophecies foretold the Great Apostasy, the rise of Trinitarianism.[51]

Most revealing of Newton's secret theological endeavors was the treatise that he began in the early 1680s and called, in one of its incomplete, chaotic manuscripts, *Theologiae gentilis origines phi-*

losophicae [The philosophical origins of gentile theology].[52] Even though he did not carry it even close to a finished form, it was easily the most important theological work he ever undertook. When Newton began to compose the original version of the final book of the *Principia* in 1685, he introduced it with materials from the *Origines,* and when he started revisions of the *Principia* in the 1690s, for the second edition, he inserted lengthy passages from the *Origines* in the so-called classical scholia in Book 3, which he never in fact published.[53] He continued to draw upon it for the rest of his life, for example, in the conclusion to Query 31 of the *Opticks* and in parts of the General Scholium added to the second edition of the *Principia.* The posthumous *Chronology* was a version of the *Origines,* thoroughly sanitized to conceal its unconventional nature.

The *Origines* presented a sketch of human history that Newton based on the mythical lore of the peoples of the ancient Near East, which he accepted as remembered accounts both of their origins and by extension of the origins of all mankind. Inevitably, the deluge occupied a prominent place in the narrative; Newton argued that mankind is descended from the single family that survived the flood. Nevertheless, in contrast to the works of orthodox Christian history on which he drew, Newton did not make the history of mankind center on the Jewish people, and he did not draw it solely from the Bible. To Newton, the historical books of the Old Testament were not divinely inspired. Rather they contained the collective memory of the Jewish people, which was one, but only one, of the sources from which to learn the early history of mankind. Newton treated these books from the Old Testament on a par with the historical lore of the other peoples, which Newton used to complete the Jewish account and to correct it.

Newton's primary interest did not lie in the political history of mankind but in its religious history. The "gentile theology" of his title referred to a religion of twelve gods shared by all the ancient peoples. It had originated in Egypt at an early time and had spread from the Egyptians to the rest of the ancient peoples (by which he meant primarily the Babylonians and the Assyrians), who had worshipped the same twelve gods under different names. To Newton, gentile theology was the incarnation of superstition and idolatry. The twelve gods were mankind's deified ancestors, and since all mankind descended from the one family that had survived the deluge, every version of the twelve gods, whatever the names applied to them, embodied the same originals. To an audience familiar with the Old Testament, the originals of the twelve gods were Noah, his children, and his grandchildren. This gentile theology was not the original religion of mankind, however; Noah and his family had worshipped the one true God, the creator of the universe. But mankind, ever prone to superstition and idolatry, had corrupted the pure religion. God sent a prophet, Moses, to restore it among a people chosen at this time. The Jews corrupted it into idolatry once more. God sent other prophets, and as many times as the pure religion was restored, it was corrupted anew. When he finally despaired of the Jews, God sent a prophet to the Gentiles, who was, of course, Jesus. The Great Apostacy of the fourth century, the triumph of Trinitarianism, with its worship of a man as God, repeated the earlier apostacies.

We could easily let the prominence of Noah and his children in the *Origines* mislead us in its interpretation. How readily their names remind us of an earlier age when everyone, including Galileo and Bellarmino, had accepted the literal truth of the Bible without serious question. We must not allow ourselves to miss the radical thrust so alien to the earlier age that Newton sought to express in the idiom available to him. To the *Origines* the historical books of the Old Testament were merely human documents that one should use in concert with other human documents. Newton treated the coming of Christ, not as the epochal event in human history, which divided the whole into two eras, but as one more in a series of efforts to destroy idolatry and to restore the true worship of God. The concept of an original pure religion, the true worship of God, meant that in Newton's eyes Christian revelation had added nothing. Thus he projected a chapter Eleven, which he did not in fact compose: "What the true religion of the children of Noah was before it began to be corrupted by the worship of false Gods. And that the Christian religion was not more true and did not become less corrupted."[54]

During his old age, one of Newton's major theological enterprises concerned the exposition of the original true religion of the children of Noah. Unencumbered with a complex theology, it was a religion that consisted of two precepts, to love God and to love one's neighbor. It is perhaps characteristic of Newton that he never filled in the details of what the second precept might mean. As for the first, he made it abundantly clear, both in the *Origines* and later, that mankind learns to recognize the existence of God and our duties toward Him from the study of nature. Thus the *Origines* argued that everywhere in the ancient world one found evidence of a particular form of temple, what he called a prytanaeum, in which mankind had worshipped the one true God. Both the Jewish temple and the temples of Vesta in the Roman world had the form of prytanaea, which represented God's creation, with a fire in the center and seven lamps to symbolize the planets.

> The whole heavens they recconed to be ye true & real temple of God & therefore that a Prytanaeum might deserve ye name of his Temple they framed it so as in the fittest manner to represent the whole systeme of the heavens. A point of religion then wch nothing can be more rational. . . . So then twas one designe of ye first institution of ye true religion to propose to mankind by ye frame of ye ancient Temples, the study of the frame of the world as the true Temple of ye great God they worshiopped. . . . So then the first religion was the most rational of all others till the nations corrupted it. For there is no way (wthout revelation) to come to ye knowledge of a Deity but by ye frame of nature.[55]

Geocentric astronomy accompanied the spread of false religion. Here was part of the meaning of Newton's title, *The philosophical origins of gentile theology.* So also he held that true religion sprang from true philosophy. As he said in the conclusion of the Thirty-first Query, in a passage that stemmed from the *Origines,*

> And if natural Philosophy in all its Parts, by pursuing this Method, shall at length be perfected, the Bounds of Moral Philosophy will be also enlarged. For so far as we can know by natural Philosophy what is the first Cause, what Power he has over us, and what Benefits we receive from him, so far our Duty towards him, as well as that towards one another, will appear to us by the Light of Nature. And no doubt, if the Worship of false Gods had not blinded the Heathen, their moral Philosophy would have gone farther than to the four Cardinal Virtues; and instead of teaching the Transmigration of Souls, and to worship the Sun and Moon, and dead Heroes, they would have taught us to worship our true Author and Benefactor, as their Ancestors did under the Government of Noah and his Sons before they corrupted themselves.[56]

For all his piety, Newton had effectively inverted the traditional order of intellectual authority and redefined the relation of science to Christianity. The change was symptomatic of the great transformation of Western civilization from a Christian to a secular, scientific culture, a change that is still proceeding but one for which the seventeenth century functioned as the pivot. Many factors contributed to the displacement of Christianity from the focus of Western civilization, but to me none seems more central than the rise of modern science. Newton's agonized religious quest, which included centrally a sustained effort to define the locus of intellectual authority, a quest that continued through virtually the whole of his adult life and occupied more of his total consciousness than any of his other activities, was part of this transformation. Even though others who wished to proceed far beyond anything he desired would appear on the stage before his death, I have not found anyone in the scientific community who fully confronted this unavoidable issue before Newton. In still another of its broadest themes, he was a culmination of the scientific revolution.

The publication of the *Principia* in July 1687 was an event of monumental importance. As we look about us today, I suspect that none of us would care to say that the progress of modern science, which has followed both upon the Scientific Revolution as a whole and upon Newton's masterpiece, has been an unmixed blessing. Equally I shall be forthright in stating my conviction that, contrary to what some critics imply, it has been far indeed from an unmixed curse. It will be a sad day indeed for humanity if we ever forget how to appreciate the colossal achievement in understanding that modern science embodies. To me, a historian of science, it presents itself as the supreme intellectual odyssey of the human species. We do well in this year 1987 to honor the man who more than any other summed up the Scientific Revolution, when the odyssey began, in the magnificent book published three hundred years ago.

Notes

1. Halley to Newton, 5 April 1687. *Correspondence of Isaac Newton,* ed. H. W. Turnbull, et al., (Cambridge: Cambridge University Press, 1959–77), 2:473.

2. Halley, "*Philosophia Naturalis Principia Mathematica,* Autore Is. Newton," *Philosophical Transactions* 16 (1686–87): 291–6.

3. Gregory to Newton, 2 Sept. 1687. *Correspondence,* 2:484.

4. See especially his pathbreaking work, *Etudes galiléennes* (Paris: Hermann, 1939). Koyré's most frequent phrase in this book is "the geometrization of space," not of nature. Even if the exact phrase is not there, however, the concept, which to me implies more than the geometrization of space, assuredly is. See also E. A. Burtt, *The Metaphysical Foundations of Modern Physical Science,* 2d ed. (London: Routledge & Kegan Paul, 1932).

5. The commentary is translated into English in William Wallace, *Galileo's Early Notebooks: The Physical Questions* (Notre Dame: University of Notre Dame Press, 1977). See also Wallace, *Galileo and His Sources. The Hertiage of the Collegio Romano in Galileo's Science* (Princeton: Princeton University Press, 1984), and Adriano Carugo and Alistair Crombie, "The Jesuits and Galileo's Ideas of Science and of Nature," *Annali dell'Istituto e Museo di Storia della Scienza di Firenze* 8 (1983): 3–68.

6. Wallace, *Galileo's Early Notebooks,* p. 247.

7. Ibid., p. 212.

8. Galileo, *The New Sciences,* trans. Stillman Drake (Madison: University of Wisconsin Press, 1974).

9. For a fuller discussion of these issues see my *Force in Newton's Physics*

(London: Macdonald, 1971), and Alan Gabbey, "Force and Inertia in Seventeenth-Century Dynamics," *Studies in History and Philosophy of Science* 2 (1971): 1–67.

10. See my articles, "Hooke and the Law of Universal Gravitation: A Reappraisal of a Reappraisal," *British Journal for the History of Science* 3 (1967): 245–61; and "Circular Motion in 17th Century Mechanics," *Isis* 63 (1972): 184–87.

11. John Herivel, *The Background to Newton's 'Principia'* (Oxford: Oxford University Press, 1965), pp. 257–74. As Herivel's notes state, some of the Definitions and Hypotheses in the first draft, including Hypothesis 3, the parallelogram of forces, were added after the initial composition.

12. Ibid., pp. 304–7, 315–17.

13. Ibid., pp. 306, 317. The way Herivel presents these two statements is somewhat misleading as to their location in the manuscript.

14. *The Mathematical Principles of Natural Philosophy,* trans. Andrew Motte, rev. Florian Cajori (Berkeley: University of California Press, 1934), p. 1. Professor I. B. Cohen emphasized this concept of measure in a paper delivered at the 13th International Congress on the History of Science in Moscow in 1971. I have not been able to locate this valuable discussion in print.

15. Herivel, *Background,* p. 294.

16. Brian Ellis, "The Origin and Nature of Newton's Laws of Motion," *Beyond the Edge of Certainty,* ed. Robert G. Colodny (Englewood Cliffs, N.J.: Prentice-Hall, 1965), pp. 29–68.

17. Proposition 53, Book 1. In *Mathematical Principles,* pp. 158–60. See also Lemma 10 and Proposition 6, Book 1 (pp. 34–35, 48–49), both important to later demonstrations, which assume that force is continuous and that the distances traversed because of its action vary as the square of the time.

18. See, for example, Proposition 1, Book 1 (pp. 40–43).

19. Herivel, *Background,* p. 315.

20. *Principia,* pp. 385–96.

21. *Principia,* p. 543. This exact statement appeared only in the second edition, in 1713.

22. MS. Dd.9.46, Cambridge University Library.

23. I. Bernard Cohen, *The Newtonian Revolution* (Cambridge: Cambridge University Press, 1980), pp. 61–68.

24. No diagram of this original version of Proposition 66 (or 36 as it was then numbered) survives. I have reconstructed one, altering the diagram of the ultimate Proposition 66 to agree with the text.

25. Add. MS. 3965.3:11, Cambridge University Library, Cf., *Principia,* p. 190.

26. In my opinion, Stillman Drake has produced irrefutable evidence from Galileo's manuscripts that he used experiments to explore his science of motion. Drake, *Galileo at Work: His Scientific Biography* (Chicago: University of Chicago Press, 1978), pp. 74–104, 123–33.

27. *Discourse on Bodies in Water,* trans. Thomas Salusbury, ed. Stillman Drake (Urbana: University of Illinois Press, 1960), p. 4.

28. *Expériences nouvelles touchant le vide* (Paris: Margat, 1647), and *Récit de la grande expérience de l'équilibre des liqueurs* (Paris: Savreux, 1648). See also the later *Traitez de l'équilibre des liqueurs de de la pesanteur de la masse de l'air* (Paris: Desprez, 1663).

29. *On the Motion of the Heart and Blood in Animals,* trans. Robert Willis, ed. Alexander Bowie (Chicago: Regnery, 1962), pp. 96–104.

30. See I. Bernard Cohen, *Franklin and Newton* (Philadelphia: American Philosophical Society, 1956), for the classic investigation of the experimental tradition that followed from the *Opticks.*

31. *Opticks* (London: Smith and Walford, 1704), p. 1.

32. J. E. McGuire and Martin Tamny, *Certain Philosophical Questions, Newton's Trinity Notebook* (Cambridge: Cambridge University Press, 1983), p. 434.

33. In the *Opticks,* for example, a version of it is the first experimental evidence cited. *Opticks,* pp. 13–14.

34. Add. MS. 3975:6, Cambridge University Library. Figure 3, together with the following five diagrams of optical experiments, is reproduced from *The Optical Papers of Isaac Newton,* and figure 10 from *The Correspondence of Isaac Newton,* by the kind permission of the Cambridge University Press.

35. Ibid., p. 6.

36. Ibid., p. 3.

37. Newton to Oldenburg, 11 June 1672; *Correspondence,* 1:184.

38. Ibid., p. 184.

39. Ibid., p. 183.

40. *The Optical Papers of Isaac Newton,* ed. Alan E. Shapiro (Cambridge: Cambridge University Library, 1984–), 1:108.

41. McGuire and Tamny, "Of Vision," *Certain Philosophical Questions,* p. 386. Newton still remembered this phenomenon forty years later and recorded it in Query 16, *Opticks* (reprint ed., New York: Dover Publications, 1952), p. 347.

42. Newton to Oldenburg, 11 June 1972; *Correspondence,* 1:182.

43. *Optical Papers,* 1:118–24.

44. Richard S. Westfall, "Isaac Newton's Coloured Circles twixt Two Contiguous Glasses," *Archive for History of Exact Sciences* 2 (1965): 181–96.

45. Wallace, *Galileo's Early Notebooks,* p. 53.

46. The record of the Inquisition. *Le opere di Galileo Galilei,* ed. Antonio Favaro (Firenze: Barbèra, 1890–1909), 19:320–23.

47. A translation of the "Letter" is found in Stillman Drake, *Discoveries and Opinions of Galileo* (Garden City, N.Y.: Doubleday, 1957), pp. 175–216.

48. John Ray, *The Wisdom of God Manifested in the Works of the Creation,* 7th ed. (London: Innys, 1717), pp. 403–5.

49. Newton to Bentley, 10 Dec. 1692; *Correspondence,* 3:233.

50. See, for example, a sheet with twelve propositions about the nature of Christ composed in the period 1672–75, Yahuda MS. 14:25, Jewish National and University Library, Jerusalem; printed in my biography of Newton, *Never at Rest* (New York: Cambridge University Press, 1980), pp. 315–16.

51. Newton's early interpretation of the Book of Revelation, from the early 1670s, is in Yahuda MS. 1.

52. The primary manuscript of the *Origines* is Yahuda MS. 16.2. The rest of Yahuda MS. 16 as well as 17 also belongs to the *Origines,* and there are bits and pieces of it scattered throughout Newton's theological papers. The sole discussion of this treatise, which has only recently been identified among the theological papers, which have themselves only recently become available to scholars, is my article, "Isaac Newton's *Theologiae Gentilis Origines Philosophicae,*" in W. Warren Wagar, ed., *The Secular Mind* (New York: Holmes & Meier, 1982), pp. 15–34.

53. What is called *The System of the World* was the first draft of the final book; a translated version is published at the end of the standard English edition of the *Principia,* pp. 549–626. Extensive passages from the classical scholia are published from the manuscripts in J. E. McGuire and P. M. Rattansi, "Newton and the 'Pipes of Pan,'" *Notes and Record of the Royal Society* 21 (1966): 108–43.

54. Yahuda MS. 16.2:45v.

55. Yahuda MS. 41:6–7. Newton inserted the parenthetical comment above the line, as an afterthought.

56. *Opticks,* pp. 405–6.

The *Principia,* the Newtonian Style, and the Newtonian Revolution in Science

I. BERNARD COHEN

Newton's *Principia* presents itself to us under a quadripartite title, in which the "principia" are modified by "philosophiae" in a special sense.[1] For the "principia" are not principles in general but "principia mathematica,"[2] and the "philosophia" in question is specifically "philosophia naturalis." In order to set these four terms in context, let me first state that Newton's treatise, like Gaul, "est omnis divisa in partes tres"—in a double sense. In formal construction there are three "Books." In logical analysis there are three tasks: to devise and develop mathematical principles; to use these principles for the construction of a natural philosophy; and, finally, to apply this system of "natural philosophy," erected on "mathematical principles," to the "system of the world."

Newton's ultimate goal was to construct a complete philosophy of nature along the lines that had proved so successful in the *Principia.* As Newton exclaimed in his Preface of 1686, "Would that we could derive the other phenomena of nature from mechanical principles by the same kind of reasoning!"[3] He suspected that many natural phenomena might "possibly depend on certain forces by which the particles of bodies, by causes not yet known, either are impelled toward one another and cohere in regular figures, or are repelled from one another and recede." Not only did Newton envisage that a kind of particle-mechanics might yield a "Principia" of the structure and properties of matter and of chemical dynamics; he also attempted again and again—without much success—to construct a system of physical optics in the style of the *Principia.*[4]

I have never found an extended discussion by Newton of the significance of the notion of "principles" ("principia"), a term that he took over from Descartes, associating it, as Descartes had done, with "philosophy" ("philosophia").[5] Descartes had put forward his own radical science of motion and his system of the world in a general *Principia philosophiae,* a work Newton studied carefully during his formative years, the mid-1660s.[6] Two decades later, in a

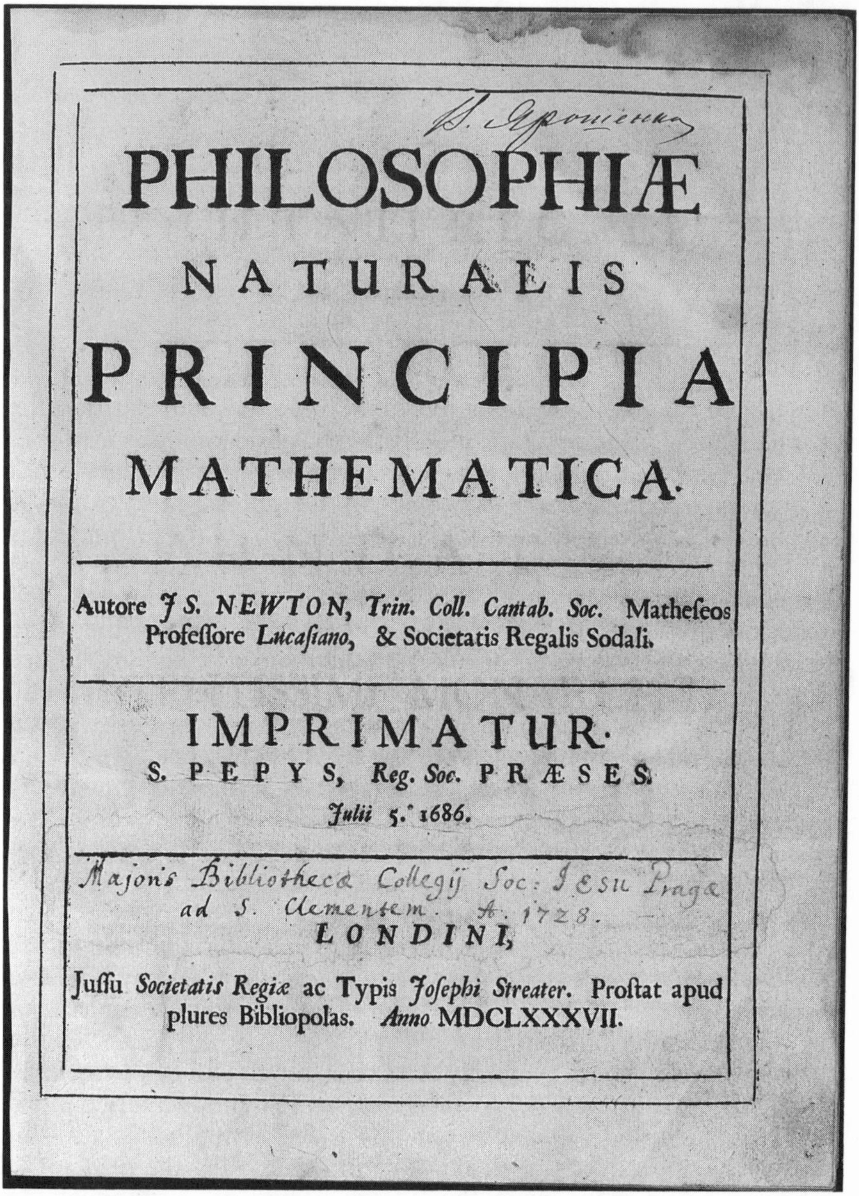

PHILOSOPHIÆ
NATURALIS
PRINCIPIA
MATHEMATICA.

Autore *JS. NEWTON*, *Trin. Coll. Cantab. Soc.* Matheseos Professore *Lucasiano*, & Societatis Regalis Sodali.

IMPRIMATUR.
S. PEPYS, *Reg. Soc.* PRÆSES.
Julii 5. 1686.

Majoris Bibliothecæ Collegij Soc: Jesu Pragæ ad S. Clementem. A. 1728.

LONDINI,

Jussu *Societatis Regiæ* ac Typis *Josephi Streater*. Prostat apud plures Bibliopolas. *Anno* MDCLXXXVII.

Figure 1. Title page of the first edition of Newton's *Principia* (1687). Note the emphasis given to the words *PHILOSOPHIAE* and *PRINCIPIA*. By permission of the Houghton Library, Harvard University.

PHILOSOPHIÆ

NATURALIS

PRINCIPIA

MATHEMATICA.

AUCTORE

ISAACO NEWTONO,

EQUITE AURATO.

EDITIO SECUNDA AUCTIOR ET EMENDATIOR.

CANTABRIGIÆ, MDCCXIII.

Figure 2. Title page of the second edition of Newton's *Principia* (1713). By permission of the Houghton Library, Harvard University.

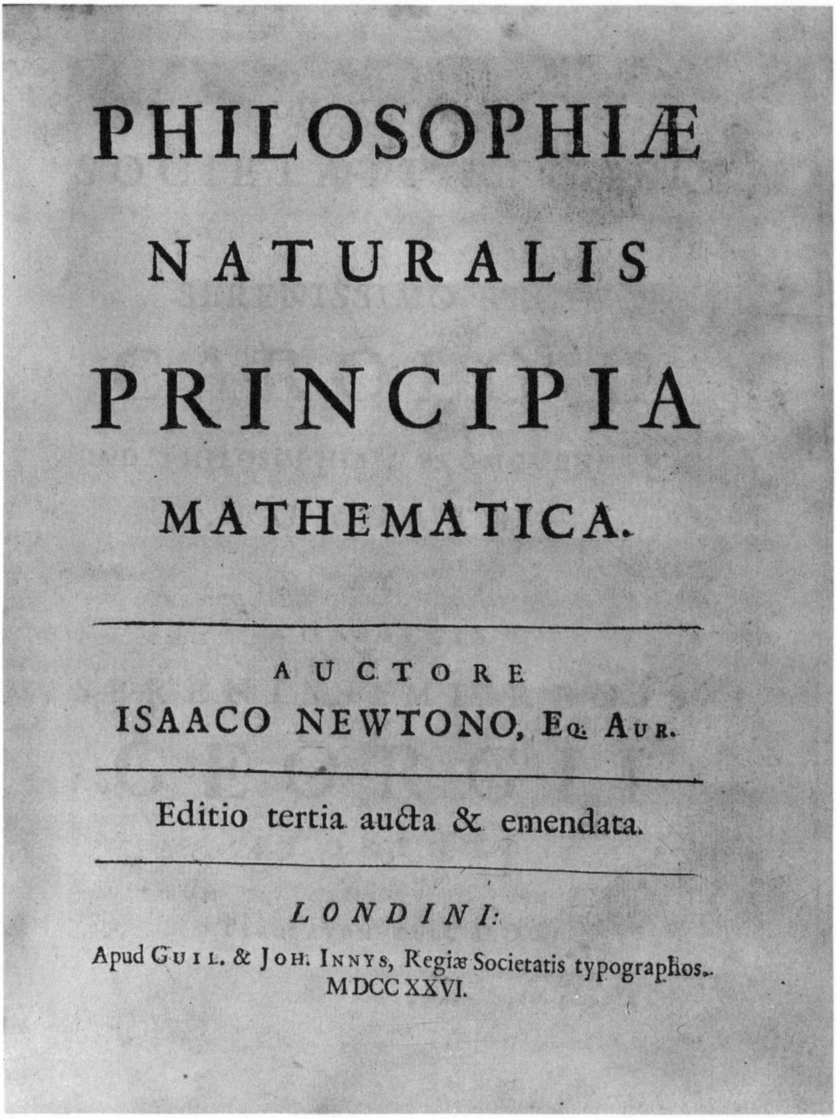

PHILOSOPHIÆ

NATURALIS

PRINCIPIA

MATHEMATICA.

A U C T O R E

ISAACO NEWTONO, Eq. Aur.

Editio tertia aucta & emendata.

L O N D I N I:

Apud Guil. & Joh. Innys, Regiæ Societatis typographos.
MDCCXXVI.

Figure 3. Title page of the third edition of Newton's *Principia* (1726). Here the words "PHILOSOPHIAE" and "PRINCIPIA" are not only emphasized by the size of the type, but are also printed in red. By permission of the Houghton Library, Harvard University.

PHILOSOPHIÆ
NATURALIS
PRINCIPIA
MATHEMATICA.

AUCTORE

ISAACO NEWTONO,

EQUITE AURATO.

EDITIO ULTIMA

Cui accedit ANALYSIS *per Quantitatum* SERIES, FLUXIONES *ac* DIFFEREN-
TIAS *cum enumeratione* LINEARUM TERTII ORDINIS.

AMSTÆLODAMI,
SUMPTIBUS SOCIETATIS.

M. D. CCXXIII.

Figure 4. In the title page of the "pirated" edition of 1723, the stress is put on the words "PHILOSOPHIAE" and "MATHEMATICA"—a noun in the genitive singular and an adjective in the nominative plural, a pair that taken together does not parse. Newton did not produce a "mathematical philosophy" but rather a "natural philosophy" based on "mathematical principles." By permission of the Houghton Library, Harvard University.

mighty burst of intellectual creativity, Newton changed Descartes's program into a more narrowly focussed set of "mathematical principles" ("principia mathematica") of a "natural philosophy" ("philosophia naturalis") and boldly declared the revolutionary force of his innovation by transforming the title of Descartes's *Principia philosophiae* into his own *Philosophiae naturalis Principia mathematica*. We shall see, below, some instances of textual borrowings and transformations of Descartes's *Principia* that occur in Newton's *Principia*, but it may be noted here that in private memorandums Newton even wrote of "my *Principia philosophiae*," using for his own book the actual title of his illustrious predecessor's.[7] It is, I believe, worthy of note that on the title page of the first edition of Newton's great work (see fig. 1), the two boldest words are *Principia* and *Philosophiae*, with lesser emphasis on the qualifiers *mathematica* and *naturalis*, a feature also immediately apparent in the second and third authorized editions (see figs. 2–3). But in the pirated Amsterdam reprint of the second edition (1723), the stress is changed to *Philosophiae* and *mathematica* (see fig. 4), a combination that not only is unfaithful to Newton's intentions but stresses two words that do not even parse when taken together![8] Andrew Motte's English translation, published two years after Newton's death, puts the whole emphasis on *principles,* with a secondary stress on *mathematical* (see fig. 5). In the so-called Cajori edition[9] produced in the United States, the emphasis of the original (see fig. 6) is lost completely. This version does not even make it clear whether the title includes "system of the world" as part of the "mathematical principles."

The meanings of the term "principle" that were current in Newton's day—as revealed, for example, in the dictionary of Newton's disciple, John Harris, the *Lexicon Technicum* of 1704—included a maxim, an axiom, or a "good practical rule of actions," or a "first principle" that is "self evident," or (in chemistry particularly) a "first constituent" or "component particle" of bodies.[10] Sometimes, however, according to Harris, "principles" merely had the sense of "Rudiments or Elements," as in such expressions as "the *Principles* of *Geometry, Astronomy, Algebra,*" where what is meant is "the Doctrine or *Rules of those Sciences.*" In the *Principia,* Newton seems to have been using "principles" in this latter general sense and was evidently more concerned with the qualifying difference of "mathematical" in "mathematical principles" than with specifying a particular meaning for "principles." Harris's stress on principles as axioms and as rules is exemplified in the final *Principia,* where Books 1 and 2 are preceded by the celebrated "Axioms

THE

MATHEMATICAL

PRINCIPLES

OF

Natural Philoſophy.

By Sir *ISAAC NEWTON.*

Tranſlated into *Engliſh* by ANDREW MOTTE.

To which are added,

The Laws of the MOON's Motion, according
to Gravity.

By JOHN MACHIN *Aſtron. Prof. Greſh.* and
Secr. R. Soc.

IN TWO VOLUMES.

LONDON:
Printed for BENJAMIN MOTTE, at the *Middle-
Temple-Gate,* in *Fleetſtreet.*
MDCCXXIX.

Figure 5. Title page of Motte's translation of the *Principia,* 1729.

Sir Isaac Newton's

MATHEMATICAL
PRINCIPLES

OF NATURAL PHILOSOPHY AND HIS
SYSTEM OF THE WORLD

Translated into English by Andrew Motte in 1729.
The translations revised, and supplied with an
historical and explanatory appendix, by

FLORIAN CAJORI

LATE PROFESSOR OF THE HISTORY OF MATHEMATICS EMERITUS
IN THE UNIVERSITY OF CALIFORNIA

UNIVERSITY OF CALIFORNIA PRESS
BERKELEY, CALIFORNIA
1934

Figure 6. The title page of the revised version of Motte's translation, published in 1934, stressed "MATHEMATICAL PRINCIPLES." Note the ambiguity with regard to "Mathematical Principles of . . . his System of the World."

or Laws of Motion" ("Axiomata, sive Leges Motus")[11] and Book 3 is introduced by the "Rules of Proceeding in Natural Philosophy" ("Regulae Philosophandi").[12]

There are some different types of sources that enable us to determine whether Newton intended by "principia" anything more than the general sense of elements or basic concepts, laws, and rules of a science. One such source is the *Opticks,* first published in 1704 but largely composed earlier. A second is the *Principia* itself, while a third is Newton's correspondence. In both the *Principia* and the *Opticks,* Newton seems to have been using "principle" in the customary scientific sense as a fundamental truth (empirically grounded or guaranteed), or a general law (or law of nature), or a natural force or power that is a cause of a phenomenon, or the basis of some kind of form of natural or mechanical action. Examples might be "Archimedes' principle" or "the principle of inertia," but not—for Newton—the basis of the functioning of a machine, as in the more recent "principle of the internal combustion engine."

In Query 31 of the *Opticks,* appearing for the first time in the Latin *Optice* in 1706, Newton writes that "a sponge sucks in Water" and "the Glands in the Bodies of Animals . . . suck in various juices from the Blood" by "the same principle."[13] In the same Query 31, he declares that "the *Vis Inertiae* is a passive Principle," one "by which Bodies persist in their Motion or Rest." This latter principle could not—of and by itself—ever produce or originate any motion in bodies; therefore, "Some other Principle was necessary for putting Bodies into Motion."

A little later on, he contrasts such passive principles with "active Principles," of which he finds examples in gravity and the cause of fermentation and of cohesion of bodies.[14] "These Principles," he writes, "I consider . . . as General Laws of Nature, . . . their Truth appearing to us by Phaenomena." He concludes that "to derive two or three general Principles of Motion from Phaenomena, and afterwards to tell us how the Properties and Actions of all corporeal Things follow from those manifest Principles, would be a very great step in Philosophy." The rubric of "principles" would thus embrace the three laws of motion and the law of universal gravity, along with varieties of forces and natural causes.

In the *Principia* Newton uses the term "principle" in these same meanings. In the *Opticks* we have seen three forces exhibited as "principles"—gravity and the cause of fermentation and of cohesion—but in the Preface to the *Principia,* Newton lists certain "natural powers" *(potentia naturalia)* in relation to "gravity, levity, elastic force, the resistance of fluids, and the like forces, whether

attractive or repulsive." Since he developed an exact science of the
motions resulting from such forces and the forces required to produce motions, he has—he says—presented this work as "mathematical principles of philosophy." He then declares, in Book 3, that
he will use the "mathematically demonstrated" propositions of
Book 1 to "derive from the celestial phaenomena, the forces of
gravity with which bodies tend to the Sun and the several planets."
From "these forces" he has deduced the motions of the planets,
comets, the moon, and the sea. And it is at this point that he
expresses the hope that "we could derive the rest of the phaenomena of Nature by the same kind of reasoning from mechanical
principles," since he has been led to suspect "that they all depend
on certain forces." Here it would seem that Newton was equating
principles and forces, with special reference to mechanical principles and natural forces; an example given is the cause "by which
the particles of bodies . . . are either mutually impelled towards
each other and cohere in regular figures, or are repelled and recede
from each other."

Our belief that Newton may have used "principle" in the sense of
force or law of force is reinforced by the opening sentence of the
scholium following the Laws of Motion. Here Newton refers to the
Laws (or "Axioms") and their corollaries as "principles" that "have
been received by mathematicians" and "confirmed by an abundance of experiments."[15] It is this same sense that appears in the
statement of Book 3, Proposition 22, that "all the motions of the
moon . . . follow from the principles which we have laid down."
Again, the term "principles" is used to imply a set of natural forces
and also their laws when, in Book 3, proposition 13, he declares
that he has demonstrated "the principles" upon which the planetary motions depend and now "from those principles" will "deduce
the motions of the heavens a priori."

A final example may serve to emphasize the restricted sense in
which Newton uses "principles" in the text of the *Principia*. In the
introduction to Book 3, in which Newton displays the System of the
World according to universal gravity, he refers explicitly to "principles of philosophy." These "principles are," he says, "the laws and
conditions of certain motions and forces." In Books 1 and 2, he
continues, he has illustrated these principles by reference to certain
aspects of "philosophy" (i.e., physics): "the density and resistance
of bodies, spaces void of all bodies, and the motion of light and
sounds." Now, "from the same principles" he will "demonstrate
the frame of the System of the World."

We may learn more of what Newton intended by the term "principle" from a letter to Roger Cotes, written during the preparation of the second edition of the *Principia*. Newton, replying (28 March 1713) to an objection raised by Cotes, compared his "Experimental Philosophy" with "Geometry." In geometry, he wrote, "the word Hypothesis is not taken in so large a sense as to include the Axiomes and Postulates"; so, "in experimental Philosophy it is not to be taken in so large a sense as to include the first Principles or Axiomes which I call the laws of motion." These "Principles," he continued, are "deduced from Phaenomena and made general by Induction," which provides "the highest evidence that a Proposition can have in this philosophy.[16] Once again principles are equated with laws of force and motion.

These examples from the *Principia,* reinforced by those from the *Opticks,* might indicate that the title of the *Principia* was intended to be a bit more specific than some vague Mathematical Elements of Natural Philosophy. Perhaps Newton had in mind a title implying that his subject was Mathematical Laws (or Conditions) of the Forces in Natural Philosophy. Or, possibly, a Mathematical Treatment of the Forces in Natural Philosophy. We must, however, take care lest we overly stress the significance of the actual occurrences of the word *principia* within the textual framework of the book. We must keep in mind that when Newton wrote the *Principia* the title was not the one familiar to the world today. In fact, at first he called the work *De Motu Corporum*. It then consisted of two "books." *De Motu Corporum liber primus* more or less corresponded to Book 1 of the *Principia* as we know it.[17] But *De Motu Corporum liber secundus,* while dealing with the same subject matter—the System of the World—as Book 3 of the *Principia,* differed radically in form since Newton rewrote the text in a mathematical manner, purposely to make it difficult to read.

In the final redaction, the rewritten and recast text that replaced *De Motu Corporum liber secundus* was named *De Mundi Systemate liber tertius,* and a wholly new *liber secundus* was composed, containing the final sections of the original *liber primus* and much more besides. Thus the final treatise consisted of three "books" rather than the original two:

De Motu Corporum liber primus.
De Motu Corporum liber secundus.
De Mundi Systemate. Liber tertius.

Clearly, this work could not be *De Motu Corporum libri tres* but needed some more encompassing title. By April 1686, when the completed manuscript of Book 1 was presented to the Royal Society,[18] the work had already been given its full title, *Philosophiae Naturalis Principia Mathematica*. We do not know exactly when Newton decided to extend the work from two to three "books" and to introduce the new title. Since the new title was associated with the manuscript of Book 1 sent to the Royal Society for publication, we know for certain that this title had been chosen by April 1686. On 7 June 1686, Halley still believed that the work would consist of two "books" as originally planned. He learned that there would be three "books" only a few weeks later, from a letter (20 June 1686) of Newton's.[19]

I believe, however, that there is good reason to suppose that the new title was adopted when Newton wrote the new Book 2. The reason is that Book 2 is strongly anti-Cartesian. So evident is this fact that Lagrange went so far as to say that the whole purpose of Book 2 was to destroy the Cartesian theory of vortices. In the scholium to Proposition 53, at the very end of Book 2, Newton even proved that a planetary system of Cartesian vortices contradicts Kepler's law of areas. It is inconceivable that Newton could have written so formidable an attack on the physics of celestial vortices without being aware of the author of this theory, René Descartes, in whose *Principia Philosophiae* Newton had long before encountered both the theory of vortices and their dramatic illustration in unforgettable diagrams.[20]

The terminal event of composing the new Book 2—giving the final death blow to the physics of Descartes's *Principia Philosphiae*—must have been loaded with affect. Consciously or unconsciously, Newton would have recognized that he was replacing Descartes's *Principia* with his own, that instead of general principles in the manner of Descartes, his own *Philosophia* was based on mathematics, and that the principles were transformed from mere *principia* to *principia mathematica*. We have seen that for Newton such principles were also found on experiment or on experience; that is, they were based on nature rather than invented by the imagination. Thus the *Philosophia* of Descartes became the *Philosophia naturalis* of Newton. We need look no further for Newton's intention.

* * *

An examination of the structure and content of Newton's *Principia* shows that the designation of "mathematical" has a double sense. The first is the obvious use of technical mathematics—

geometry, trigonometric relations, proportions, algebra and infinite series, the calculus or "fluxions"—in derivations, proofs, and solutions to problems.[21] The second introduces the ontological level of discourse, notably of the first two Books, in relation to what I have called "the Newtonian style."[22]

Let me begin with a few words about mathematics in the first sense. Newton's qualifying adjective *"mathematical"* is definitely not to be taken as synonymous with "exact" or with "quantifiable" or "measurable." Newton's *Opticks* of 1704, for example, one of the seminal scientific works for the eighteenth century, although quantitative and exact, is not mathematical. Whereas the *Principia* proceeds, proposition after proposition, by using ordinary and recognizable mathematical tools, the *Opticks* makes constant use of what Newton explicitly declares is "Proof by Experiments."[23] Early in his career, Newton had hoped to reduce the subject of optics to "mathematical principles" in the Newtonian style of the *Principia,* and he declared his hope that the science of colors might "become mathematical." But, as he explained, "Optiques and many other mathematical sciences depend as well on Physical Principles as on Mathematical Demonstrations." The foundation of "the Proposition of colours," he said, was from *"Experiments* and so but *Physicall,"* with the result that "the Propositions themselves can be esteemed no more than *Physicall Principles* of a science."[24] But he had "good reason to believe," he averred, that "the *Science of Colours* will be granted *Mathematicall* and as certain as any part of *Optiques"* because of the possibility that a mathematical system can be erected that will compute (predict) and demonstrate the phenomena of colors. In the sense in which mathematics—rather than measurement, calculation, number, or mere quantification—is a characteristic feature of the *Principia,* Newton's vision of a mathematical system of optics remained a dream[25] or wish and never became reality. In the event, as Zev Bechler and Alan Shapiro have delineated, each of his attempts ended in failure.[26]

When Newton eventually published his *Opticks* (1704), the conclusion consisted of a series of "Queries" or unanswered questions, following a section of experiments on colored fringes, what we would designate today as "diffraction phenomena." He said he had intended to repeat these experiments "with more care and exactness," adding yet further "observations."[27] He admitted frankly that he had not adequately investigated the final topic of his book; in this sense his conclusion in the form of "Queries," as a guide to "a farther search to be made by others," must be read as a kind of public confession of the treatise's inadequacy, incompleteness, and even failure. We may see this feature of the book revealed in a number of facts. Newton kept the text private for years until finally goaded into publication. Furthermore, the work was written and published in the vernacular language, English, rather than at once disseminated in the international language of the scientific world, Latin. As an aside, I note that once the book had been published,

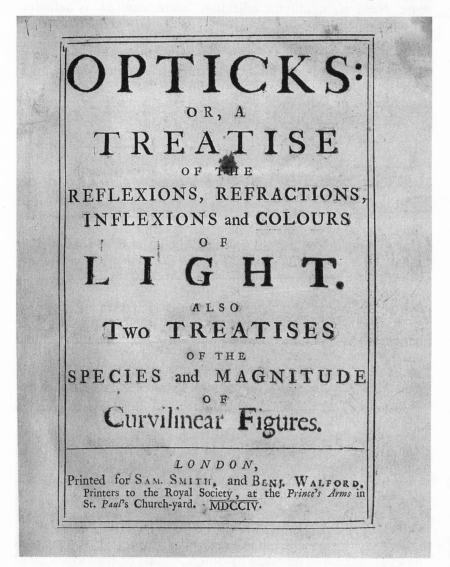

Figure 7. Title page of Newton's *Opticks* (1704). Note that there is no author's name. By permission of the Houghton Library, Harvard University.

Newton wanted credit for his positive contributions—which were many—and commissioned a Latin version that was published in 1706.[28]

I believe that it was this sense of failure that lay behind what is otherwise a wholly unexplained aspect of the *Opticks,* that it appeared without the name of the author on the title-page (see fig. 7), as if a not quite legitimate offspring and hence not fully acknowledged by its parent. Even the preface was not signed "Is. Newton," as had been the case for the *Principia,* seventeen years earlier, but merely bore the initials "I. N."[29]

Failure or not, the *Opticks* was to become one of the most influential books of the Enlightenment. It showed that mathematics, admittedly the key to rational mechanics[30] and celestial dynamics, was not the only path for the development of the sciences. Newton's *Opticks* became a veritable handbook for experimenters of the eighteenth century, a "vade mecum" of the experimental art and a primary textbook for those who saw that progress in science could be made by direct questioning of nature in the laboratory. For these investigators—developing the new experimental sciences of such varied subjects as plant physiology, electricity, heat, and chemistry—Newton's *Opticks* became the primary guide.[31] The expanded "Queries"[32] introduced into the later editions came to constitute a research program for the century. In a sense, then, as I discovered and documented early in my career, there arose two separate and quite distinct kinds of Newtonian scientific traditions: one based on the mathematical style of the *Principia,* whose adherents include Clairaut, d'Alembert, Lagrange, and Laplace, the other based on what I have called the speculative experimental science of the *Opticks,* whose adherents include Stephen Hales, Benjamin Franklin, and A.-L. Lavoisier.[33]

It is a measure of Newton's genius that even the occasion of a confession of failure could lead to positive advances in the sciences. This occurred not only with respect to his *Opticks,* but with respect to his *Principia* as well. The famous Scholium Generale, with which the *Principia* concludes in the later editions,[34] was essentially a declaration of Newton's failure to explain and to understand how a force of universal gravity could extend out through hundreds on hundreds of millions of miles. He would not, he said, conflate hypotheses or fictional explanations with true science; HYPOTHESES NON FINGO, he declared, "I do not feign hypotheses."[35] But in the light of its influence, this General Scholium has been read as a battle cry against hypotheses, as a declaration of a new set of goals for science: that it is enough ("satis est" are Newton's words) to predict and to retrodict the observed phenomena of nature.[36]

* * *

Like most classics of science, Newton's *Principia* is a work more honored than read. Even scholars who are interested in the Enlightenment or in the Scientific Revolution do not usually get much beyond the introductory and concluding prose sections, that is, the definitions and axioms plus the concluding general scholium. Largely unread are the mathematical portions, comprising about 95 percent of the whole work, the great oceans of mathematical derivations and ancillary materials of the first two "books" of *principia mathematica* or "mathematical principles." Only the most dedicated Newtonian scholars are hardy enough to follow Newton through the stages of his mathematics, proposition by proposition, section by section, "book" by "book."

To read through some five hundred pages of dense mathematics is, in any case, a daunting exercise. The ardors of mathematics set up *chevaux-de-frise* for all but the hardiest of philosophers, historians, and even historians of science. Being trained as a mathematician eases the task of reading the *Principia* but, even so, there are formidable difficulties of two major sorts. First, there is the intrinsic difficulty of the subject matter: rational mechanics and celestial mechanics. Second, Newton's algorithms seem distant and alien to any post-Newtonian reader. Ever since the invention of the calculus (by Newton himself and by Leibniz), and especially after the adoption of the Leibnizian algorithm, the mathematics of the *Principia* has appeared to us strange and difficult, an alien kind of exact science that in form is neither an exact copy of Greek mathematics nor an example of any kind of mathematics with which we are familiar. William Whewell, the nineteenth-century philosopher and historian of the "inductive sciences," has given a dramatic expression of the gulf separating the mathematics in the *Principia* from what post-Newtonians know familiarly as the language, methods, and algorithms of mathematical discourse. We gaze, Whewell wrote, "with admiring curiosity" on "the ponderous instrumental synthesis," which was "so effective in his hands," as if we were contemplating "some gigantic implement of war, which stands idle among the memorials of ancient days, and makes us wonder what manner of man he was who could wield as a weapon what we can hardly lift as a burden."[37] I may add that Whewell was in an exceptional situation to make such a judgment because he not only knew his mathematics and rational mechanics but had edited a Latin version of the central core of the *Principia*,[38] explaining the concepts and the proofs. He had also produced two versions of major portions of the *Principia* in English paraphrases with recastings of the proofs and developments into the algorithm of the calculus.[39]

I have mentioned some of the difficulties that arise in reading Newton's *Principia*. There is another: Newton purposely made the work hard to read. In the opening of Book 3, he admits to the reader that he had "composed an earlier version . . . in popular form, so that it might be more widely read"; yet he later cast this aside and "translated the substance of the earlier version into propositions in a mathematical style."[40] The reason he gave was that he did not want his exposition of the system of the world to be read and discussed by anyone who had not "first mastered the principles." In this way, he said, he sought "to avoid lengthy disputations" with "those who have not sufficiently grasped the principles" and who "will not perceive the force of the conclusions" and "lay aside the preconceptions to which they have become accustomed over many years."[41] As is well known, however, Newton did not still his critics, who faulted him for introducing the allegedly "occult" force of attraction. Diderot especially criticized Newton for thus having put a "veil" between "mankind and nature." It would have taken a month for a great mind like Newton's, Diderot said, to make everything plain; this month would have saved three years of intensive labor and exhaustion to a thousand "good minds." Nature, he concluded, has enough of a veil; we should not double the veil by adding mystery.[42] At least one thinker was recorded as having said that he would prefer to remain in Cartesian error than to have to learn enough mathematics to master Newton's *Principia*.[43]

It is certainly a paradox that a largely unintelligible book—its pages closed to all but the most skilled and dedicated mathematicians—would dominate the intellectual character of the Enlightenment and become the most generally influential work of its age. So important was the message of Newton's new physics, however, that a gallant company of interpreters took on the assignment of producing explanations for nonmathematical readers. The primary expositions of the Newtonian natural philosophy were produced by Henry Pemberton, a medical doctor who had prepared the third edition of the *Principia* under Newton's direction, and Colin MacLaurin, a brilliant mathematician who is immortalized in the infinite series that bears his name.[44] These two masterful works were translated into French and Italian and their message was thus broadcast throughout Europe.

One of the finest introductions to the Newtonian natural philosophy was written by Voltaire, under the tutelage of the "divine Emilie," the Marquise du Chastellet, who translated Newton's *Prin-*

cipia into French.[45] Important expositions of Newtonian physics were given in the Boyle Lectures, inaugurated by Richard Bentley in the 1690s.[46] On another level, the principles of Newtonian science were displayed in an attractive manner by experiments and demonstrations—a form of exposition pioneered by William Whiston and J. T. Desaguliers. A popular work, written in Italian and reprinted in many editions in French and English translations, was Francesco Algarotti's *Newtonianismo per le dame*.[47] In short, there were exegeses of Newton's *Principia* and his *Opticks* for almost every imaginable level of reader. And, of course, there came into being a series of admirable textbooks on Newtonian rational mechanics and astronomy by such authors as Keill, Gregory, 'sGravesande, La Caille, and Musschenbroek.[48]

Because the *Principia* remains generally unread, save by a small company of Newton specialists, its mathematical quality is all too often misrepresented. A common and quite erroneous over-simplified opinion holds that the *Principia* is written in the style of Greek geometry. In fact, a superficial glance does seem to reveal a work written on the Greek model. As in Euclid there are preliminary Definitions and Axioms.[49] The propositions are numbered and appear to be geometrical. They generally do not at first show themselves to be developed as sequences of algebraic or functional relationships subjected to the processes of differentiation and integration, nor are they usually seen to be written in either the Newtonian algorithm of dotted letters or the Leibnizian algorithm of *d*'s.

But "looks," as we all know, "are deceiving." Some propositions, notably those in Sections 4 and 5 of Book 1, are indeed purely geometric and synthetic, somewhat in the Greek manner.[50] But most are not. Section 1 of Book 1 sets the tone for the whole treatise: it is a discussion of "first" and "last" ratios, in actuality a short tract on the theory of "limits." The first lemma of this initial Section 1 of Book 1 announced that the mathematics of the *Principia* is that of the moderns and not of the ancients. "Quantities, and also ratios of quantities," Newton declares, "which in any finite time constantly tend to equality, and which before the end of that time approach so closely to one another that their difference is less than any given quantity, became ultimately equal." Together with Lemma 2 of Book 2, this collection of "the first eleven lemmas of Book 1" comprises—as Newton explained in an unpublished Preface—the "elements" of the "method of synthesis for fluxions and moments," Newton's own name for the calculus.[51] Those who have actually read Newton's *Principia* have always known that this work was "completely filled with the calculus," as the Marquis de l'Hospital (apparently generously but perhaps grudgingly) admitted, to Newton's great satisfaction.[52] Many propositions of the *Principia* indicate that they depend on the calculus by such phrases as "granting the quadrature of curves," that is, taking as granted the existence of certain integrals, or the ability to find the area of ("under") curves. In other propositions that explicitly use infinitesimals, appearing as little o's, Newton casts out those of higher orders in obtaining the final result.[53] And there are numerous examples of what are patently calculus applications to infinite series. In Lemma 2 of Book 2, Newton gave explicit rules for differentiating algebraic polynominals term by term, even giving the derivative of $A^m B^n$ correctly as

$$mA^{m-1}B^n + nB^{n-1}A^m.$$

We may easily see the quality of the *Principia* by a careful examination of two propositions: Proposition 1 and Proposition 11 of Book 1. Proposition 1 states that "The areas which bodies in orbit describe by radii drawn to an unmoving center of force lie in unmoving planes and are proportional to the times." In the proof of Proposition 1, Newton proceeds in a series of clearly demarcated steps. First he uses the simplest elementary geometry to demonstrate that a body moving with only linear inertial motion will sweep out equal areas in equal times, with respect to any point not in the line (see fig. 8). Here, at the beginning of the first proposition of the first book of the *Principia*, the reader was presented with an

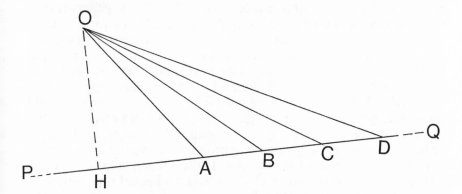

Figure 8. Inertial motion and the law of areas. In pure inertial motion a body will move in equal time intervals through equal spaces, AB = BC = CD = ... If O is any point not on the line AB, then if OH is erected perpendicular to AB, and the lines AB, BC, and CD, . . . are drawn, then the triangles AOB, BOC, COD, . . . will be of equal area since they have equal bases and a common altitude (OH). (From *The Birth of a New Physics* [see n.57], p. 161.)

absolutely stunning, breaktaking announcement that there is a close logical connection between the law of inertia and Kepler's law of areas. Until the publication of Newton's *Principia* in 1687, these two laws were discussed in relation to wholly different domains: Descartes's law of inertia was part of the general science of motion, but Kepler's law of areas was considered only as a property of planetary orbits or of the orbits of planetary satellites. Newton, however, revealed in a single master stroke that linear inertial or uniform motion has the property of area conservation. And this is only a preliminary to the major part of Proposition 1, which advances from simple linear motion to the generalities of motion along a curve.

In the *Principia* Newton does not boastfully call attention to his important discovery, which links the observational subject of astronomy with the abstract principles of rational mechanics. This mode of presentation is characteristic of the tone of the whole volume; Newton does not point out to the reader the seemingly endless parade of novelties that follow one another in rapid succession. Even the greatest of all, the law of the inverse square and the principle of universal gravity, are presented in a kind of understatement. We do, however, have convincing contemporary evidence concerning the importance of Newton's discovery of the physical significance of Kepler's law of areas and its relation to the law of inertia. The brilliant researches of Domenico Bertoloni Meli have revealed how deeply impressed Leibniz was by Newton's extraordinary feat. In his manuscript annotations on the first edition of the *Principia,* Leibniz characterized Newton as "rem generalissime considerans, nam communem omnibus proprietatem invenit, quam observavit Keplerus," thus, of course, demonstrating his own recognition that Newton had made a great generalization since he had found the property observed by Kepler (for the heavenly bodies) to be a common property of all bodies. In another manuscript Leibniz declared: "Sed eadem tempora pulcherrimo Kepleri invento, (quod Neutonus generale reddidit) sunt areis proportionales [*sic*]." That is, "But the same times, by a very beautiful discovery of Kepler's (which Newton made general) are proportional to the areas."[54]

In Newton's proof of Proposition 1, the second step is to postulate that there is a sequence of successive and equal time-intervals. In the first time-interval the body would go from A to B; in the second, from B to c. But, when the body arrives at the point B it is struck by an impulsive or instantaneous blow or force directed toward a point S (see fig. 9). This alters the path so that the body goes, during the second time interval, from B to C rather than from

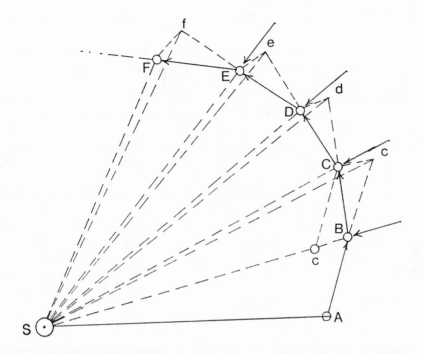

Figure 9. The meaning of the law of areas, according to the first three propositions of Book 1 of the *Principia*. A body moves initially with pure inertial motion. During some time interval T it will go from A to B. If there is no force acting, then during the next equal time interval T, the body will move along the same straight line from B to *c*, where AB = B*c*. But, if at point B the body is struck with an impulsive force, directed toward the point S, its path during the second time interval T will not be from B to *c* but rather from B to C, determined from the parallelogram rule for combining velocities or displacements. Newton proves, from simple and straightforward geometrical considerations, that the area of the new triangle BSC is equal to the area of the triangle BS*c*. Since he has already shown that the area of triangle ASB is equal to the area of triangle AS*c*, it follows that the area of triangle ASB is equal to the area of triangle BSC. Now, after another time interval T, when the body is at C, it receives another blow and the previous argument is repeated. Thus Newton proves that in the polygonal path ABCD . . . produced by a regular succession of impulsive forces directed toward a given point, the law of areas still holds. In the limit, the succession of impulsive forces becomes a continuously acting force and the polygon becomes a continuous curve (From *The Birth of a New Physics* [see n. 57], p. 162).

B to *c*. By the simplest geometry Newton proves that the area of triangle SAB is equal to the area of triangle SBC. At the point C, the body receives another impulse toward S and its resultant path is along CD rather than C*d*. Such a series of impulsive forces directed at S, following each other at regular intervals, produces the polygonal path ABCDEF . . . , in which the triangles SAB, SBC, SCD,

SDC, SDF . . . all have the same area; that is, area is still conserved.

Then, in the third step, Newton introduces the new mathematics
as he shifts the level of discourse from the finite to the infinitesimal.
For he now lets "the number of triangles be increased and their
width decreased indefinitely," whereat "their ultimate perimeter
will be a curved line" and "the centripetal force by which the body
is constantly drawn back from the tangent of this curve will act
uninterruptedly." In the limit, as the centripetal force becomes a
continually acting force and the polygon becomes a smooth curve,
the area law still holds.[55]

Newton next inverts the demonstration to prove that the area law
implies a centripetal force. We may note that not only has Newton
introduced a proof in a mode quite alien to Greek geometry; he has
also given a physical explanation of Kepler's law of areas: the area
law is a necessary and sufficient condition for a centripetal force
acting on a body with an initial component of inertial motion.[56] In
this master stroke Newton has brought Kepler's law of areas into
the domain of rational mechanics.

*   *   *

We may see the stages of Newton's mathematical procedure even
more strikingly by going step by step through Proposition 11, a
typical Newtonian problem that dramatically delineates how Newton's geometry-cum-limits differs from classical models. In Proposition 11, the problem is to find the law of force for Keplerian
planetary motion. A planet P moves in an elliptic orbit (see fig. 10)
according to the law of areas, reckoned with respect to the sun S
located at a focus of the ellipse. We are to determine the quantitative measure of the force F directed to S. The solution is that the
force F is inversely proportional to the square of the distance SP or
$F \propto 1/SP^2$.

We may observe that this is the problem whose solution had
eluded Hooke, Wren, and Halley in the mid-1680s. It was this
problem that impelled Halley to journey from London to Cambridge in August 1684 to ask Newton about forces and elliptical
orbits. When Newton told Halley he had calculated the force in an
elliptical orbit and had found it to vary as the inverse square of the
distance, Halley pressed Newton to write up the solution.[57] After
Newton had done so, Halley then urged Newton to send the written
account of his work for inclusion in the records of the Royal Society
in order to preserve his priority. Halley was fully aware that Newton
had solved the outstanding problem of the age: to find the force

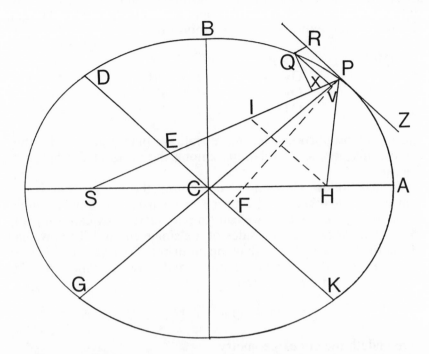

Figure 10. Diagram for Book 1, Prop. 11, of the *Principia:* to find the centripetal force, directed toward a focus, for a body moving in an elliptical orbit. This diagram is lettered in the same way in all editions. Note that the body moving in orbit is labelled P (for *P*laneta) and that the central body is S (*Sol*).

acting to produce the motions observed in the solar universe. Once Newton got started on his account he actually wrote a short tract for the Royal Society, which survives in a number of drafts or versions, one of which may still be read in the records of the Royal Society. Newton went on to enlarge this work and produced his *Principia.*[58] So it may be seen that Proposition 11 is far from trivial; it is a turning point in the road toward the eventual *Principia* and the major step in the making of modern exact science.

Newton's proof of Proposition 11 is strictly mathematical. Like many proofs in the *Principia,* it is divided into three distinct stages. The first is purely geometrical—in the manner of Greek geometry, reminiscent of Apollonius or Archimedes—as we saw was the case in the first stage of Proposition 1. Using the properties of the ellipse and certain features of the construction, Newton develops a series of proportions that—for simplicity—may be tabulated anachronistically in a set of four symbolic statements. Newton did not actually use these symbolic notations in the *Principia*. He generally stated the proportions in words. There is no injustice to his thought, however, in our expressing these proportions symbolically.

$L \times QR$	:	$L \times Pv$	: :	AC	:	PC
$L \times Pv$	:	$Gv \times Pv$	: :	$2BC^2$	:	$AC \times Gv$
$Gv \times Pv$	:	Qv^2	: :	PC^2	:	CD^2
Qx^2	:	QT^2	: :	CD^2	:	BC^2

These four proportions, let me repeat, are purely geometric: they express the geometric relations among the parts of the figure or diagram.

We may follow Newton by combining the ratios, that is, by multiplying together all the terms in each of the four columns. Before doing so, however, we can simplify matters by canceling any terms that appear on both sides of the single colon.[59] Thus we can cancel the following terms that appear in both column 1 and column 2: $L \times Pv$, $Gv \times Pv$. From columns 3 and 4 we can cancel AC, PC (in PC^2), BC^2 (in $2BC^2$), CD^2:

$$L \times QR \times Qx^2 : Qv^2 \times QT^2 : : 2PC : Gv.[60]$$

Here endeth the Greek geometry.

Now Newton shifts through some twenty centuries to advance from ancient to modern mathematics. He allows PQ to be "indefinitely diminished." In other words, to find out what happens when P approaches Q, he introduces a limiting process.

Of course, in Greek geometry there were some examples of the use of limiting processes. Notable among them was the so-called "method of exhaustion," found in the writings of Archimedes. But there is a very fundamental difference between Newton's use of limits and the practices of Archimedes. Archimedes essentially was concerned to find areas or volumes in cases where ordinary geometric methods would not yield the desired result. He would begin with one or more simple geometric figures and then alter them in a sequence in order to discover a limiting relation.[61] Newton, however, was interested in a quite different process, one in which he sets up a complex proportion from a geometric figure, different only in the form of expression from our method of setting up an equation. He then proceeds to the limit, finding out what the fate of his algebraic statement becomes.

In Newton's proof of Proposition 11, in the limit, as P approaches

Q, Qx^2 becomes equal to Qv^2 and at the same time 2PC becomes equal to Gv.[62] Hence, we may substitute Qv^2 for Qx^2 and Gv for 2PC to get[63]

$$L \times QR \times Qv^2 : Qv^2 \times QT^2 :: Gv : Gv$$

a relation that is true only in the limit, only on the infinitesimal level. It follows (after canceling out the factor QV^2) that

$$L \times QR : QT^2 :: Gv : Gv$$

or that the ratio

$$L \times QR : QT^2$$

has the value unity. That is,

$$L = \frac{QT^2}{QR}$$

or

$$\frac{QR}{QT^2} = \frac{1}{L} .$$

This completes the second stage, in which Newton has worked up in a geometrical manner, or by the methods of synthetic geometry, a proof that is a special case of his method of fluxions.

In the third and final stage of the demonstration, Newton makes use of his measure of a force (from Proposition 6, Coroll. 1 and 5),[64] which in this case yields the result that the force F, directed from P to S, has the value

$$F \propto \frac{QR}{QT^2} \times \frac{1}{SP^2}$$

By substitution of our previously obtained result, that

$$\frac{QR}{QT^2} = \frac{1}{L}$$

it follows that

$$F \propto \frac{1}{L \times SP^2} \, .$$

Since for any given ellipse L (the latus rectum) is a constant, Newton has proved that the force is inversely proportional to the square of the distance SP, from the sun to a planet. Q.E.I.

*   *   *

Again and again in the *Principia,* Newton proceeds in these three stages. In short, rather than derive a general algorithm for the calculus, which he could then universally apply, he tends to develop each theorem or proposition as a more or less independent problem. Why did Newton proceed in this manner? There are at least three possible reasons. The first is that if he had written the *Principia* in the language of his newly invented fluxions, there would have been a double barrier to its being read: the difficulty and novelty of Newton's concepts and methods (that is, the subject matter of rational mechanics) plus the difficulty and novelty of a new mathematical language that would have to be mastered before the book could be read.

A second problem arises from the chronology of Newton's inventions. We usually think of Newton's fluxions, or derivatives, in terms of his brilliant symbolism of dotted letters in which, for example, $\dot{x}$ stands for the first derivative of x (what we know as dx/dt), $\ddot{x}$ for the second derivative (d^2x/dt^2), and so on. But D. T. Whiteside has shown that Newton did not invent this particular symbolism until well into the 1690s, quite some time after the *Principia* had been published.[65]

Newton himself gave a third reason. In writing the *Principia,* he claimed, he had made extensive use of his fluxions. But he had then rewritten the propositions as synthetic demonstrations in what he called "the manner of the Ancients." In other words, he said, the propositions of the *Principia* had largely been "invented by Analysis," but since "the Ancients (so far as I can find) admitted nothing into Geometry before it was demonstrated by Composition [i.e., synthesis]," Newton had rewritten the whole work "to make it more Geometrically authentic & fit for the publick."[66] Anyone who is mathematically trained, he added, could easily reverse the order and find the "method of Analysis" by which the proposition had been "invented." By this means, he concluded, the "Marquess de l'Hospital was able to affirm that this Book was 'presque tout de ce Calcul' almost wholy [sic] of the infinitesimal calculus [or Analysis]."[67] We may agree with the judgment of the Marquis de l'Hospital, but I do not believe that Newton's mode of discovery or invention was in any major way different from his mode of presentation in the *Principia*.[68]

* * *

Newton's extraordinary prowess was revealed to readers of Book 1 of the *Principia* in a group of propositions that includes Propositions 40–42. Here Newton supposes "a centripetal force of any kind" and seeks "to find the trajectories in which bodies will move and also the times of their motions in the trajectories so found." A subsidiary problem is, with this same "law of central force," to find "the motion of a body setting out from a given place with a given velocity along a given straight line."[69] In 1687, when the *Principia* was published, readers were not prepared for problems of such generality as Propositions 41 and 42.

Even more astonishing were the propositions in Section 11 of Book 1, in which Newton considers the action of bodies mutually attracting one another. Hooke, Halley, and Wren had struggled in vain to solve the problem of a single planet moving around a fixed sun. We know how amazed Halley was to learn in 1684 that Newton had proved that if a planet moves in an orbit that is a Keplerian ellipse, the force directed toward the sun must be in the ratio of the inverse square of the distance. In the *Principia,* however, Newton took a giant's step further by considering a problem on a much higher level, that of any two bodies mutually pulling on each other. Such a system is much closer to the external physical world of nature than the imagined mathematical construct of a single body or mass point moving about a fixed center of force, the basic problem posed by Halley on that famous visit in August 1684. The result of Newton's analysis of the two-body problem was very profound. In the study of the one-body-plus-center-of-force system, Newton had revealed the physical significance of Kepler's first two laws. He had also shown under what limited or artificial circumstances of an idealized or imagined system these laws of Kepler were accurately or exactly true.[70] In his study of a two-body system, Newton advanced from this restricted realm of a mathematical Keplerian universe to one more nearly like that of physical nature. He not only showed that in a two-body system Kepler's laws are not strictly true; he went on to find the new laws, the actual alteration needed in the laws if they were to be valid in a natural system.[71]

But even this gigantic intellectual leap forward was not the limit of his creative innovation. It was, rather, only a preliminary step toward the grand climax on a higher level of generality, one that is still breathtaking in the splendor of his concept and achievement. For Newton at once moves on from Propositions 57–65 on two bodies to the great three-body problem. Before the *Principia* only a very few scientists had even conceived that such a problem might exist. And no one had the vaguest notion that it could be solved mathematically.

We know today that there is no analytic solution to the general problem of three bodies mutually attracting one another. Newton explores a limited case of this problem in an extensive series of twenty-two corollaries to Book 1, Proposition 66. Here a body like the moon (P) moves in orbit around a body like the earth (T) under the perturbing action of a body like the sun (S).[72] In order to see the magnitude of this achievement we have only to recall that for some fifteen hundred years the irregularities in the motion of the moon had been studied by purely geometric models, circles on circles, without reference to cause. Newton invented celestial mechanics as we understand the term today and completely altered the study of the moon's motion, shifting it from a set of problems in celestial geometry to an analysis of the forces producing the motions and their effects.[73] In a review of the second edition of the *Principia* in the *Acta Eruditorum*,[74] this breakthrough was hailed as one of Newton's most notable achievements, a judgment shared by later astronomers.

Newton's lunar theory was not wholly successful[75] and the full solution of the problems he posed had to await the work of later mathematicians such as Clairaut and Euler.[76] But it was Newton who had set the problem, who had shifted the level of discourse from ad-hoc contrivances to an analysis of physical causes and their effects. A few years ago, when I was writing an introduction to a reprint of a pamphlet containing Newton's theory of the moon's motion,[77] I consulted the textbook of celestial mechanics by Forrest Ray Moulton that I had used as a graduate student. To my astonishment and joy I found that in presenting the problems of the moon's motion, Moulton essentially followed the geometric development introduced by Newton in the *Principia*.[78]

The diagram for Proposition 11 has a very interesting feature. The orbiting body (see fig. 10) is denoted by P, the center of force by S. Readers would be at once aware that ultimately this mathematical demonstration was intended to illuminate an astronomico-physical problem: the orbital notion of a planet P (*Planeta*) around the Sun

S(*Sol*). In the first edition, this same pair of letters (S, P) appears in a number of propositions that are intended to be applied later in the *Principia* to orbital motions of a planet round the sun. In proposition 3, Newton explores the motion of a body that describes areas proportional to the times with respect to a moving body. In the second and third editions, these bodies are denoted by L and T, since their motions are evidently intended to be like those of the moon L (*Luna*) with respect to a moving earth T (*Terra*). This alteration was suggested to Newton by N. Fatio de Duillier[79] so that Proposition 3 would be brought into harmony with its neighbors.[80] And it is the same for Propositions 56–58 on the perturbations of the moon's motion by the sun; in the first edition the central body was S and the circulating P, as in the earlier propositions, while the external perturbing body was Q. But in the later editions, the central body becomes T, and the perturbing body S, for *Terra* and *Sol*. Newton did not, however, change P to L for the circulating body. In this context we must keep in mind that in the *Principia* Newton, following Galileo, still called satellites "planetae," as in a reference to "planetas circumjoviales." His usage was generally to distinguish such secondary or circumplanetary "planetae" from primary "planetae" or circumsolar "planetae." In the first edition there was a reference to but one moon of Saturn, discovered by Huygens and called the "planeta Hugenianus"; in the second and third editions Newton referred (e.g., in Book 3, Phen. 2) both to Saturnian satellites ("Satellitum Saturniorum") and to circum-Saturnian planets ("planetas circumsaturnios"), and he also wrote of "planetis qui saturnum comitantur" (Proposition 1).

* * *

The crowning glory of the *Principia* was to disclose the properties and actions of the force of universal gravity and the effects of this force in producing the motion of planets and their moons and of comets, and in being the cause of the phenomena of tides. Book 1 was devoted almost entirely to an exploration of forces of attraction. How could Newton reconcile his apparent advocacy of a force acting at a distance with the reigning or "received" philosophy—the so-called "mechanical philosophy,"[81] in which there was no place for any kind of force other than one acting through contact? We know that a major criticism leveled against the *Principia* was that it had abandoned sound philosophy and had reintroduced into physics the "occult" force of attraction that had been supposedly banned.[82] Newton himself, in a famous letter to Richard Bentley, written soon after the publication of the *Principia*, had expressed his own views on this subject in no uncertain terms. That "one Body may act upon another at a Distance thro' a *Vacuum*," he wrote, "without the Mediation of anything else, by and through which their Action and Force may be conveyed from one to another" is "so great an Absurdity" that he could not believe that anyone who had "a competent Faculty of thinking" about "philosophical Matters" could ever adopt such a position.[83]

And yet, in the *Principia*, Newton introduced a force that could apparently spread out through hundreds and hundreds of millions of miles of empty space, out into invisible reaches of the heavens. At so great a distance the solar force was still powerful enough to affect the motion of a comet so as to turn it around and bring it back to the regions of our visible solar system.[84] We know that Newton repeatedly sought some explanation of how universal gravity might act. That is, he attempted to reduce universal gravity to the action of something else, to a shower of aether particles, to electrical effluvia, to variations in an all-pervading aether.[85] All of these attempted "explanations," these reductions of universal gravity to some accepted kind of mechanism, failed—because none could fulfill two major requirements: that the resultant force vary inversely as the square of the distance and that this force act mutually on every pair of bodies so as to attempt to bring them together. It is well known that by the time of the second edition of the *Principia*, Newton was so aware of the difficulty his fellow scientists shared with him in accepting universal gravity that he set forth a new position in the General Scholium, which appeared for the first time in the second edition of 1713. It was there that he said that he could not explain gravity and that he would not feign hypotheses[86] in place of sound explanation. Confessing the failure of his attempts to reduce gravity to the action of some kind of matter (and motion), so as to eliminate a force acting at a distance through empty space, he asserted that it "is enough" that this force really ("revera") exists and that it serves abundantly to account for the major phenomena of the heavens and our earth.

* * *

There is a paradox in Newton's having constructed his *Principia* on a concept of attraction that he admittedly abhorred and that therefore he was not willing to admit into natural philosophy as a primary quality or property of matter. In order to understand how Newton could have based his system of the world on a force that acts at a distance, I have attempted to systematize and to make precise Newton's own procedures as developed in the *Principia*. I have also drawn on certain private and public statements made by Newton about the way he developed his ideas. This analysis reveals a coherent Newtonian approach to natural philosophy that—adapting a phrase used by Edmund Husserl in relation to Galileo—I have called the Newtonian style.[87]

In an unpublished preface to the *Principia,* Newton stressed that his aim was not to develop all aspects of the mathematics of "magnitudes, motions, and forces," but rather "only those things which relate to natural philosophy and especially to the motions of the heavens."[88] This goal was also stated clearly in the opening section of Book 3, where Newton declares that in the previous Book 1 and Book 2 he has set forth "strictly mathematical principles of philosophy" on which "natural philosophy can be based."[89] Here we get an insight into Newton's method in the *Principia:* the development of mathematical principles that can ultimately be applied to natural philosophy—mathematical principles chosen expressly because they relate to natural philosophy, to the realm of physical science revealed to us by experiment and observation, ordered and reduced to mathematical rules by reason.

We have seen that in Proposition 1 of Book 1, Newton deals with a mass point initially moving along a straight line in pure uniform inertial motion. In Proposition 11, the mass point moves along a pure elliptical orbit with a force directed toward a mathematical point as "center." Where do such conditions obtain? Only, as Newton explains in the *Principia,* in the realms of the imagination, in a world of mental constructs, in a geometric rather than a physical space, in an imagined realm where everything occurs according to pure *mathematical* principles rather than *philosophical* (in the sense of natural philosophy) principles (that is, principles of natural philosophy) or physical laws. This is the explicitly stated realm of the first two books of Newton's *Principia.* For it is in this realm of Books 1 and 2 that Newton, as he says expressly in the *Principia,* has presented principles ("the laws and conditions of motions and of forces") that, although relating to natural philosophy, "are not, however, philosophical but strictly mathematical."[90] Newton adds the comment that natural philosophy not only includes dynamics ("motions and forces") or "rational mechanics" (the term that Newton made standard for this realm of discourse), but also embraces such "topics . . . as the density and resistance of bodies, spaces void of bodies, and the motion of light and sounds," that is, the classical domains of physics.[91]

If we reexamine closely Newton's proof of Proposition 1 and Proposition 11, we see that Newton has considered the laws of a mathematical force in a mathematical configuration or in a mathematical space. When I say that the force is mathematical, I mean that Newton is dealing with motion resulting from the action of a central force directed to a mere geometric point in space, whereas he knew that in the world of nature forces do not originate in empty points in geometric space, but in bodies—as in the case of a magnet pulling on a piece of iron or a rubbed or electrified piece of amber pulling on a bit of straw or chaff, and so on. The moving body, in both Proposition 1 and Proposition 11, furthermore, is essentially a mass point, a dimensionless unspecified object in a free space, that is, a mathematical construct rather than a physical object in the world of observed nature.[92] In short, Newton is exploring laws of motion for a mathematical construct, rather than elaborating a phenomenologically based natural philosophy.

Newton's mathematical construct derives in the first instance from nature simplified, that is, from considering bodies reduced to mass points. But the end product is a mathematical construct, a creation of the mind, in which Newton is perfectly free to consider whatever kinds of motions he pleases, subject to any type of force that he may imagine—because he is dealing with a mathematical construct and not with a physical situation. Specifically, it is this lack of the constraint of the world of physical experience, of the rules and principles and even limitations of experimental natural philosophy, that enables Newton to consider a kind of force that acts as an attraction and that might even be repugnant to him as a physical entity. He is not concerned with the philosophical action of a physical force in the realm of natural philosophy, but only with a mathematical construct. I have called this aspect of Newton's thinking the first stage or stage one of the method of the *Principia:* stage one of what I have called the "Newtonian style."

Of course, the construct in question has been designed by Newton to be applied eventually to a specific end-use in natural philosophy, and so the construct has certain elements similar to the situation of the world of physics, the realm of natural philosophy revealed to us by our external senses, by experiment, and by observation. And it is clear to any reader that Newton is directing his efforts to producing rules and laws that can eventually be applied in natural philosophy. For example, in Proposition 11 and other early propositions in Book 1, Newton has the mass point move in a closed orbit, just as the planets do about the sun and as planetary satellites do about the parent body. Furthermore, in Proposition 11 this mass point moves in an elliptical orbit according to the law of areas, just as Kepler said the observed planets do and their satellites do as well. We have mentioned that Newton has even called the moving body P for "planeta" and the center of force S for "sol." But it may be noted that Newton also introduces conditions that do not originate in the world of observed reality, such as motion in an ellipse with a force directed toward a center, rather than a focus. And it is made obvious that Newton is dealing with the realm of mathematical and imagined constructs when he introduces orbital motion in a hyperbola, a situation he knew does not occur in nature.

Newton could not help being aware that in this stage one his construct is so oversimplified that it differs extremely from nature. For, in the most elemental of nature's systems, in the heavenly universe, we would find not just one but at least two bodies: for instance, the sun and the earth or the earth and its moon. And, as I have mentioned, Newton tells us that he is fully aware that forces originate in bodies and not in mathematical points. Accordingly, a comparison of Newton's mathematical construct with the world of natural philosophy reveals that certain essentials have been omitted. This confrontation with the external physical world of phenomena I call stage two. In the case of the early propositions concerning orbital motion, this confrontation or stage two results in a revision of the mathematical construct of stage one and its replacement by a new mathematical construct. The consequence is that the mathematical laws or rules found in stage one must be modified to accord with the new revised mathematical construct.

That this is indeed Newton's procedure is shown by an examination of the text of the *Principia*. For Newton says at the beginning of Section 11 that he has, in the earlier sections of Book 1, been dealing exclusively with "the attractions of bodies toward an immovable center," even though "most likely there is no such thing existent in nature." In the world of phenomena, he says, "attractions are made toward bodies" and not toward mathematical centers. Accordingly, he will now consider the motion of a body or mass point that moves under the action of a force directed not toward a mathematical point but toward a second body or mass point. In any such system, however, the third axiom or law of motion requires that each of the two bodies act upon the other. In such a system there are two forces, not just one.

When Newton, in Section 11, introduced the action of each of two bodies on the other, he knew that he had to make it absolutely plain to the reader that he was concerned here (and *here* means in Book 1 of the *Principia*) with the realm of mathematics, with artificial or imagined mathematical constructs and not with physics. Thus, he says, he is using the word "attraction" for simplicity of discourse. But in the realm of physics, as Alexandre Koyré pointed out, this word "attraction" with its overtones of action at a distance and "occult" forces was bound to incite the hostility of natural philosophers who were committed to the new mechanical philosophy.[94] It would, perhaps, have somewhat lessened the criticism if Newton had eschewed the word "attraction" and had rather written of "the mutually interacting centripetal forces arising from and acting upon bodies" or used some other circumlocution. He writes that he has introduced the common expression "attraction" because he no longer can use "centripetal force"; the reason is that there is no longer a single center of force but as many such centers as bodies. He insists that he is concerned with mathematics and not physics. In the "language of physics," he observes, these forces "might more truly be called impulses." But, for now, he is "putting aside any debates concerning physics"; he is addressing himself exclusively to "mathematical readers."

In a scholium at the end of Section 11, Newton expounds in even greater detail the elements of the Newtonian style. He asserts that he is concerned "with any endeavor whatever of bodies to approach one another." In this context, it does not matter "whether that endeavor occurs as a result of the action of the bodies either mutually drawn toward one another or acting on one another by means of spirits emitted." Nor is it of any import whether this endeavor "arises from the action of aether or of air or of any medium whatever—whether corporeal or incorporeal—in any way impelling toward one another the bodies floating therein." It is the same for "impulses," he adds, since he is "considering in this book [i.e., Book 1] not the species of forces and their physical qualities" but only "their quantities and mathematical proportions."

Then Newton himself sets forth three separate and distinct levels of inquiry. "Mathematics," he writes, "requires an investigation of quantities of forces and their proportions that follow any conditions that may be supposed." This is Newton's own statement of what I have designated stage one of the Newtonian style. He continues: When we later "come down to physics, these [mathematical] proportions must be compared to the phenomena." This is a clear and unambiguous description of what I have called stage two. Only then, and "finally," will it be possible to argue philosophically "concerning the physical species, physical causes, and physical proportions of these forces."[95]

When Newton develops the mathematical laws of a two-body system in the revised stage one, he discloses two very important aspects of the *Principia*. The first is the demonstration of the true significance or physical meaning of each of Kepler's laws. The second is the revelation that these laws are strictly or exactly true, or mathematically true, only in the realm of the imagination, in an imagined construct in mathematical space, one in which a single mass point moves about a mathematical center of force. In the *Principia* Newton then shows what modifications must be made to the simplified form of Kepler's laws.[96]

After developing the mathematical properties of a two-body system, in the revised stage one, Newton advances to yet a new stage two, where it is seen that the physical world of our solar system consists of more than two bodies, each of which must interact with all the others in a way that we have seen earlier in the case of a two-body system. Thus, we proceed to yet a new stage one, in which we have three mutually interacting bodies. Ultimately this contrapuntal alternation between the mathematical properties of an imagined construct in a stage one and the observed physics of a stage two leads Newton to develop a system more and more like that of the world we see around us. After dealing with a system of more than two bodies, Newton proceeds to bodies that are not mass points but have physical shapes. First he discusses the attraction of uniform or homogeneous spheres, then of spheres made of successive homogeneous or uniform shells, that is, bodies with properties more and more like those observed in physical objects in the external world such as planets and moons. Eventually, Newton will introduce shapes other than spherical.[97]

Finally, Newton advances to what I have called stage three. He applies to the physical world of our earth and of the solar system the rules he has developed mathematically in his imagined sequence of constructs. This application to the realms of the physical universe is—as I have mentioned—the goal of the whole exercise. The crowning glory of the *Principia* is the elaboration of the Newtonian System of the World, functioning on the principles of rational mechanics in which there is a force of universal gravity directly proportional to the product of the masses of bodies and inversely proportional to the square of the distances between them. This third stage occupies Book 3 of Newton's *Principia* and is, of course, the summit of his life's work.

* * *

In conclusion, we may note the extraordinary power of the Newtonian style. It enables Newton to be freed from physical and philosophical considerations and restraints in the early development of "mathematical principles," which he intends to apply later to "natural philosophy." It even enables him to discuss and to develop the mathematical properties and laws of a kind of force that—considered of and by itself, as an entity of physics—he finds repugnant. We know how strong a restraining force the abhorrence of action at a distance actually was. After the *Principia* was published, Christiaan Huygens wrote a letter to Leibniz in which he carefully explained that he too might have produced many, if not all, of the mathematical results of Newton's *Principia,* had he not been inhibited by the prejudice that did not permit him even to think of a force acting at a distance.[98] We may, incidentally, thus imagine the nature of the blow dealt to Huygens's conceptions by Newton since, at the conclusion of Book 2, Newton proved that the Cartesian vortices—an alternative to a system based on gravitational attraction—could not exist. The problem for Huygens was extremely severe since he could not accept the alternative of merely believing in action at a distance; the attempt to produce a non-Cartesian aether occupied much of Huygens's later creative energy. At the present, however, we are more interested in the example of Huygens as an illustration of the restraint on creative activity produced by the received mechanical philosophy. In contrast, the Newtonian style displays a powerful liberating quality.

For Newton himself, as we have seen, the very success of the Newtonian style posed a fundamental question: how to account for a force that does exist and that does explain so much. His successors, Lagrange and Laplace, started out with the assumption that a force such as gravity does indeed account for the phenomena, and they proceeded to develop mathematical rational mechanics and celestial mechanics in the Newtonian style without being overtly concerned with how such a force may act.

Laplace began his monumental *Traité de mécanique céleste* with a statement about Newton's "discovery of universal gravitation." Mathematicians, Laplace continued, "have, since that epoch, succeeded in reducing to this great law of nature all the known phenomena of the system of the world."[99] In his mathematical development of the principles of rational mechanics and his applications of those principles to Newtonian universal gravity there is no hesitation, no qualm expressed, in making use of the concept of a force acting at a distance. But we know that in the nineteenth century many physical scientists were deeply troubled by this concept and that the philosophical problem that so disturbed Newton and his contemporaries eventually led to a series of revolutions, of a profundity that rivals the shattering effect of the Newtonian revolution in science.

Notes

1. The full title is *Philosophiae Naturalis Principia Mathematica,* that is, *Mathematical Principles of Natural Philosophy.*

2. This part of the title has become celebrated in our century as a result of its reappearance in the *Principia Mathematica* of Bertrand Russell and Alfred North Whitehead (Cambridge: at the University Press, 1910–13), both of whom had studied portions of Newton's *Principia* as undergraduates in Trinity College, Cambridge. I suspect, however, that they were using this title only in part in obvious imitation of Newton, but in even greater part as a supposed Latin equivalent of "Principles of Mathematics," rather than as an indication of "Principles" that were themselves "mathematical" (the sense of Newton's title). The reason is that their *Principia Mathematica* was, to a considerable degree, a reworking (making use of the technical tools of mathematical or symbolic logic) of an earlier work by Russell alone, entitled *Principles of Mathematics* (Cambridge: at the University Press, 1903).

3. This Preface appeared in all three editions of the *Principia.* There is a further discussion of these topics in a "Conclusio" written for the first edition (1687) but never used. This has been published, with a commentary, in A. Rupert Hall and Marie Boas Hall, eds., *Unpublished Scientific Papers of Isaac Newton* (Cambridge: at the University Press, 1962), §IV.3.

4. On this topic see Alan Shapiro, "Experiment and Mathematics in Newton's Theory of Color," *Physics Today* 18 (1974): 184–222; and the publications of Zev Bechler, "Newton's Search for a Mechanistic Model of Colour Dispersion," *Archive for History of Exact Sciences* 11 (1973): 1–37; "Newton's Law of Forces Which Are Inversely as the Mass: A Suggested Interpretation of His Later Efforts to Normalize a Mechanistic Model of Optical Dispersion," *Centaurus* 18 (1974): 184–222.

5. There do not appear to have been many books of the seventeenth century that contained the word "principia" in the title. Since Newton did study Descartes's *Principia* and did forge some of his most basic ideas in response to those which Descartes expressed in his *Principia,* it is pointless to look beyond Descartes for the source of Newton's "principia."

6. Our present understanding of the significance of Descartes's *Principia* for Newton began with the discovery and publication (with an important interpretative commentary) by Rupert and Marie Hall (see n. 3 above) of an unknown and unpublished essay, beginning "De gravitatione et aequipondio fluidorum . . . ," apparently written by young Newton soon after first reading Descartes's *Principia.* See further Alexandre Koyré, "Newton and Descartes" in his *Newtonian Studies* (Cambridge: Harvard University Press, 1965), pp. 53–200.

For further information on Newton and Descartes, see I. Bernard Cohen: "'Quantum in se est'. Newton's Concept of Inertia in Relation to Descartes and Lucretius," *Notes and Records of the Royal Society* 19 (1964): 131–55; I. Bernard Cohen, "Newton and Descartes," pp. 607–34 of vol. 2 of Giulia Belgioioso, Guido Cimino, Pierre Costabel, & Giovanni Papuli, eds., *Descartes: Il Metodo e i Saggi*

(Rome: Istituto della Enciclopedia Italiana, 1990). Firm evidence concerning Newton's reading of the Latin edition of Descartes's correspondence was given in my
The Newtonian Revolution (Cambridge, London, New York: Cambridge University Press, 1980), pp. 332–33.

For a major additional source of information concerning Newton's reading of
Descartes and a valuable essay on the significance of Cartesian ideas in the
development of Newton's thought, see J. E. McGuire and Martin Tamny, eds.,
Certain Philosophical Questions: Newton's Philosophical Notebook (Cambridge,
London, and New York: Cambridge University Press, 1983), pp. 127–94. On Des-
cartes and Newton, and on Descartes as a scientist, see William R. Shea, *The
Magic of Numbers and Motion: The Scientific Career of René Descartes* (Canton
[Mass.]: Science History Publications, 1991).

7. He writes in this fashion to such an extent as to leave little doubt that
Newton's *Philosophiae Naturalis Principia Mathematica* was a direct transformation of Descartes's *Philosophiae Principia* (actually *Principia Philosophiae*). Newton's MSS contain this phrase and also a number of variations based on one, two,
or three words of the title of the *Principia;* see, especially, C.U.L. MS Add. 3965.
In the Preface to the first edition of the *Principia,* Newton wrote of "philosophiae
principia mathematica." In the anonymous review *(Recensio libri)* of the *Commercium Epistolicum,* Newton wrote about himself in the third person, "By the help
of the new *Analysis* Mr. *Newton* found out most of the Propositions in his
Principia Philosophiae," *Philosophical Transactions* 29 (1714–16): 206. The use of
the possessive pronoun ("his") rather than the definite article ("the") indicates
Newton's awareness that this was not the only *Principia Philosophiae*.

Examples of direct borrowing from Descartes abound in the *Principia*. In the
scholium (on space and time) following the Definitions, Newton refers no less than
seven times to variant forms of a "conatus recedendi ab axi motus," an "endeavor
of receding from the axis of motion." These are variations on Descartes's "conatus
recedendi a centro." We may be doubly surprised to find this Cartesian phrase
appearing so explicitly in Newton's *Principia*. First, the notion of "conatus" was
strictly Cartesian and as such has no place in Newtonian rational mechanics,
which is a system of dynamics based on force. Second, this "conatus recedendi" is
related to a pre-Newtonian kind of analysis of curved motion; Newton showed that
such a centrifugal tendency arises from a body's component of inertial motion,
which urges it along a tangent to the curve and thus makes it seem to be
endeavoring to recede from the center or from an axis of rotation. A striking
example of the appearance of "conatus recedendi a centro" occurs in Book 1,
Proposition 66, arguably the most original portion of the *Principia*. For it is here
that Newton proceeds from the relatively simple problem of a body in orbit,
attracted by a centrally directed force, to the difficult and complex problem of two
mutually interacting bodies, each exerting a force on the other.

Another instance of direct borrowing from Descartes's *Principia* occurs in
Newton's *Principia* in the previously mentioned scholium on time and space. Here
Newton says that "true and absolute motion cannot be defined as transfer from the
vicinity of bodies which are regarded as being at rest." The latter part of this
sentence—per translationem e vicinia corporum, quae tanquam quiescentia spec-
tantur—is taken almost word for word from Descartes, appearing (in variant form)
in four different sections of the latter's *Principia*.

8. On the various editions of the *Principia,* see Appendix 8 of *Isaac Newton's
Principia . . . the Third Edition with Variant Readings,* ed. Alexandre Koyré, I.
Bernard Cohen, and Anne Whitman (Cambridge: Harvard University Press; Cam-

bridge: at the University Press, 1972).

9. There is good reason to suppose that Florian Cajori had very little to do with the actual preparation of the final text that bears his name. At the time of his death the text of his edition was far from completed. This topic is discussed further in the introduction to a new translation of the *Principia,* and the so-called *System of the World,* made by I. Bernard Cohen and Anne Whitman (Berkeley and Los Angeles: University of California Press, 1992).

10. There are "principia," in a sense very close to that of chemistry, in the literature of alchemy.

11. The "Axiomata, sive Leges Motus" follow the "Definitiones" in the portion of the *Principia* before the official opening of Book 1.

12. In the second edition (1713), three "Regulae Philosophandi" were separated out by Newton from a set of "Hypotheses" that had appeared at the opening of Book 3 in the first edition (1687). In the third edition (1726), the "Regulae" were increased in number from three to four. On the evolution of the "Regulae" from "Hypotheses," see I. B. Cohen, "Hypotheses in Newton's Philosophy," *Physics* 8 (1966): 163–84; for textual details of this transformation see the edition cited in n. 8 above. Also, I. B. Cohen, *Introduction to Newton's 'Principia'* (Cambridge: Harvard University Press; Cambridge: at the University Press, 1971, 1978), §II.1, 2; §V.6.

13. See n. 27 below.

14. See McGuire and Tamny (cited in n. 6 above), pp. 275–95.

15. A somewhat different sense, that of a mathematical rule, appears in the final scholium in the opening Section 1 of Book 1, on the method of "first and last ratios." Having "demonstrated the principles," he writes, we may now apply them with impunity. Of a very different sort is "the thinking principle" of a man ("in persona hominis seu principia ejis cogitante") that is mentioned in the fourth paragraph of the concluding General Scholium.

16. *Correspondence* (cited in n. 24 below), 5:391–97.

17. See my *Introduction* (cited in n. 12 above).

18. Ibid., p. 130.

19. *Correspondence* (cited in n. 24 below), 2:434, 437.

20. *De Gravitatione et aequipondis fluidorium,* in *Unpublished Scientific Papers* (see n. 3 above), pp. 89–156.

21. One of the outstanding features of the *Principia* was Newton's extensive and original use of the methods of infinite series, as Edmond Halley noted in his review (*Philosophical Transactions* 186 [1687]: 291–97). Newton's general use in the *Principia* of the limiting process is, in effect, what François de Gand has felicitously called Newton's "non-algorithmic calculus." But there are some instances in the *Principia* of an algorithm, although these do not make use of the dot-notation, which was developed only in the 1690s, after the *Principia* had been completed and published; see, e.g., Lemma 2, Book 2.

22. The concept of a Newtonian style is developed fully in my *Newtonian Revolution* (cited in n. 6 above) and at the end of the present chapter.

23. This phrase occurs in theorem after theorem.

24. The *Correspondence of Isaac Newton,* 7 vols. (Cambridge: at the University Press, 1959–77), 1:187.

25. Newton did produce many interesting examples of mathematicization of optical problems, for which see vols. 4 and 6 of D. T. Whiteside's edition of *The Mathematical Papers of Isaac Newton* (Cambridge: at the University Press, 1971, 1974); some of his results are summarized in my *Newtonian Revolution* (see n. 6

above), pp. 134–41. There is, however, one truly mathematical proof (in words) in the *Opticks*, Book 1, Part 1, Proposition 6; this single example differs greatly from the rest of the work.

26. See n. 4 above.

27. A facsimile edition of the *editio princeps* of the *Opticks* (1704) was published by Editions Culture et Civilisation, Brussels, in 1966. A reprint of the fourth edition is available from Dover Publications (New York, 1952; revised edition, 1979). The quotation comes from page 132 (second numeration) of the 1704 edition, page 339 of the Dover reprint. The late Henry Guerlac devoted his last years to producing a scholarly "variorum" edition of the *Opticks*, not quite completed at the time of his death.

28. Newton commissioned Samuel Clarke (of the "Leibniz-Clarke correspondence") to produce the Latin version.

29. It would not have been hard to guess who the author was. Not only did he sign the preface ("Advertisement") with the initials "I. N."; he also identified himself by stating that "Part of the ensuing Discourse about Light was written at the desire of some Gentlemen of the Royal Society, in the year 1675," and was "sent to their Secretary, and read at their Meetings."

30. This term became current in the exact sciences following Newton's introduction of it in the *Principia;* Preface to the 1687 edition, "mechanica rationalis."

31. This aspect of Newton's influence is displayed in I. B. Cohen: *Franklin and Newton* (Philadelphia: American Philosophical Society, 1956; Cambridge: Harvard University Press, 1966), ch. 6.

32. The elaboration of the Queries, edition by edition, has been traced by Henry Guerlac (see n. 27 above).

33. See n. 31 above.

34. It appeared for the first time in the second edition (1713).

35. The rendering "I do not feign hypotheses" was first suggested by Alexandre Koyré.

36. For example, Ernst Mach wrote that

> Newton's reiterated and emphatic protestations that he is not concerned with hypotheses as to the causes of phenomena, but has simply to do with the investigation and transformed statement of *actual facts,*—a direction of thought that is distinctly and tersely uttered in his words Hypotheses non fingo, (I do not frame hypotheses)—stamps him as a philosopher of the *highest* rank.

Quoted from *The Science of Mechanics,* trans. Thomas J. McCormack, 6th ed. (La Salle: The Open Court Publishing Company, 1960), pp. 236–37.

37. William Whewell: *History of the Inductive Sciences,* 3d ed., 2 vols. (New York: D. Appleton & Co., 1865), 1 : 408 (Book 7, §II.4).

38. William Whewell, ed., *Newton's Principia: Book I. Sections I.II.III. in the Original Latin, with Explanatory Notes and References* (London: John W. Parker, 1846).

39. *An Introduction to Dynamics, Containing the Laws of Motion and the First Three Sections of the Principia* (Cambridge: J. and J. J. Deighton, 1832); *The Doctrine of Limits, with its Applications; Namely, Conic Sections, the First Three Sections of Newton, the Differential Calculus* (Cambridge: J. and J. J. Deighton; London: John W. Parker, 1838).

40. This more popular book was entitled *De motu corporum liber secundus* at a time when Newton intended the whole work to consist of only two books. This "liber secundus" was published posthumously in an English version as *A Treatise*

on the System of the World (London, 1728; 2d ed., 17–31), facsimile reprint, with an introduction by I. B. Cohen (London: Dawsons of Pall Mall, 1969). An edited Latin version was published in London in 1729. For information on this work, see the introduction to the facsimile reprint and my *Introduction to Newton's Principia* (see n. 11 above), § IV.6. A new edition of this work, edited by I. Bernard Cohen and Anne Whitman, is scheduled for publication in 1993 by the University of California Press.

41. *Principia*, Book 3, Introduction.

42. See *Franklin and Newton* (n. 31 above), p. 124; the quotation occurs in Diderot's essay "De l'Interprétation de la nature."

43. This was reported by Newton's disciple, J. T. Desaguliers; see my *Franklin and Newton* pp. 246–47.

44. For the various editions and translations of these two works, see *A Descriptive Catalogue of the Grace K. Babson Collection of the Works of Isaac Newton* (New York: Herbert Reichner, 1950; supplement, 1955); also Peter Wallis and Ruth Wallis, eds., *Newton and Newtoniana 1672–1975* (Folkestone: Dawson, 1977).

45. See the two works cited in n. 44 above.

46. These are reprinted in I. B. Cohen and Robert E. Schofield, eds., *Papers & Letters of Isaac Newton on Natural Philosophy* (Cambridge: Harvard University Press, 1958; revised ed., 1978), §IV. 3.

47. See the two works cited in n. 44 above.

48. A discussion of the various books disseminating the Newtonian natural philosophy is given in my *Franklin and Newton* (see n. 31 above).

49. But there are no "Postulata" such as accompanied the Euclidean "Definitiones" and "Axiomata."

50. These two purely geometrical chapters had been written independently of the *Principia* and were included on the weak grounds that they dealt with ellipses (and other conic sections) and so belonged to a work on planetary orbits. See my *Introduction* (n. 5 above), §IV.4.

51. But there are many exceptions that would be revealed by a close examination.

52. *Analyse des infiniment petits* (Paris, 1696), Preface.

53. For example, Book 2, Proposition 10.

54. These manuscripts are transcribed in full and analyzed in Bertoloni Meli's doctoral dissertation, "The Formation of Leibniz's Techniques and Ideas about Planetary Motion in the years 1688 to 1690" (Cambridge University, 1988). See his "Leibniz's Excerpts from the *Principia Mathematica*," *Annals of Science* 45 (1988): 477–505.

It may be observed that in the second quoted passage, Leibniz probably wrote "proportionales" for "proportionalia" because he was thinking of the form found in Book 1, Proposition 1, of the *Principia,* where "proportionales" agrees with "areae"; cf. Leibniz's notes on the *Principia* transcribed in Bertoloni Meli's dissertation, p. 111, ll. 299–301: "Areae . . . sunt temporibus proportionales et vicissim."

Bertoloni Meli's dissertation (pp. 59, 133) makes special note of Leibniz's recognition of the importance of Newton's generalization of Kepler's area law.

55. This limiting process is a common one in the *Principia*. It, in fact, enables Newton to proceed from a second law of motion based on instantaneous or impulsive forces to a law based on continuously acting forces. See my "Newton's Second Law and the Concept of Force in the *Principia*" in Robert Palter, ed., *The Annus Mirabilis of Sir Isaac Newton, 1666–1966* (Cambridge: The MIT Press, 1970), pp. 143–85. It may be observed that Newton introduced into physics the

technical term "vis centripeta" or "centripetal force," which—he said—he had named in honor of Christiaan Huygens, who had used the term "vis centrifuga" or "centrifugal force."

56. This single statement encompasses two propositions (1 and 2) of Book 1, a result that Newton extended (Proposition 3) to a system with a moving center of force, rather than a static one.

57. For details, see my *Introduction* (cited in n. 12 above), §III.1. The records of Halley's conversation are contained in some notes by John Conduitt, who intended to produce a biography of Newton. Conduitt got the story from the mathematician A. de Moivre, who in turn got it from Halley himself. Did Halley ask Newton what the force would be in an elliptical orbit or did he ask what the orbit would be if the force were to vary as the inverse-square of the distance? This problem has been brought to the attention of Newtonian scholars by the work of Robert Weinstock, esp. "Dismantling a Centuries-Old Myth: Newton's *Principia* and Inverse-Square Orbits," *American Journal of Physics* 50 (1982): 610–17, although most scholars do not agree with his final conclusions. For a discussion of this question, see my *The Birth of a New Physics*, revised and updated edition (New York: W. W. Norton & Company, 1985), supplement 13.

58. See my *Introduction* (n. 11 above), §III.2,3.

59. This canceling operation shows us an advantage of expressing these relations as ratios.

60. Actually, Newton does not ever write the equivalent of the last ratio (the one compounded after four primary ratios). He compounds the ratio and takes the limit in a single step. In the limit ("punctis Q & P coeuntibus"), Qv^2 will equal Qx^2, that is, "Qv *quad.* ad Qx *quad.*" will be a "ratio aequalitatis." Newton writes: "$L \times QR$ fit ad QT *quad.* ut $AC \times L \times PCq \times CDq$, seu $2CBq \times PCq \times CDq$ ad $PC \times GV \times CDq \times CBq$, sive ut $2PC$ ad GV." That is, since $Qv^2 = Qx^2$ it is the case that the ratio $L \times QR \times Qx^2$: $Qv^2 \times QT^2$ will become equal to the ratio $= L \times QR$: QT^2. And hence, $L \times QR : QT^2 :: AC \times L \times PC^2 \times CD^2 : PC \times Gv \times CD^2 \times CB^2$

$$:: 2CB^2 \times PC^2 \times CD^2 : PC \times Gv \times CD^2 \times CB^2$$

where in the second row, the quantity $2CB^2 \times PC^2 \times CD^2$ replaces its equal $AC \times L \times PC^2 \times CD^2$. Now by canceling the equivalent terms (CB^2, PC, CD^2) in columns 3 and 4, the final result is

$$L \times QR:QT^2 :: 2PC : Gv.$$

61. See "The *Method* of Archimedes" in T. L. Heath, ed., *The Works of Archimedes* (New York: Dover Publications, n.d.), suppl. A convenient, succinct presentation of Archimedes' method may be found in Marshall Clagett, *Greek Science in Antiquity* (New York: Collier Books, 1963).

62. See n. 60 above.

63. In Newton's actual procedure, the results of the limiting process are first applied to get $Qv^2 = Qx^2$ (as in n. 60 above) and then, "Sed punctis Q & P coeuntibus aequantur 2PC & Gv" (that is, with Q and P coming together, 2PC and Gv will be equal).

64. Newton's "dynamical" measure of a force constitutes a remarkable and original concept, one that gives him a powerful tool for the analysis of curved orbits.

65. Newton publicly introduced this new notation in *Tractatus de quadratura curvarum*, one of the two mathematical tracts (in Latin) published as supplements to the *Opticks* (1704).

66. These and some rather similar autobiographical statements of Newton's may be found in my *Introduction* (n. 12 above), suppl. 1. The same point is made on p. 206 of the anonymous review of the *Commercium epistolicum* (1712), written by

Newton himself and published in the *Philosophical Transactions* 29 (1715): 173–224. See A. R. Hall, *Philosophers at War: The Quarrel between Newton and Leibniz* (Cambridge: Cambridge University Press, 1980).

67. See n. 52 above.

68. There is not a scrap of evidence to support Newton's claim (repeated by some others) that he had originally found the propositions (or corollaries or lemmas) of the *Principia* by essentially any other method (e.g., by using fluxions) than that displayed in the printed presentation.

69. It is explicit in these propositions that the quadrature of curvilinear figures is possible, that is, that the integral calculus enables areas under such figures to be determined. Newton's exact working is "concessis figurarum curvilinearum quadraturis" or "if the quadrature of curves be granted." We know that at one time Newton proposed to insert into the *Principia,* near these propositions, a classification of algebraic curves whose quadrature could be effected. He also planned to append to the text of the *Principia* his treatise *De quadratura* on the subject of integration. For details see my *Introduction* (n. 11 above) and D. T. Whiteside's edition of Newton's *Mathematical Papers*.

70. That is, for a "system" composed of a single mass point moving about a mathematical center of force.

71. See I. B. Cohen, "Newton's Theory vs. Kepler's Theory: An Example of a Difference Between a Philosophical and Historical Analysis of Science" in Yehuda Elkana, ed., *The Interaction Between Science and Philosophy* (Atlantic Highlands, N.J.: Humanities Press, 1974), pp. 299–338.

72. These corollaries and Newton's investigations of the moon's motion in terms of perturbations have been studied by Craig Waff: "Universal Gravitation and the Motion of the Moon's Apogee: The Establishment and Reception of Newton's Inverse-Square Law," (unpublished doctoral dissertation, The Johns Hopkins University, 1975). We eagerly look forward to Waff's completion of his revisions so as to produce a published book on a much needed topic.

73. See my edition of *Isaac Newton's Theory of the Moon's Motion* (Folkestone: Dawson, 1975).

74. *Acta Eruditorum,* Leipzig, March 1714, p. 140: "Indeed, the computation made of the lunar motions from their own causes, by using the theory of gravity, the phenomena being in accord, proves the divine force of intellect and the outstanding sagacity of the discoverer."

75. See D. T. Whiteside, "Newton's Lunar Theory: From High Hope to Disenchantment," *Vistas in Astronomy* 19 (1975): 317–28.

76. See, inter alia, Philip P. Chandler II, "Newton and Clairaut on the Motion of the Lunar Apse," (unpublished doctoral dissertation, University of California, San Diego, 1975).

77. See n. 73 above.

78. F. R. Moulton, *Introduction to Celestial Mechanics* (New York: The Macmillan Co., 1914, 1962), p. 337.

79. For details see our edition of Newton's *Principia* with variant readings (n. 8 above) and my *Introduction* (n. 12 above).

80. Fatio's notes for a projected edition of Newton's *Principia* are to be found in the library of the Royal Society.

81. On the "mechanical philosophy," see R. S. Westfall, *Force in Newton's Physics* (London: Macdonald; New York: American Elsevier, 1971), pp. 81–88, 213–14, 264–65, 327–29, 331–32.

82. See A. Koyré, *Newtonian Studies* (Cambridge: Harvard University Press, 1965), ch. 3, appendixes A, B, C.

83. Newton's *Correspondence*, 3:254. On this exchange of letters see Newton's *Papers* (n. 46 above), §IV.

84. In Book 3, Newton explored the orbits of comets, having concluded that comets are a "sort" of planet. He devoted most of his attention to parabolic orbits, observing that in the region of the solar system around the sun (where the planet is visible) the difference between a parabolic and an elliptic shape of the orbit is very small.

85. I have listed the variations in Newton's concept of the aether in my "The *Principia,* Universal Gravitation, and the 'Newtonian Style,' " in Zev Bechler, ed., *Contemporary Newtonian Research* (Dordrecht, Boston, London: D. Reidel Publishing Company, 1982), pp. 21–108.

86. See n. 35 above.

87. E. Husserl, *The Crisis of European Science and Transcendental Phenomenology: An Introduction to Phenomenological Philosophy,* trans. David Carr (Evanston: Northwestern University Press, 1970), pp. 23–59. See my *The Newtonian Revolution* (cited in n. 6 above), pp. 132–33.

88. U.L.C. MS Add. 3968.9, fol. 109r/109v; see D. T. Whiteside's version in *Mathematical Papers of Isaac Newton,* 8:452–57.

89. In Book 1, Section 11, of the *Principia,* Newton declares that the propositions in Book 1 are "purely mathematical" and that he has put aside "all physical considerations."

90. *Principia,* Book 3, Introduction.

91. Ibid.

92. If this were not the case, Newton would have had to introduce considerations based on size and shape. For instance, in the next stage he will have the circulating "body" and the central "body" attract each other. If these bodies had to have physical dimensions, for example, be physical spheres, then Newton would have had to deal with the rather difficult question (addressed only much later) of the attraction of a sphere and so on.

93. Newton to Bentley, 15 Feb. 1692/93. See Newton's *Papers & Letters* (n. 83 above), pp. 302–03 and Newton's *Correspondence,* 3:254.

94. Alexandre Koyré had sketched for us the criticism of Newton for specifically having used the term *attraction* with its many overtones; see n. 82 above.

95. *Principia,* Book 1, Section 11, concluding scholium.

96. See n. 71 above.

97. *Principia,* Book 1, Sections 12, 13.

98. See my *Newtonian Revolution* (n. 6 above), p. 80.

99. *Celestial Mechanics,* trans. Nathaniel Bowditch, vol. 1 (Boston, 1829; reprinted., New York: Chelsea Publishing Company, 1966), Preface by the Author, p. xxiii.

"The Unity of Truth": An Integrated View of Newton's Work

BETTY JO TEETER DOBBS

Introduction

The search by this investigator for an integrated view of Newton's work began with one very questionable advantage—questionably advantageous in the opinion of most scholars, that is, because the starting point was Newton's alchemy. Newton's alchemy had almost always been considered the most peripheral of his many studies, the one furthest removed from his important work in mathematics, optics, and celestial dynamics. Most students of Newton's work preferred to ignore the alchemy, or, if not to ignore it, then to explain it away as far as possible. But one may, perhaps pardonably, remain unconvinced that a mind of the caliber of Newton's would have lavished so much attention upon any topic without a serious expectation of learning something significant from his study of it. Indeed, working one's way through Newton's alchemical papers, one becomes increasingly aware of the meticulous scholarship and the careful quantitative experimentation Newton had devoted to alchemical questions over a period of many years. Clearly, *he* thought his alchemical work was important. So one is forced to question what it meant to him: if Newton thought alchemy was an important part of his life's work, then what was that life's work? Was it possible that Newton had a unity of purpose, an overarching goal, that encompassed *all* of his various fields of study?

Blinded by the brilliance of the laws of motion, the laws of optics, the concept of universal gravitation, the methodological success, scholars have seldom wondered whether the discovery of the laws of nature was all Newton had in mind. Both Enlightenment and

From Betty Jo Teeter Dobbs, *The Janus Faces of Genius: The Role of Alchemy in Newton's Thought* (New York and London: Cambridge University Press, 1991). Reprinted by permission.

post-Enlightenment Newtonians have often missed the religious nature of Newton's ultimate quest and have taken the stunningly successful byproducts for his primary goal. But Newton wished to look through nature to see God, and it was not false modesty when in old age he said he had been only like a boy at the seashore picking up now and again a smoother pebble or a prettier shell than usual while the great ocean of truth lay all undiscovered before him.[1]

Newton's quest was immeasurably large; it generated questions starkly different from those of modern science. For him, the most important questions were never answered, but in reconstructing them lies the best chance of grasping the singleness of Newton's mind and the unique thrust of his genius. His questions were not ours. They encompassed fields of knowledge that seem to modern scholars to have no relevant points of contact. But the evidence for the unity of Newton's thought emerges when his alchemical papers are considered in conjunction with his other manuscript remains and with his published works. The same may be said for the changes in his explanatory "mechanisms" over the long decades of his search. There was consistency in Newton's thinking, and a demonstrable pattern in its changes. The consistency lay in his overwhelming religious concern to establish the manner of God's acting in the world. The pattern of change resulted from his slow fusion and selective disentanglement of systems of thought that were often essentially antithetical to each other: Neoplatonism, Cartesian mechanical philosophy, Stoicism; chemistry, alchemy, atomism; biblical, patristic, and pagan religions.

It is the purpose of this paper to argue that Newton did have a unity of purpose and that his overarching goal did indeed encompass all his various fields of study. One can hardly expect to do justice so briefly to a complex and very active career that spanned more than six decades. But one may approach the problem first by a discussion of Newton's methodology, then by a brief review of the significance of his alchemical studies. Finally, with the example of Newton's changing ideas on the concept of gravity, one may attempt to evaluate his success—or lack of it—in achieving his goal.

Newton's Methodology

In certain ways Newton's intellectual development is best understood as a product of the late Renaissance. Thanks to the revival of ancient thought, to humanism, to the Reformation, to developments in medicine/science/natural philosophy prior to or contemporary with his early period of most intense study (1660–84), he had access to an unusually large number of systems of thought. Each system had its own set of guiding assumptions, so in that particular historical milieu some comparative judgment between and among competing systems was perhaps inevitable. But such judgments were difficult to make without a culturally conditioned consensus on standards of evaluation, which was precisely what was lacking. The formalized skepticism of Pyrrhonism had been revived along with other aspects of antiquity, but in addition one may trace an increase in a less formal but rather generalized skepticism at least from the beginning of the sixteenth century, as competing systems laid claim to truth and denied the claims of their rivals. As a consequence, Western Europe underwent something of an epistemological crisis in the sixteenth and seventeenth centuries. Among so many competing systems, how was one to achieve certainty? Could the human being attain truth?[2]

Newton, however, was not a skeptic. As Quinn has suggested, Newton may profitably be compared to such twentieth-century thinkers as G. E. Moore and Bertrand Russell, each intent on truth, none skeptical of human capacity to obtain it. In fact each expected to save humanity from skepticism and usher in a millennium.[3]

To save humanity from skepticism was the ambition of many a thinker in the seventeenth century. Quinn has reported a serious conversation between Descartes and John Dury in which it was agreed that the emergence of skepticism constituted the profound crisis of their period and that a way needed to be found to counter it with epistemological certainty. Descartes chose mathematics, Dury the interpretation of biblical prophecy, as the most promising response to the crisis.[4] The point, of course, is that no one knew then what would ultimately be established as effective. A modern scientist might be inclined to think Descartes chose the better part, but in fact the natural philosophy that Descartes claimed to have established with mathematical certainty was soon overthrown by Newton's. Descartes's mathematico-deductive method was not adequately balanced by experiment, observation, and induction; Newton's was. Perhaps the most important element in Newton's methodological contribution was that of balance, for no *single* approach ever proved to be effective in settling the epistemological crisis of the Renaissance and early modern period.

Newton had perhaps been convinced of the necessity of methodological balance by Henry More, who had worked out such a procedure within the context of the interpretation of prophecy.[5]

Since every single approach to knowledge was subject to error, a more certain knowledge was to be obtained by utilizing each approach to correct the others: the senses to be rectified by reason, reason to be rectified by revelation, and so forth. The self-correcting character of Newton's procedure is entirely similar to More's and constitutes the superiority of Newton's method over that of earlier natural philosophers, for others had certainly used the separate elements of reason, mathematics, experiment, and observation before him.

But Newton's method was not limited to the balancing of those approaches to knowledge that still constitute the elements of modern scientific methodology. Nor is there any reason to assume that he would deliberately have limited himself to those familiar approaches even if he had been prescient enough to realize that those were all the future would consider important. Newton's mind was equipped with a certain fundamental assumption that was fairly common in his time and indeed constituted one answer to the problem of skepticism: the assumption of the unity of truth.[6] Not only did Newton respect the idea that truth was accessible to the human mind, but also he was very much inclined to accord to several systems of thought the right to claim access to some aspect of the truth. For Newton then the many competing systems he encountered tended to appear complementary rather than competitive. The mechanical philosophy that has so often been seen as the necessary prelude to the Newtonian revolution probably did not hold a more privileged or dominant position in Newton's mind than did any other system. The mechanical philosophy was one system among many that Newton thought to be capable of yielding at least a partial truth.

For Newton, true knowledge was all in some sense a knowledge of God.[7] Truth was one, its unity guaranteed by the unity of God. Reason and revelation were not in conflict but were supplementary. God's attributes were recorded in the written Word, but were also directly reflected in the nature of nature. Natural philosophy thus had immediate theological meaning for Newton, and he deemed it capable of revealing to him those aspects of the divine never recorded in the Bible or the record of which had been corrupted by time and human error. By whatever route one approached truth, the goal was the same. Experimental discovery and revelation; the productions of reason, speculation, or mathematics; the cryptic, coded messages of the ancients in myth, prophecy, or alchemical tract; the philosophies of Platonists, atomists, or Stoics—all, if correctly interpreted, found their reconciliation in the ultimate unity and majesty of the Deity.

In Newton's conviction of the unity of truth and its ultimate source in the divine, one may locate the fountainhead of all his diverse studies. His goal was never the relatively narrow one of the modern scientist, i.e., a knowledge of nature, but was rather the creation of a system that included not only natural principles but divine ones as well. Newton wished to penetrate to the divine principles beyond the veil of nature, and beyond the veil of human record and received revelation as well. His goal was a unified truth—ultimately a knowledge of God and all His works—and for achieving that goal Newton marshaled the evidence from every source available to him: mathematics, observation, reason, revelation, historical record, myth, the tattered remnants of ancient wisdom. With the post-Newtonian diminution of interest in divinity and heightened interest in nature for its own sake, scholars have too often read the Newton method narrowly, selecting from the breadth of his studies only mathematics, experiment, observation, and reason as the essential components of his scientific method. For a science of nature, a balanced use of those approaches to knowledge suffices, or so it has come to seem since Newton's death, and one result of the restricted interests of modernity has been to look askance at Newton's biblical, theological, chronological, and alchemical studies; to consider his pursuit of the *prisca sapientia* as irrelevant. None of those was irrelevant to Newton, for his goal was considerably more ambitious than a knowledge of nature. His goal was truth, and for that he utilized every possible resource. Can one deny that his willingness to consider alternative systems of all sorts contributed to the flexibility and creativity he evinced in his still admired "scientific" work?

Thus Newton's method was not limited to the balancing of the elements of modern scientific methodology. Newton's balancing procedure included also the knowledge he had garnered from theology, revelation, alchemy, history, and the wise ancients. It has been difficult to establish this fact because Newton's papers largely reflect a single-minded pursuit of each and every one of his diverse studies, as if in each one lay the only road to knowledge. When he wrote alchemy, he wrote as an alchemist, as Sherwood Taylor long ago observed.[8] But when he wrote chemistry, his concepts conformed to those of contemporary chemists.[9] When he wrote mathematics, no one doubted him to be a pure mathematician.[10] When he adopted the mechanical philosophy, he devised hypothetical aethereal mechanisms with the best.[11] When he undertook to interpret prophecy, his attention to detail implied that nothing else mattered.[12] In only a few of his papers may one observe his attempt to balance one apparently isolated line of investigation with another.

There are two conclusions to be drawn from these observations on Newton's methodology. One is that, so far as one can tell from the manuscripts he left behind, Newton never rejected the guiding assumptions behind any system of thought a priori. He seems to have withheld judgment on the validity of every one of the several approaches to knowledge that he pursued, at least at first, and often for a considerable period of time, assuming that there was some portion of the truth to be obtained by following that avenue to the end. The other conclusion to be drawn is that, inevitably, his various lines of investigation came into conflict at some points. When that happened, he was forced to make judgments, for, according to the doctrine of the unity of truth, partial truths should be reconcilable. If they are not, it indicates that errors have crept in and must be rectified by bringing the evidence from yet another line of investigation to bear upon the problem. At that point Newton did his most creative synthetic work and was most apt to challenge and reject the guiding assumptions behind the putative error he had discovered. But even when rejecting some guiding assumptions, he was apt to be quite selective and specific about what he discarded and to draw the line well short of a rejection of the entire system.

The Significance of Newton's Alchemy

One may begin to comprehend the centrality of Newton's alchemical studies to his life's work by considering some of the many different ways in which alchemy addressed his other concerns. Alchemy offered Newton access to spiritualistic guiding principles in the natural world and provided hints for the experimental examination of them and of non-mechanical causation in general. An effective demonstration of such principles at work in the organization of matter would have been a demonstration of divine activity in nature. Through the argument from design, a proof of divine interaction with the material world would not only have been an experimental proof of the existence of God (and thus a resounding blow struck against both skepticism and atheism) but also a decisive refutation of the materialistic and deistic implications of Cartesian mechanism that had so alarmed the Cambridge Platonists.

Of the approximately one million words on alchemy that Newton left in manuscript, some of the most enlightening are in a set of papers now belonging to the Smithsonian Institution and designated as Dibner MSS 1031 B.[13] Newton left the manuscript untitled and undated, but he probably wrote it about 1672, when he had been studying alchemy for some four years, and its incipit makes an appropriate title—"Of Natures obvious laws & processes in vegetation." Newton concentrated in this manuscript on the vegetation of metals, but he was convinced that the processes of vegetation were similar in all three kingdoms of nature: the animal kingdom, the vegetable kingdom, and the mineral kingdom. He said that metals vegetate after the same laws as those observed in the vegetation of plants and animals, and that vegetation is the effect of a latent spirit that is the same in all things.[14] The manuscript continues through twelve closely written pages, with many corrections and alterations as Newton pondered the issues involved in nature's obvious laws and processes in vegetation, and especially in the vegetation of metals—which is what one usually calls alchemy.

The problems that Newton hoped to solve through his study of alchemy were especially intense in the seventeenth century. The mechanical philosophy had been established earlier in the century by Descartes and other philosophers and by Newton's time was widely accepted by the community of natural philosophers as the most promising way of approaching the study of the natural world. The mechanical philosophy was predicated upon the existence of atoms or corpuscles—discrete particles of matter—that were in constant motion. All events in the natural world were to be explained by matter and motion, by the tiny particles of a passive, inert matter that transferred motion by pressure on, or impact with, other small particles. But because of the general passivity of matter in the mechanical philosophy, certain problems arose regarding cohesion and life. How can passive little billiard balls of matter cohere and stick together in organized forms? How can those tiny passive particles of a common universal matter combine to produce the immense variety of living forms to be found in the world? The cohesion and differentiation of living forms seems intuitively to be qualitatively different from anything that the mechanical motion of small particles of matter might produce.[15]

It was Newton's conviction that the passive particles of matter could not organize themselves into living forms. Their organization required divine guidance, the latent spirit that he said was present in all things, a vitalistic and active spirit that could guide the particles of passive matter into all the beautiful forms of plants and animals and minerals that God had ordained. It was a vital agent, or a fermental virtue as he sometimes called it, or, in the Smithsonian manuscript, the vegetable spirit.[16] Since alchemical literature claimed to provide information regarding that spirit of life, and to offer laboratory techniques for its isolation and manipulation, it is small wonder that Newton pursued his study of alchemy so indefatigably. One must insist, however, that he never saw alchemy and the alchemical worldview as *substitutes* for the mechanical philosophy and the mechanical worldview. On the contrary, Newton expected alchemy to *supplement* the mechanical philosophy in areas where mechanisms failed to provide satisfactory explications of the natural world.

In addition to regarding the vegetable spirit as a much needed *natural* explication for the peculiarities of life in the three kingdoms of nature, Newton was confident that the vital agent was in some sense divine, for it was so defined against passive matter as an "active" principle, in the time-hallowed traditions of both Platonism and Stoicism.[17] Inspired by his interest in this divine active principle, he sought in alchemy the source of all the apparently spontaneous processes of fermentation, putrefaction, generation, and vegetation—that is, everything associated with normal life and growth, such as digestion and assimilation, for the word "vegetation" was originally from the Latin *vegetare,* to animate or enliven. Such processes produced the endless variety of living forms and simply could not be explained by the mechanical action of inert corpuscles. Mechanical action could certainly account for many classes of phenomena, but it could never account for the process of assimilation, in which food stuffs are turned into the bodies of different animals, vegetables, and minerals, Newton said.[18] Nor could it account for the sheer variety of forms in this world, all of which has somehow sprung from a common lifeless matter.[19] An effective demonstration of the operation of a divine active principle in the natural phenomena of life would constitute "scientific" evidence for the existence of a Deity.

Such a Deity, furthermore, would have been an intimately involved providential one, one who noted falling sparrows and numbered hairs in traditional Judaeo-Christian fashion, and not a remote, distant, deistic one. One root of Newton's extreme interest in alchemy probably stemmed from the critique of Descartes's version of the mechanical philosophy that had been promulgated by the Cambridge Platonists, and especially by Henry More. Since Descartes had insisted that the only acceptable natural explanations were constituted by matter and motion, it soon became apparent to More that Descartes's universe was a closed mechanical system in which the Deity had no means for exercising His providential care, that Cartesianism had inherent materialistic, deistic, even atheistic tendencies. More argued for the reinstatement of spiritual, nonmechanical principles in the natural world, and just such a program was adopted by Isaac Newton in his alchemical studies.[20] So the demonstration of the operations of the vegetable spirit would not only have provided a "scientific" corrective for the mechanical philosophy but would have solved also the theological problem posed by Newton's Cartesian heritage.[21]

Gravity: A Test Case

One may now turn to Newton's conceptualization of gravity as a case in point, in particular to what he thought to be the "cause" of gravity. Clearly Newton had set himself a gigantic task in his effort to reconcile the partial truths from his many lines of investigation, and in the case of gravity one finds a very lengthy struggle on Newton's part to "get it right." Some of his efforts were much more successful than others, as one might expect.

For about twenty years Newton appears to have been a fairly orthodox mechanical philosopher on the matter of gravity. Although some Newton scholars would disagree with that interpretation, there exists a lengthy series of papers in which Newton devised various speculative systems that involved dense, plenistic, Cartesian-type aethers that "caused" gravitational effects. Part of the evidence for this interpretation is fresh material from Newton's alchemical papers, which, taken in conjunction with much better known material from other classes of Newton's papers, makes a strong case for Newton's acceptance of some type of mechanical gravitational aether for about twenty years: from the mid-1660s until late 1684 or early 1684/85. There were variations in his proposed systems, but there was also a strong continuity in that all of them involved the impact of pressure of fine particles of matter, imperceptible to the senses but acting mechanically as the material "cause" of gravity.[22]

In 1684, however, when Newton derived mathematically the general area law for bodies revolving around a center of force (Theorem 1 in the various versions of *De motu* and then Proposition 1, Theorem 1 of Book 1 of the *Principia*), he found that his mathematical demonstration matched Kepler's area law rather precisely.[23] The area law had first emerged in Kepler's work as an "approximation," based on an Aristotelian doctrine that made force proportional to velocity, though Kepler later realized that it was a deductive consequence of his hypothesis that an immaterial virtue emanating from the sun constituted the physical cause of planetary motion about the sun. In the Keplerian hypothesis the component of a planet's velocity at right angles to the radius vector was inversely proportional to the distance from the sun, so that the planet moved most rapidly at perihelion and most slowly at aphelion, which in turn implies the area law in which the radius vector sweeps out equal areas in equal times no matter where the planet is in its orbit. Combined with Kepler's even later discovery of the elliptical orbit for Mars, the area law received empirical verification (in that combined form), though for technical reasons the area law was little used by post-Keplerian astronomers, even by those who accepted elliptical orbits.[24] Newton's new dynamical analysis, which included an inertial component, rendered the Aristotelian and Keplerian analyses obsolete; nevertheless, this convergence of mathematical and observational lines of investigation must have been immensely satisfying to Newton, for the verification of the first two Keplerian laws had been effected with the very best observational data assembled by Tycho Brahe.

But however satisfactory that may have been, within the context of the mechanical philosophy a problem immediately arose. The Keplerian area law that matched empirical observation should *not* fit so closely with the exact area law derived mathematically by Newton if the heavens were filled with the aethereal medium postulated by the mechanical philosophers. Unless the medium is somehow disposed to move with exactly the same variable speed that the planetary body exhibits, the planet should encounter enough resistance from the medium to cause an observed deviation from the mathematical prediction, just as projectiles in the terrestrial atmosphere are observed to deviate from mathematical prediction. Newton undertook some new pendulum experiments to test aethereal resistance, found it to be nil or almost so, and briefly devised a new version of the gravitational aether that was a "nonresisting" medium. But very shortly thereafter, probably early in 1684/85, he rejected the plenistic and corporeal nature of the medium "utterly and completely" and opted for the idea "that by far the largest part of the aetherial space is void, scattered between the aetherial particles."[25] The result was that Newton made a dramatic break with impact physics and with material, mechanical causation for gravity.

The question sometimes arises as to why Newton was willing to abandon aethereal vortex explanations for celestial motions when many of his brightest contemporaries were not.[26] It was probably his voluntarist theology that enabled Newton to be so flexible.[27] As a voluntarist, Newton was convinced that God's absolute will was His primary attribute and that He could will to do anything that did not involve a contradiction. Man's puny mind is not the measure of what God may will to do. So although mechanically acting aethereal vortices and impact physics might well appear to be the only reasonable and rational explanation of celestial motions (as those concepts seemed to Leibniz and Huygens), Newton's voluntarist stance insisted upon God's unlimited freedom to order the heavens by any means whatsoever, even by non-mechanical means. Then also one must include Newton's alchemical studies in his mental repertory, especially with respect to nonmechanical causation. In fact, in the Smithsonian alchemical manuscript of the early 1670s, one finds a succinct restatement by Newton of voluntarist theology in the context of arguments for nonmechanical causal principles in "vegetation."[28] The "reasonableness" and "rationality" of mechanical principles fell, one may suggest, before the combined onslaught of voluntarism, mathematics, observation, and Newton's conviction that nonmechanical causal principles operated in "vegetable chemistry" or alchemy.

Having concluded that the largest part of the cosmic aether was empty space and could not serve as a mechanical "cause" for gravity as he had once thought, Newton published in the *Principia* the "mathematical principles" of gravity only—"not defining in this treatise the species or physical qualities of forces, but investigating the quantities and mathematical proportions of them."

> I here use the word *attraction* in general for any endeavor whatever, made by bodies to approach to each other, whether that endeavor arise from the action of the bodies themselves, as tending to each other or agitating each other by spirits emitted; or whether it arises from the action of the ether or of the air, or of any medium whatever, whether corporeal or incorporeal, in any manner impelling bodies placed therein towards each other. In the same general sense I use the word impulse.[29]

Whether a medium "corporeal or incorporeal," whether bodily contact of the aether or the air or "spirits emitted," might be a "cause" of gravity he would not then say. But in the post-*Principia* period Newton struggled on with each of the options still open to him. The corporeal medium, involving bodily contact with air particles or the extra fine corpuscles of aether, had been excluded and was never to return. But there remained as possibilities the "incorporeal medium" and the "spirits emitted," the latter of which Newton apparently conceived as quasi-material inhabitants of the grey area between the complete incorporeality of God and the full solidity of body. The search for a "cause" for gravity was not over.

The possibility of the "incorporeal medium" as the "cause" of gravity comprised Newton's first tentative solution. Even in his student notebook of the 1660s Newton had recognized the literal omnipresence of God, the omnipresence that comes to subsume universal gravity for a period of time after the *Principia*. Violent motion in a vacuum meets with no resistance, Newton had said in his notebook.

> It is true God is as far as vacuum extends, but he, being a spirit and penetrating all matter, can be no obstacle to the motion of matter; no more than if nothing were in its way.[30]

In conversation with David Gregory in 1705 Newton developed the notion that God was the cause of gravity, emphasizing His omnipresence. Newton had posed the question, *"What the space that is empty of body is filled with."* Gregory recorded that Newton believed God to be present in the literal sense, supposing "that as God is present in space where there is no body, he is present in space where a body is also present."

> But if this way of proposing this his notion be too bold, he thinks of doing it thus. *What Cause did the Ancients assign of Gravity.* He believes that they reckoned God the Cause of it, nothing els, that is no body being the cause; since every body is heavy.[31]

In the General Scholium to the second edition of the *Principia* (1713) Newton emphasized the penetrating quality of the "cause" of gravity.

> This is certain, that it [gravity] must proceed from a cause that penetrates to the very centres of the sun and planets . . . that operates not according to the quantity of the surfaces of the particles upon which it acts (as mechanical causes used to do).[32]

Glossed with the passage from the student notebook, the penetrating "cause" of the General Scholium likewise becomes the omnipresent Deity.

Many other examples might be drawn from his unpublished papers and published works to demonstrate that, when the mechanical aether failed him, Newton had alternative sets of assumptions ready to hand that provided a nonmechanical divine "cause" for gravity.[33] But the assumption of the "incorporeal medium"—an omnipresent, all-penetrating Deity as the "cause" for gravity—conflicted with another basic assumption fundamental to Newton's theology, so it was not quite a satisfactory solution to his problem.

In addition to being a voluntarist, Newton was an Arian in theology, refusing to accord equal status to all three Persons of the orthodox Trinity. The Supreme Deity in Arianism was defined in an elevated and transcendental manner, whereas the Christ was considered to be a Being created in time and not a part of the Godhead from all eternity. One concomitant tenet of Arianism was that the Supreme God did not ordinarily interact directly with the world He had made but on the contrary utilized an agent (Christ) to put His will into effect in the world. Newton's treatment of the Supreme God as the "incorporeal medium" causing gravity, and so interacting with matter continuously and intimately, could not be fully reconciled with Arian belief. The transcendent Arian Deity should not in fact be responsible for the operation of gravity from moment to moment.[34]

Newton's later theological papers show him engaged in a search of scripture for information on the nature of Christ's body before and after the Incarnation.[35] Given Newton's belief in the Christ as God's cosmic agent in His interactions with the world, it is apparent that Newton hoped to find biblical evidence for a substance that was not composed of ordinary matter like everything else (and hence one that would not constitute a frictional medium) yet not wholly incorporeal either; a hybrid substance between spirit and matter that could serve as a "cause" of gravity.

The Bible did not of course offer much insight, but Hauksbee's electrical experiments did. Hauksbee's work provided evidence on the "spirits emitted" by bodies.[36] Those demonstrations gave Newton a glimpse of a different type of material substance operating in short-range phenomena, an electric "spirit" that he promptly began to utilize as a speculative explanation for life and "vegetation" in micromatter.[37] Newton also boldly extrapolated that microcosmic spirit to a macrocosmic one in the so-called "aether" queries added to the 1717/18 edition of the *Opticks,* and appearing in all subsequent editions, Queries 17–24. "And is it not (by its elastick force) expanded through all the Heavens?" he asked in Query 18.[38] Newton was probably in part responding to criticisms by continental philosophers on his lack of a physical substrate for gravity, but both the timing of his new aether and its nature can be more fully explained if Hauksbee's electrical experimentation and Newton's theological commitments are taken into account. It was the hybrid substance between corporeal body and incorporeal spirit for which he had been searching, the cosmic intermediary between God and matter.

The new elastic gravitational aether differed substantially from the old Cartesian type, to which Newton never returned. It was composed of active, elastic particles of a stuff that was not like ordinary matter. It did not act by impact, nor was it dense; there still remained in the new aether much space that was empty of body of any sort. It was Newton's second tentative solution to the problem of the "cause" of gravity, and it rested in the last editions of the Opticks in uneasy juxtaposition with his first solution of God's literal omnipresence.

Conclusion

Thus that Newton had several alternative sets of guiding assumptions available to him is readily apparent. If, as argued above, he treated all his lines of investigation as if they probably had some validity and would lead to at least a partial truth if followed consistently, he was presumably not very dogmatic about any of them and so was extraordinarily flexible, willing to try out different tentative solutions. His willingness to dispose of the mechanical aether as the "cause" for gravity while he was writing the Principia proved to be one of his greatest triumphs. But perhaps in the end his effort to find a satisfactory balance among so many different conflicting assumptions may have proved frustrating to him and impossible of achievement. So he never quite achieved his goal of establishing a complete system of the world that incorporated both natural and divine principles, and his "mathematical principles" came to stand alone.

Nevertheless, to return in conclusion to the central argument: Newton himself had a unity of purpose in his life's work. Newton engaged in the study of many topics that appear bizarre to the modern mind, but, given the stature of Newton's intelligence, it seems to be both good psychology and good history to assume that he knew what he was doing—or at least that he thought he knew what he was doing—in the context of his own time. There has been in the past an immense amount of confusion regarding Newton's attitude toward a "cause" for gravity, partly because of the misdating of one key manuscript (De gravitatione) but primarily because

Newton's pursuit of alchemy, theology, and other forms of ancient wisdom have been arbitrarily and artificially separated out from his "real" scientific interests by almost all post-Newtonian scholarship. Once the breadth of Newton's methodology and the scope of his ambition are grasped, however, the difficulties fade away. It is useful, and indeed necessary, to draw upon both his alchemical and his theological papers in order to explicate his developing thought on a concept central to his scientific work. Thus, only by assuming the unity of Newton's own purpose may one obtain an integrated and internally self-consistent view of the complex corpus of Newton's life's work.

Notes

1. Cf. David Brewster, *Memoirs of the Life, Writings, and Discoveries of Sir Isaac Newton,* 2 vols. (Edinburgh: Thomas Constable and Co.; Boston: Little, Brown, and Co., 1855), 2: 407–8.

2. Richard H. Popkin, *The History of Scepticism from Erasmus to Spinoza* (Berkeley: University of California Press, 1979); Charles G. Nauert, Jr., *Agrippa and the Crisis of Renaissance Thought,* Illinois Studies in the Social Sciences, no. 55 (Urbana: University of Illinois Press, 1965); Walter Pagel, *Paracelsus: An Introduction to Philosophical Medicine in the Era of the Renaissance* (Basel and New York: Karger, 1958); idem, *Joan Baptista van Helmont: Reformer of Science and Medicine* (Cambridge, London, New York, New Rochelle, Melbourne, and Sydney: Cambridge University Press, 1982); Paul Oskar Kristeller, *Renaissance Thought and Its Sources,* ed. Michael Mooney (New York: Columbia University Press, 1979); Ernst Cassirer, *The Platonic Renaissance in England,* trans. James P. Pettegrove (Austin: University of Texas Press, 1953).

3. Arthur Quinn, "On Reading Newton Apocalyptically," in *Millenarianism and Messianism in English Literature and Thought 1650–1800. Clark Library Lectures 1981–82,* ed. Richard H. Popkin, Publications from the Clark Library Professorship, University of California, Los Angeles, no. 10 (Leiden, New York, København, and Köln: E. J. Brill, 1988), pp. 176–92.

4. Ibid., p. 179.

5. Richard H. Popkin, "The Third Force in Seventeenth Century Philosophy: Scepticism, Science, and Biblical Prophecy," *Nouvelle République des Lettres* 1 (1983): 35–63.

6. For a general statement on the concept of the unity of truth in the Renaissance, see Kristeller, *Renaissance Thought* (n. 2), pp. 196–210; for similar views contemporary with Newton, see Arthur Quinn, *The Confidence of British Philosophers: An Essay in Historical Narrative* (Studies in the History of Christian Thought, vol. 17, edited by Heiko A. Oberman, in cooperation with Henry Chadwick, Edward A. Dowey, Jaroslav Pelikan, Brian Tierney, E. David Willis; Leiden: E. J. Brill, 1977), especially pp. 8–20. For numerous variants of the attempt to achieve "right reason" by balancing experience, reason, and revelation in the generation just prior to Newton's, see Lotte Mulligan, " 'Reason,' 'right reason,' and 'revelation' in mid-seventeenth-century England," in *Occult and Scientific Mentalities in the Renaissance,* ed. Brian Vickers (Cambridge, London, New

York, New Rochelle, Melbourne, and Sydney: Cambridge University Press, 1984), pp. 375–401.

7. For a sketch of the Christian background to Newton's position, see Seyyed Hossein Nasr, *Knowledge and the Sacred. The Gifford Lectures, 1981* (New York: Crossroad, 1981), pp. 4–6, 12–29; for a sweeping explication of traditional belief in "the cosmos as theophany," a tradition to which Newton adhered, see ibid., pp. 189–220.

8. Frank Sherwood Taylor, "An Alchemical Work of Sir Isaac Newton," *Ambix* 5 (1956): 59–84.

9. B. J. T. Dobbs, "Conceptual Problems in Newton's Early Chemistry: A Preliminary Study," in *Religion, Science, and Worldview: Essays in Honor of Richard S. Westfall* (Cambridge, London, New York, New Rochelle, Melbourne, and Sydney: Cambridge University Press, 1985), pp. 3–32.

10. Derek T. Whiteside, "Isaac Newton: Birth of a Mathematician," *Notes and Records of the Royal Society of London* 19 (1964): 53–62.

11. J. E. McGuire and Martin Tamny, *Certain Philosophical Questions: Newton's Trinity Notebook* (Cambridge, London, New York, New Rochelle, Melbourne, and Sydney: Cambridge University Press, 1983).

12. See, for example, Newton's notes in his copy of Henry More's *A Plain and Continued Exposition of the several Prophecies or Divine Visions of the Prophet Daniel, Which have or may concern the People of God, whether Jew or Christian; Whereunto is annexed a Threefold Appendage, Touching Three main Points, the First, Relating to Daniel, the other Two to the Apocalypse* (London: Printed by *M. F.* for *Walter Kettilby* at the *Bishop's Head* in Saint *Paul's* Church-Yard, 1681), now BS 1556 M 67 P 5 copy 2, Bancroft Library, University of California at Berkeley.

13. Dibner MSS 1031 B, Dibner Library of the History of Science and Technology, Special Collections Branch, Smithsonian Institution Libraries, Smithsonian Institution, Washington, D.C. The history of this manuscript is traced in B. J. T. Dobbs, *Alchemical Death & Resurrection: The Significance of Alchemy in the Age of Newton. A Lecture Sponsored by the Smithsonian Institution Libraries in Conjunction with the Washington Colloquium for the Humanities Lecture Series: Death and the Afterlife in Art and Literature. Presented at the Smithsonian Institution, February 16, 1988* (Washington, D. C.: Smithsonian Institution Libraries, 1990) nn. 4 and 5. The entire manuscript is also reproduced in facsimile in an appendix to that volume.

14. Dibner MSS 1031 B (n. 13), fol. 1r.

15. B. J. T. Dobbs, *The Foundations of Newton's Alchemy, or "The Hunting of the Greene Lyon"* (Cambridge, London, New York, and Melbourne: Cambridge University Press, 1975), pp. 100–105; idem, "Newton's Alchemy and His Theory of Matter," *Isis* 73 (1982): 511–28.

16. Dibner MSS 1031 B (n. 13), fol. 4r. See also, Dobbs, "Newton's Alchemy and His Theory of Matter" (n. 15).

17. B. J. T. Dobbs, "Newton's Alchemy and His 'Active Principle' of Gravitation," in *Newton's Scientific and Philosophical Legacy,* ed. P. B. Scheuer and G. Debrock, International Archives of the History of Ideas, no. 123 (Dordrecht, Boston, and London: Kluwer Academic Publishers, 1988), pp. 55–80; idem, "Newton and Stoicism," in *Spindel Conference 1984: Recovering the Stoics, The Southern Journal of Philosophy* 23 Supplement (1985): 109–23; idem, "Newton's Alchemy and His Theory of Matter" (n. 15).

18. Dibner MSS 1031 B (n. 13), fol. 5v–6r.

19. Ibid.

20. See the references in n. 15 and the literature cited in them.

21. For a more extended discussion of the theological nexus of Newton's alchemical work, see B. J. T. Dobbs, "Newton as Alchemist and Theologian," in *Standing on the Shoulders of Giants, A Longer View of Newton and Halley: Essays Commemorating the Tercentenary of Newton's Principia and the 1985–86 Return of Comet Halley,* ed. Norman J. W. Thrower (Berkeley, Los Angeles, and Oxford: University of California Press, 1990), pp. 128–40.

22. Isaac Newton, "Of Gravity & Levity," in McGuire and Tamny, *Certain Philosophical Questions* (n. 11), pp. 362–65, 426–31 (written in the mid-1660s); idem, "Of Natures obvious laws & processes in vegetation," Dibner MSS 1031 B (n. 13), fol. 3v–4r (written ca. 1672); idem, "An Hypothesis explaining the Properties of Light, discoursed of in my severall Papers," in *The Correspondence of Isaac Newton,* ed. H. W. Turnbull, J. P. Scott, A. R. Hall, and Laura Tilling, 7 vols. (Cambridge: Published for the Royal Society at the University Press, 1959–77), 1:362–82 (Newton to Oldenberg, 7 Dec. 1675); idem, two different systems in a letter to Boyle: *Correspondence,* 2:288–96 (Newton to Boyle, 28 Feb. 1678/79); idem, use of gravitational vortices in correspondence with Burnet: *Correspondence,* 2:319–34 (Newton to Burnet, 24 Dec. 1680; Burnet to Newton, 13 Jan. 1680/81; Newton to Burnet, ? Jan. 1680/81); idem, use of gravitational vortices in correspondence with Flamsteed: *Correspondence,* 2:340–56, 358–67 (Newton to Crompton for Flamsteed, 28 Feb. 1680/81; Flamsteed to Crompton for Newton, 7 March 1680/81; Newton to [?Crompton] ? April 1681; Newton to Flamsteed, 16 April 1681); idem, reference to the material fluid of the heavens that carried the planets in their courses, in University Library, Cambridge, Add. MS 3964.14, fol. 613r, cited in Derek T. Whiteside, "Before the *Principia:* The Maturing of Newton's Thoughts on Dynamical Astronomy 1664–1684," *Journal for the History of Astronomy* 1 (1970): 5–19, especially pp. 14 and 18, n. 42 (dated by Whiteside ca. 1682).

23. B. J. T. Dobbs, "Newton's Rejection of a Mechanical Æther for Gravitation: Empirical Difficulties and Guiding Assumptions," in *Scrutinizing Science: Empirical Studies of Scientific Change,* ed. Arthur Donovan, Larry Laudan, and Rachel Laudan, Synthese Library Studies in Epistemology, Logic, Methodology, and Philosophy of Science, vol. 193 (Dordrecht, Boston, and London: Kluwer Academic Publishers, 1988), pp. 69–83.

24. Curtis Wilson, "Kepler's Derivation of the Elliptical Path," *Isis* 59 (1968): 5–25; idem, "How Did Kepler Discover His First Two Laws?" *Scientific American* 226, no. 3 (1972): 93–106; idem, personal communication, 30 Jan. 1989.

25. Dobbs, "Newton's Rejection of a Mechanical Æther" (n. 23). The quotations are from *De Gravitatione et aequipondio fluidorum et solidorum in fluidis scientiam,* published in Isaac Newton, *Unpublished Scientific Papers of Isaac Newton, A Selection from the Portsmouth Collection in the University Library, Cambridge,* chosen, ed., and trans. A. Rupert Hall and Marie Boas Hall (Cambridge: Cambridge University Press, 1962), pp. 112–13 and 146–47 (Halls' translation). The Halls suggested a date of 1664–68 for the manuscript, and subsequent scholars have tended to follow the Halls' dating with relatively minor variation, no one suggesting a date as late as 1684 or 1684/85. My reasons for redating it to the later period are given at length in *The Janus Faces of Genius.*

26. Brian S. Baigrie, "The Vortex Theory of Motion 1687–1713: Empirical Difficulties and Guiding Assumptions," *Scrutinizing Science: Empirical Studies of Scientific Change* (n. 23), pp. 85–102.

27. J. E. McGuire, "Force, Active Principles, and Newton's Invisible Realm," *Ambix* 15 (1968): 154–208.

28. Dibner MSS 1031 B (n. 13), fol. 4v.

29. Isaac Newton, *Sir Isaac Newton's Mathematical Principles of Natural Philosophy and His System of the World* (1729, trans. Andrew Motte), ed. Florian Cajori (reprint of the 1934 ed.; 2 vols.; Berkeley and Los Angeles: University of California Press, 1962), 1:192; *Isaac Newton's Philosophiae naturalis principia mathematica. The Third Edition (1726) with Variant Readings,* assembled and ed. by Alexandre Koyré and I. Bernard Cohen with the assistance of Anne Whitman, 2 vols. (Cambridge: Harvard University Press, 1972), 1:298.

30. McGuire and Tamny, *Certain Philosophical Questions* (n. 11), p. 409, original English on p. 408.

31. *David Gregory, Isaac Newton and Their Circle: Extracts from David Gregory's Memoranda 1677–1708,* ed. Walter George Hiscock (Oxford: Oxford University Press for the Author, 1937), p. 30.

32. Newton, *Principia* (n. 29): (Motte-Cajori), 2:546; (Koyré-Cohen), 2:764.

33. Stoic or neo-Stoic doctrines probably provided some alternatives. Cf. Dobbs, "Newton and Stoicism" (n. 17); idem, "Newton's Alchemy and His 'Active Principle' of Gravitation" (n. 17).

34. Cf. Dobbs, "Newton's Alchemy and His Theory of Matter" (n. 15).

35. See, for example, Jewish National and University Library, Jerusalem, Yahuda MS Var. 1, Newton MSS 15.3, fol. 66v and 15.5, fol. 96r.

36. Henry Guerlac, "Francis Hauksbee: Expérimentateur au Profit de Newton," in Henry Guerlac, *Essays and Papers in the History of Modern Science* (Baltimore and London: The Johns Hopkins University Press, 1977), pp. 107–19 (reprinted from *Archives Internationale d'Histoire des Sciences* 16 [1963]: 1113–28); J. L. Heilbron, *Physics at the Royal Society during Newton's Presidency* (Los Angeles: William Andrews Clark Memorial Library, University of California, Los Angeles, 1983).

37. Cf. Dobbs, "Newton's Alchemy and His Theory of Matter" (n. 15), especially pp. 525–26 and the literature cited there. In his own interleaved and annotated copy of the second edition of the *Principia* Newton described this spirit as "electrici & elastici," a description that eventually found its way by an incompletely known route into the standard modern English translation of the *Principia:* I. Bernard Cohen, *Introduction to Newton's 'Principia'* (Cambridge: Harvard University Press, 1971), pp. 26–27, and Newton, *Principia* (n. 29): (Motte-Cajori), 2:547; (Koyré-Cohen) 2:765.

38. Isaac Newton, *Opticks, or a Treatise of the Reflections, Refractions, Inflections & Colours of Light* (based on the 4th London ed. of 1730; New York: Dover, 1952), pp. 347–54, quotation from p. 349.

Newton's Chemical Experiments: An Analysis in the Light of Modern Chemistry

PETER SPARGO

Although Newton's intense and extended interest in alchemy and its voluminous literature is well known, and has provoked considerable comment and debate among Newtonian scholars, the fact that he carried out an extensive series of several hundred chemical experiments, some of which were complex and perhaps unique, is even today much less well known and has been the subject of comparatively little study and discussion.

Newton's practical laboratory work in chemistry was only accorded the most superficial treatment by his first two major biographers, Sir David Brewster[1] and Louis Trenchard More.[2] This is not difficult to understand as both were professional physicists without the technical competence (or, one suspects, the inclination!) to assess either the scientific importance or the historical significance of Newton's work in this area. In the case of Brewster one can feel some sympathy, for in having only a short period in which to work through the huge mass of Newton's papers, then stored intact at Hurstbourne Park in Hampshire, it is understandable that the relatively slim records of Newton's excursion into practical chemistry should have gone virtually unnoticed among the torrent of manuscripts on theology, alchemy, chronology, and mathematics. More, however, not only had much more time at his disposal, for by the early 1930s the scientific portion of Newton's papers, including the material describing the chemical experiments, had been deposited by the Earl of Portsmouth in the University Library, Cambridge, and therefore could be examined at leisure, but he also had the 1888 catalogue of Newton's papers in which the section on the chemical experiments, prepared by G. D. Liveing of Cambridge, was well done and which should have alerted More to the importance of Newton's work in this area.[3]

The first serious study of Newton's chemical experiments in terms of their chemical content and significance was carried out more than twenty years later by Marie Boas and A. Rupert Hall, the former of whom was a trained chemist.[4] Published in 1958 this pioneer paper—which the authors freely admit only reflected a preliminary study of Newton's chemical experiments—highlighted some of the numerous practical problems that a study of the experiments entails, e.g., materials, processes, and apparatus.

The Halls' work was followed in 1975 by that of Betty Jo Teeter Dobbs.[5] Although this outstanding study was primarily concerned with an analysis of part of the corpus of Newton's chemical papers, it nevertheless also addressed the question of what chemical processes underlay some of the most important of Newton's chemical experiments. Even if one does not concur with all of the conclusions reached by Professor Dobbs it nevertheless remains the finest piece of work in this field and its value has been acknowledged by other eminent Newton scholars such as Westfall,[6] who has made extensive use of it in subsequent work, and Figala,[7] who commented on it at length in a major essay review. An inevitable shortcoming of Dobbs's work—and that through no fault of the author—is that it only examines a portion of Newton's chemical experiments and makes little attempt to view them as a whole. It also says comparatively little about the practical problems encountered, as well as the skills evidenced, by Newton in carrying out his work in his tiny laboratory set against the wall of the chapel in Trinity College, Cambridge.

During his lifetime Newton published three works that indicated clearly his great interest and competence in what we might loosely call practical chemistry. The first, entitled *Scala Graduum Caloris*, defined a temperature scale containing a number of "Degrees of Heat," varying from the "Heat of the winter air when water begins to freeze" (Degree of Heat 0), to "The heat of a small coal fire not urged by bellows" (Degree of Heat 5). The scale, which was published anonymously in the *Philosophical Transactions of the Royal Society* of May 1701, shows clear evidence of familiarity with a number of laboratory processes. The second is a paper *De Natura Acidorum*, which Newton wrote in about 1691 and which was published in English in 1710 by John Harris in his *Lexicon Technicum*.[8] Once again this piece shows clear evidence of familiarity with many laboratory processes. Finally, many of the queries appended to the second English edition of the *Opticks*, and particularly Queries 30 and 31, are essentially chemical in nature.[9]

Our concern in this paper, however, is not with these three

published pieces, but rather with the long series of chemical experiments, still unpublished, which occupied so much of Newton's time. The descriptions of these experiments are to be found in two places. The first is a small leather-bound notebook that commences with quotations from various authors—and in particular Robert Boyle—but most of which is occupied with detailed descriptions of several hundred chemical experiments.[10] The second is a set of loose sheets also containing detailed descriptions of chemical experiments.[11] Many of the experiments described occur in both the notebook and the loose sheets. Both are now located in the Cambridge University Library, having been donated to the University by the Earl of Portsmouth in 1888.

The earliest dated experiment in the manuscripts was carried out on 10 December 1678 while the last took place in February 1695/96, not long before Newton forsook Cambridge for London to enter the service of the Royal Mint. The dated experiments therefore extend over a period of some eighteen years. However, as we shall see below, Newton had in fact purchased chemicals and apparatus as early as 1669 (and perhaps even as early as 1665) so that his period as a practising chemist must have extended over a much longer timespan than is dealt with in these two manuscripts. Unfortunately, there is no known manuscript record of his chemical work in the period 1669 to 1678.

Newton's Experiments

In examining the records of Newton's chemical experiments for a paper such as this, one can either attempt an overview of the several hundred experiments as a whole, as the Halls did in their 1958 paper, or undertake a more detailed examination of a limited group of experiments. I have chosen to do the latter as not only does this reduce the task to more manageable proportions in terms of length but it also ensures the analysis of a reasonably homogeneous set of experiments.

An immediate problem encountered by anyone examining Newton's records of his chemical experiments is the fact that not only are they closely written, in the true Newton style, but they also run on from one to the next with little or no break. They are not numbered or identified in any way, apart from the occasional sets of dated experiments. In order to assist analysis I have therefore divided them up into what I believe to be discrete experiments—although this is by no means a simple task—which I have then numbered. As in this paper we are particularly interested in New-ton the laboratory chemist, I have chosen to examine only those experiments which are recorded in the so-called loose sheets, as I believe these to provide a more faithful and vivid, record of New-ton's experimental work than the somewhat more refined version contained in the leatherbound notebook into which Newton seems to have copied his results later.

The first three sets of experiments in the loose sheets are dated December 10 1678 to January 15 1679/80; January 1679/80; and February 1679/80 respectively. They have been numbered, 1, 2, and 3 in a later hand, presumably that of G. D. Liveing when he was editing this portion of Newton's papers for the 1888 Catalogue mentioned above. As the next set of experiments is dated August 1682 the three sets chosen form a naturally homogeneous group carried out during the period December 1678 to February 1679/80. According to my analysis these three sets of records describe fifty discrete experiments. Although I am naturally very aware of draw-ing too many conclusions from a relatively small sample of New-ton's experiments, I believe that for the purposes of this paper the decision is justified.

The results of the experiments are written in a small but perfectly legible hand and describe Newton's techniques in explicit terms even if, as we shall see below, they make use of names and symbols for materials that are not always sufficient to identify them in terms of their modern equivalents. The style, details, vividness of descrip-tion, and general "feel" of these experimental records give a mod-ern reader with appropriate chemical experience a curious feeling of almost being present with Newton in the laboratory. After work-ing with these papers for some time the reader begins to construct in his mind a portrait of Newton that, in my case at any rate, the mathematical, Mint, alchemical, and theological papers have never done.

Newton's Laboratory

Newtonian scholars have reason to be eternally grateful that Newton's amanuensis and assistant from 1685 to 1690, Humphrey Newton of Grantham, has provided us with a vivid description of various aspects of Newton's life. Among these are some extraor-dinarily graphic accounts of Newton's laboratory and his work there:

He very rarely went to bed till *two* or *three* of the clock, sometimes not till *five* or *six,* lying about *four* or *five* hours, especially at spring and fall of the leaf, at which times he used to employ about six weeks in his elaboratory, the fire scarcely going out either night or day, he sitting up one night and I another, till he had finished his chemical experiments, in the performances of which he was the most accurate, strict, exact. What his aim might be I was not able to penetrate into, but his pains, his diligence at these set times made me think he aimed at something beyond the reach of human art and industry. . . . On the left end of the garden [in Trinity College, Cambridge] was his elaboratory, near the east of the chapel, where he at these set times employed himself in with a great deal of satisfaction and delight. Nothing extraordinary, as I can remember, happened in making his experiments.[12]

And again:

About 6 weeks at spring, and 6 at ye fall, ye fire in the elaboratory scarcely went out, which was well furnished with chymical materials as bodyes, receivers, heads, crucibles, &c, which was made very little use of, ye crucibles excepted, in which he fused his metals; he would sometimes, tho' very seldom, look into an old mouldy book wch lay in his elaboratory, I think it was titled *Agricola de Metallis,* the transmuting of metals being his chief design, for which purpose antimony was a great ingredient. . . . His brick furnaces *pro re nata,* he made and altered himself without troubling a bricklayer.[13]

Accepting the accuracy of Humphrey Newton's account, the picture of Newton that emerges is that of a highly motivated and deeply committed practical chemist working assiduously in a private, well-equipped laboratory in the garden just below his rooms in Trinity. According to Humphrey Newton, Isaac was particularly concerned with using his furnaces in the calcination of various materials and, as we shall see below, this was a remarkably accurate picture of Newton the chemical experimenter.

The role of Humphrey Newton in the performance of Newton's chemical experiments is by no means clear, for although he claims here that he shared the performance of the experiments with Newton—or at least the night watches!—my examination of the experiments leads me to believe that their performance was in fact very largely in Newton's own hands. The vividness of description, together with the fine details so frequently recorded, could not have arisen from anybody who did not carry out the work himself. Humphrey Newton was almost certainly little more than the late-night fire watcher he claimed to be and to call him a "laboratory assistant" is to accord him altogether too grand a title.[14]

Materials Used by Newton

As we shall see below, one of the first records we have of Newton's interest in chemistry is that of the purchase in 1669 of various chemicals—the first of what must have been many such purchases over the years.

A study of the fifty experiments dealt with in this paper shows that in them Newton used thirty-eight different materials—an impressively high number for the time. The materials used, together with possible modern equivalents, are set out in table 1. Also included is the number of experiments in which each material was used.

The table is revealing in a number of ways. First, it illustrates in a most graphic way the importance to Newton of two materials, namely "antimony" (by which Newton would mean not the metal but rather the compound antimony sulfide, Sb_2S_3, which is the major constituent of the most common antimony ore, stibnite), and "sal ammoniac," now called ammonium chloride. Their frequent occurrence is hardly surprising in the light of the now undisputed close connection between Newton's chemical experiments and his extensive reading in the alchemical literature, for both occur very frequently in many alchemical writings and were presumably used in much of the laboratory work the alchemists carried out.

It must be stressed that this list is not necessarily representative of the experiments as a whole, for one of the most interesting features of Newton's experiments is the extent to which they change over the years. A preliminary examination of the whole set of experiments reveals a number of differences between the materials used in the earlier and the later experiments. For example, these early experiments do not well illustrate the nonstandard, probably unique, symbols that Newton developed to identify various materials, whether reactants or products. One such symbol, at least, is present in these experiments, for the symbol ♒ does not appear in the normal lists of alchemical symbols of the period.[15]

The question of the purity of the materials used by early chemists has been a major problem for later scholars attempting to understand early experimental work in chemistry.[16] Although Boyle was very conscious of this problem, and on various occasions took precautions to ensure that his starting materials were pure,[17] I have found no evidence that Newton was similarly aware. However, on examining these experiments carefully one frequently gets the feeling that impurities were (in our terms anyway!) a problem. Thus, to take just one example, in an experiment involving antimony, Newton mentions "green precipitated sublimate of ☿ ."[18] As there is no green antimony salt that I am aware of, this almost certainly indicates the presence of copper, or perhaps nickel, as an impurity in the antimony.

The repetition of some of Newton's experiments may well go some way towards solving this problem and I am now in the process of repeating a number of these experiments in my own laboratory.

Chemical Processes Employed by Newton

In the course of the fifty experiments on which this paper is principally based Newton made use of a wide variety of chemical processes. Most of these are identified by name by Newton himself while others are readily inferred from experimental descriptions.

The processes involved, together with the number of occasions on which they were used in these fifty experiments, are as follows:

1.	Sublimation	31
2.	Weighing	30
3.	Reacting in specified proportions	24
4.	Fusion	17
5.	Fusion or heating on a hot iron or fire shovel	14
6.	Drying of materials	9
7.	Tasting materials	8
8.	Precipitation	8
9.	Dissolution in water	8
10.	Edulcoration (i.e., freeing the soluble particles from a material by washing)	5
11.	Aqueous extraction (cold)	4
12.	Dissolution in *aqua fortis* (i.e., nitric acid)	4
13.	Digestion	4
14.	Fusion/heating on a piece of glass	3
15.	Boiling, or evaporation, to dryness	3
16.	Grinding	3
17.	Sublimation at room temperature	1
18.	Sublimation in hot sand	1
19.	Fusion/heating over a candle	1

Table 1

Newton's Name or Symbol	Probable Modern Equivalent	Occurrence in 50 Experiments
✳	Ammonium Chloride [NH_4Cl]	22
☿ (antimony symbol)	Crude Antimony Ore [Sb_2S_3]	18
☿ sublimate	Presumably Sb_2S_3	10
♈ or ▽ or ▽̶	"Aqua fortis," i.e., Nitric acid [HNO_3]	9
♄☿ iate or saturn ☿ iate	Possibly an Sb·Pb alloy	8
♀ or copper or "filings of copper"	Copper metal	6
♂ or Iron	Iron metal [Fe]	5
Iron ore	Probably iron oxide [Fe_3O_2]	5
☿ once acted on by ♈	? Antimony nitrate [$2Sb_2O_3 \cdot N_2O_5$]	4
Regulus of ☿	Antimony metal [Sb]	3
♄ impregnated	Perhaps lead sulphate or nitrate [$PbSO_4$ or $Pb(NO_3)_2$]	3
♄	Lead metal [Pb]	3
Oyle of ⊕	Sulphuric acid [H_2SO_4]	3
Calx of ☿ once wrought on	Perhaps antimony sulphate [$Sb_2(SO_4)_5$]	3
♒ vir or white ♒ or white ♒ viridis	?	3
Vitriol made with ♀	Copper sulphate [$CuSO_4$]	3
Liquor of ☿	Perhaps a solution of antimony nitrate	3
Glass of ☿	Perhaps a mixture of Sb_2O_3 and Sb_4O_6	3

Newton's Name or Symbol	Probable Modern Equivalent	Occurrence in 50 Experiments
Vinegre of ☿	? Antimony acetate	2
Vinegre	Solution of acetic acid [CH_3COOH]	2
Salt of ⊟ or Salt of tartar	Potassium hydrogen tartrate [$KHC_4H_4O_6$]	2
Reg of ♂	Iron metal [Fe]	1
Scoria of Reg of ♂	?	1
Tin	Tin metal [Sn]	1
Tinglas	Bismuth metal [Bi]	1
Bismuth	Bismuth metal [Bi]	1
Bole armonack	Ammonium chloride [NH_4Cl]	1
Fullers earth	Fuller's earth	1
Sal ♂ tis	?	1
☿	Mercury metal [Hg]	1
Vitriol of ∩ vir	?	1
Calx ☿ ij saturniate or calx ♄☿ iate	?	1
Sublimate of ☿ dissolved in ▽R and precipitated with water	?	1
Arsnick	Arsenic metal [As]	1
Net	An alloy of iron, antimony, and copper	1
⋂R	Aqua regia	1
Lead ore	Lead sulphide [PbS]	1
☿ iate vitriol of ♀	?	1

20. Extraction *"in balneo aquae ebullientis"* (i.e., over a
    water bath)   1
21. Digestion in acid   1
22. Digestion in mixture of solvents   1
23. Calcination   1
24. Condensation of a sublimate in a separate vessel
    "whelmed" over another   1
25. Dissolution in *aqua regia* (i.e., a mixture of
    concentrated nitric and hydrochloric acids)   1
26. Filtration   1
27. Distillation   1

This list reveals a quite remarkably diverse set of processes employed in a relatively small set of experiments and is indisputable evidence of the extent of Newton's competence in the laboratory. He was clearly master of the whole spectrum of contemporary laboratory processes in which, as we have seen above, he used a large number of materials.

The extent to which Newton employed the balance in his chemical experiments is also quite extraordinary—apart from Boyle I believe it to be unique for a chemist of his time. As the list above shows, in thirty of the fifty experiments analyzed he weighed the materials involved, never to less than 1 grain (i.e., 65 mg), and sometimes estimating to ¼ of a grain, or about 16 mg. The later experiments were carried out with greater accuracy than the earlier ones, presumably because Newton had acquired a better balance. Almost all the experiments in which Newton did not actually record that he had weighed his materials involved the reaction of materials in stated proportions, which is of course another form of quantification. In only six out of the fifty experiments did Newton specify neither mass nor proportions. To the best of my knowledge no contemporary chemist, including Boyle, approached this degree of quantification in chemistry—nor indeed was anyone to do so until some time later.

The extensive use of the process of sublimation fits in well with the use of *Sal ammoniac* (ammonium chloride), which was one of the materals most commonly used by the alchemists in sublimation experiments. The relative paucity of processes involving solutions, i.e., what we would today call "wet chemistry," is also noteworthy: the experiments are dominated by sublimation and fusion.

Another noteworthy feature of the experiments is the frequency and care with which Newton recorded the time taken for various experiments. The accounts of the experiments abound with phrases such as "& it fumed all away in about ⅛ of an hour,"[19] "when I had continued this heat about three hours,"[20] or "after a digestion of 7 or 8 hours."[21] As might be expected of somebody of his background and methodical nature, he was precise in all things.

Newton's propensity to taste the products of his reactions occurs throughout these records, as well as those of the rest of his chemical experiments. I have discussed this matter in detail elsewhere and here will do no more than repeat that this habit may well have been a contributory factor, albeit a small one, to what I believe to be the chemical basis of his alleged "nervous breakdown" of 1695/96.[22]

These records of his experiments contain both a touch of drama and great vividness of description. "In yᵉ distillation yᵉ retort crackt & about ¼ or ⅓ of yᵉ liquor ran out. I changed the retort and distilled on"[23] illustrates the drama that is part of the day-to-day life of any chemical laboratory, while phrases such as "a white fat clamy slime"[24] and "white like one's thumb nail"[25] show a rich and effective use of descriptive language.

Finally, one must record the acute awareness Newton showed with respect to temperature. His descriptions of his experiments abound wth phrases such as "till the glass was of a dark red heat and began to melt,"[26] "a gentle heat not much exceeding that of hot blood."[27] No doubt the role of temperature in these experiments, and the need to set up some quantitative scale to represent it, played an important part in Newton's work that culminated in his *Scala Graduum Caloris* many years later.

Apparatus and Chemicals

We know that Newton purchased chemicals and apparatus on a number of occasions. The earliest such purchase may well have been in 1665 when, as a young man of twenty-three, we read that he purchased "Glass bubbles" (almost certainly glass flasks for use in the laboratory) for four shillings. This was followed a few years later, in April 1669, by the following impressive purchases.[28]

For aquafortis, sublimate, oyle pink, fine silver, antimony, vinegar, spirit of wine, white lead, salt of tartar, ☿	. . . £2 0 0
A furnace	. . . £0 8 0
Air furnace	. . . £0 7 0
Theatrum chemicum	. . . £1 8 0

Newton was now well and truly launched as a practical chemist—an activity that, as we have seen, was to last for some twenty-seven years.

We have to wait until 1687 before Newton's manuscripts record another purchase of chemicals and apparatus:[29]

Bought of Mr Stonestreet 1687

Antimony at 4^d ye lb
Ol. Vitriol at 3^s 0^d ye lb
Crucibles at 5^d ye Nest
Sal Armoniack at 2^s ye lb
Double Aqua fortis at 5^s 4^d ye lb

Mr Stonestreet Druggist by Bow Church on ye same side ye street towards Pauls at ye sign of ye Queens head wth a rope in her breast.

Six years later, in 1693, we again find him purchasing chemicals, but this time from Mr. Timothy Langley who, Newton informs us, had succeeded Mr. Stonestreet:

Quicksilver at 5^s ye lb
salarmoniack at 1^s 9^d ye lb
Singl Aqua fortis at 2^s 8^d ye lb

(It is interesting to note the drop in price of *sal ammoniac* from 2s. to 1s 9d. per pound.)

In an undated reference Newton also mentions the "third sort of ☿ bought of Mr Box in small loafs wth small hair veins running all one way & a little soot & many colours up & down the veins & soot."[30] Unfortunately we have no further information concerning Messrs. Stonestreet, Langley, or Box.

In his descriptions of his experiments Newton frequently mentions in explicit terms the apparatus used while, in many other cases the processes described make it relatively easy to infer what items of apparatus were available to him.

The central item of apparatus in the seventeenth-century laboratory was undoubtedly the furnace. Used for heating solid materials, as well as distilling liquids in flasks and retorts, furnaces played a key role in the chemistry of the day, as indeed they have done until very recently. Although Newton's accounts of his chemical experiments contain no description of the furnaces that he used so frequently, one of the sheets in the alchemical papers sold at Sotheby's in July 1936 contains an excellent outline in Newton's hand of a variety of laboratory furnaces.[31] There are sketches of four wind furnaces, one "reverberatory" furnace and one "Athanor Piger Henricus, or Furnace Acediae for long digestions (y^e) vessel being set in sand heated wth a Turret full of Charcoale wch is contrived to burne only at the [botto]m the upper coales continually sinking downe for a supply." There is little reason to doubt that the furnaces depicted in this sketch were very similar to those which Newton himself constructed and used. (See fig. 1)[32]

Apart from furnaces the records of Newton's experiments indicate the use of a wide variety of other pieces of apparatus:

1. Retorts of various sizes ("and it setled in ye neck of ye retort")
2. Mortars and pestles ("ground them together in a glass")
3. Fire shovels ("I put on a fireshovel in ye fire")
4. Water baths ("balneo aquae ebullientis")
5. Candles (for heating small quantities of material: "and holding the glass over a candle")
6. Iron plates ("melted easily upon a hot Iron")
7. Glass flasks ("in a glass unstoppered"; "in a glass phiol")
8. Quills (for transferring small quantities of solids: "I took out of ye glass a little of it on a quill")
9. Crucibles of various sizes ("ye largest crucible") (Compare the "nest of crucibles," mentioned above, which Newton purchased from Mr. Stonestreet in 1687.)
10. As "egg-glass," i.e., a sealed glass vessel used for heating materials out of contact with the air ("the neck of ye egg-glass . . . Upon breaking the glass I found")[33]
11. Sealed glass flasks ("in a Viol glass luted")
12. Glass vessels used to condense volatile materials ("in a glas whelmed over it")
13. Cold water baths ("to water in frigido")
14. Corks ("I had put a cork in ye mouth of ye retort to stop it imperfectly during ye sublimation")
15. "Receivers," i.e., glass containers used to collect distillate from retorts ("there came over flegm into ye Receiver")
16. A heavy mortar and pestle—perhaps metal ("ye Reg. was beaten fine")
17. Filtering apparatus ("filtrated ye solution & there rested in y^e filter")

Newton also describes a particular piece of distilling apparatus that, to the best of my knowledge, is unique.[34] This is illustrated in fig. 2 and consists of three circular earthernware pots. In use the

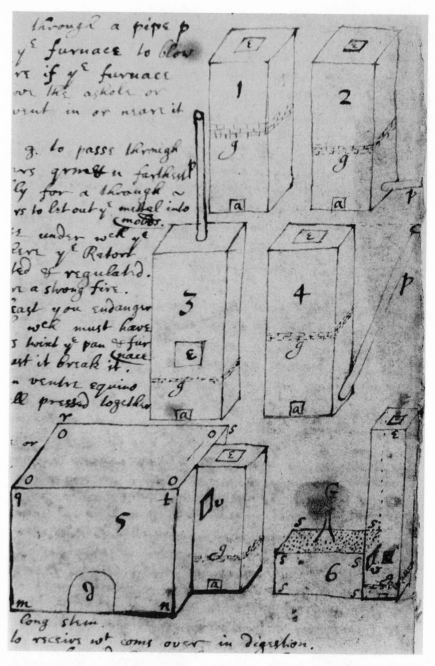

Figure 1. Courtesy of the Department of Special Collections, University of Chicago Library.

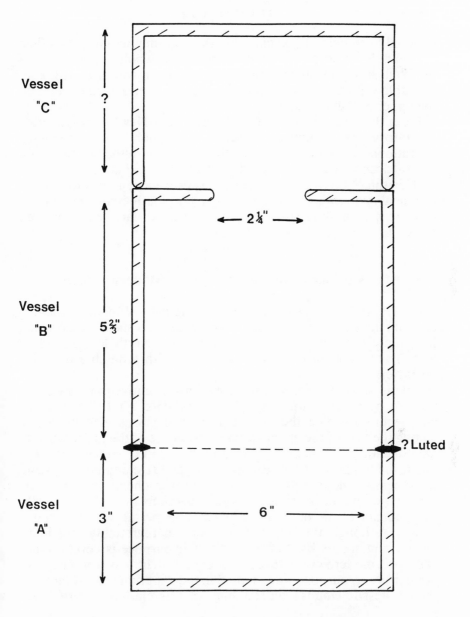

NEWTON'S SUBLIMATION APPARATUS

Figure 2

material to be heated was placed in the bottom pot, which was then presumably luted to the middle one. The whole apparatus, assembled as in the figure, was then placed on the fire. Any condensable vapors driven off from the heated material would pass into the upper pot and condense.

From the above it is abundantly clear that Newton possessed a laboratory well equipped to carry out all the standard chemical operations of his time. It is worth noting that although there are several accounts in the early 1700s of a large burning glass designed by Newton,[35] there is no record of his ever having used one in his chemical experiments—even though his south-facing laboratory and the surrounding spacious garden were in fact relatively well placed to intercept the summer sun.

Newton's Experiments in the Light of Modern Chemistry

Analyzing Newton's experiments in the light of modern chemistry is no light task, for although in some respects these experiments have a curiously modern ring about them, nevertheless a modern investigator is soon brought face to face with a number of major obstacles.

Let us start by stating that a modern scientist is at once impressed by many features of the experiments. First, the frequent and accurate use of the chemical balance places Newton firmly among the first of the chemists who was aware of the importance of quantification in chemistry. Second, Newton's ongoing concern about the temperature of reactions is impressive and implies acute awareness of the importance of this factor in chemistry. Third, his frequent noting of the time for which reactions take place. Fourth, the undeniable air of purposefulness that permeates the experiments. Although the experiments contain regrettably few interpretive statements by Newton, and their purpose is certainly not easy for a modern eye to discern, one is left in little doubt that there is clear order, purpose, and method in the experiments—something that Professor Dobbs's outstanding work is clarifying very effectively.

However, when one attempts to make sense of the chemistry in the experiments it is an altogether different matter, for here problems abound. For convenience let us number these:

1. *The use of antimony.* The extensive use of antimony and antimony sulfide is a major stumbling block for the modern in-

vestigator, for the study of this metal and its compounds, although fairly common in the great heyday of classical inorganic chemistry from ca. 1870 to 1910, has virtually disappeared from chemistry as we know it today. Most modern university chemistry courses today hardly even mention the metal and its compounds and I would guess that a high proportion of the chemists alive today have hardly ever seen the metal or its compounds, let alone studied them in detail. Added to this is the fact that the chemistry of antimony and its neighbors in Group VA of the periodic table, i.e., arsenic, bismuth, and phosphorus, is complex and demanding—any element that has oxidation states of $+3$, $+5$, and -3 deserves to be treated with respect!

2. *Unusual alloys.* Newton's experiments with antimony frequently involve the use of other metals, thereby producing what would today be called intermetallic compounds, or alloys. If two metals are involved a binary system is formed, if three a ternary system. Clearly the number of ternary systems that can be formed from the metals in the periodic table is very large and many ternary systems have never been studied. For example, one of Newton's favorite materials was "the net," an alloy consisting of antimony, iron, and copper. To the best of my knowledge the post-1800 literature on this material is totally silent. (I know however that it can exist for I have made some.)

3. *Purity of materials.* There is also the problem of the purity of Newton's materials, which has been mentioned above. In Newton's case the problem is exacerbated by the fact that he made frequent use not only of metals, but also their ores (as did so many of the alchemists) and these are notoriously variable in composition, depending upon the particular mine from which they originate.

4. *Nonstandard nomenclature.* Newton's use of nonstandard symbols and nomenclature continues to be a major stumbling block to anybody seeking to understand these experiments, although I believe that this problem will probably yield to the pressure of systematic and sustained study of the experiments.

5. *Re-use of products of a reaction.* Newton's predilection for using one or more of the products of a reaction as the starting materials for the next one simply compounds several of the problems mentioned above.

6. *Ignorance of experimental conditions.* Although, as we have seen, Newton was a meticulous observer, there are many cases in which his failure to specify factors such as temperature or the

duration of a reaction is a serious problem in understanding it today.

Thus, to summarize, to attempt an understanding of Newton's experiments in the light of modern chemistry is a difficult task with no certainty of success at the end of it. Nevertheless I have found my first skirmish with the problem stimulating and enjoyable—if frustrating—and I intend to continue the battle.

Anyone who has studied Newton's chemical experiments sooner or later is questioned as to whether there are great chemical discoveries hidden in these experimental records, for surely the mind that gave the world the *Principia* must also have made at least some discoveries of major importance when it turned its great powers to chemistry? The question is in every way a reasonable one, but I am afraid that the answer must be in the negative. Having examined these accounts of his chemical experiments many times, I am absolutely convinced that, whatever their intrinsic interest to Newtonian scholars, they contain nothing of real chemical value that would have materially altered the history of chemistry if their contents had been known at any time between their composition and the present day.[36]

A Final Note—Partly Personal

Science is carried out by scientists, and, as those of us who have been practicing scientists will know only too well, it is a process compounded of intellectual activities on the one hand and very human feelings and reactions on the other—frustration, unremitting toil, physical and mental exhaustion, the excitement of discovery, and elation that can at times verge on ecstasy. To do full justice to any figure in the history of science, both components need to be dealt with, for only by so doing can a balanced, holistic picture of the true process of scientific creativity emerge. To look only at the intellectual products of science means that we run a very real danger of seeing the flickering of the flame but missing the fire.

With the above in mind I would like to conclude this study of Newton's chemical experiments on a personal note. I grew up in Johannesburg, that great city which is the center of the world's largest gold mining industry. Many years ago, during one of my vacations as a student at the University of the Witwatersrand I worked as an assistant in the assay laboratory on a gold mine near the city. Our principal task was to determine the percentage of silver and gold in the hundreds of ore samples that were collected underground every day. Although all the newer mines had fine modern laboratories much like the ones we know today, on the older mines, such as the one where I was employed, the assay laboratories in many ways did not differ significantly from that in which Newton worked. Thus in our assaying we made use of much apparatus that has been mentioned in this paper: various types of furnace fired with solid fuel, tongs, crucibles of various shapes and sizes, fire shovels, concentrated nitric acid for parting the silver from the gold (Newton's *aqua fortis*), and mechanical balances.

That vacation employment took place thirty years ago yet I still retain the most vivid memories of the intense heat of the furnace room, the sulfureous fumes from the coal, the scorching feeling on hands, face, and forehead as a result of the radiation from the furnaces, and the great care needed in handling the brightly glowing crucibles or cupels as they were taken from the furnace. Together with this was the growing exhaustion as the day wore on, and the great thirst that was slaked with water from an unglazed earthenware bottle that cooled its contents by evaporation through the clay.

I record this not simply to indulge in nostalgia but rather to substantiate my belief that when one examines with the benefits of appropriate experience the accounts of Newton's chemical experiments and notes the frequency of heating in the furnace, the high temperatures reached, and the time for which some experiments were continued, then one begins to get a glimpse of the enormous intensity with which these experiments were pursued and how important they must have been to him.

We have already quoted from Humphrey Newton's vivid account of Newton's work in the laboratory, in which he speaks of the fires in the laboratory furnaces "scarcely going out either night or day" for weeks on end and of Newton's extraordinary "pains and diligence" in the laboratory. The records we have of Newton's laboratory work fully justify Humphrey Newton's graphic description, and I believe that in carrying out these experiments, particularly bearing in mind the conditions of great heat and discomfort that were inevitable in that cramped, poorly ventilated little laboratory, Newton was truly driven by intense forces the like of which few, if any, of us has either experienced or can even begin to understand.

Notes

1. Sir David Brewster, *Memoirs of the Life, Writings and Discoveries of Sir Isaac Newton*, 2 vols. (Edinburgh: Thomas Constable, 1855).

2. Louis Trenchard More, *Isaac Newton, A Biography* (New York and London: Charles Scribner's Sons, 1934; reprint ed., New York: Dover, 1962).

3. *A Catalogue of the Portsmouth Collection of Books and Papers Written by or Belonging to Sir Isaac Newton* . . . (Cambridge, England: Cambridge University Press, 1888).

4. Marie Boas and A. Rupert Hall, "Newton's Chemical Experiments," *Archives Internationales d'Histoire des Sciences* 11 (1958): 113–51.

5. Betty Jo Teeter Dobbs, *The Foundations of Newton's Alchemy, or "The Hunting of the Greene Lyon"* (Cambridge, England: Cambridge University Press, 1975).

6. R. S. Westfall, *Never at Rest. A Biography of Isaac Newton* (Cambridge, England: Cambridge University Press, 1980).

7. Karin Figala, "Newton as Alchemist," *History of Science* 15 (1977): 102–37.

8. John Harris, *Lexicon Technicum or An Universal English Dictionary of Arts and Sciences*, 2 vols. (London: D. Brown, 1704 and 1710), 2, Introduction.

9. Isaac Newton, *Opticks: or, A Treatise of the Reflections, Refractions, Inflexions and Colours of Light*, 2d edition (London: W. and J. Innys, 1718), pp. 349–82.

10. University Library, Cambridge; MS Add 3975.

11. University Library, Cambridge; MS Add 3973.

12. King's College, Cambridge; Keynes MS 135; also in Brewster, n.2 above, vol. 2, pp. 93–94.

13. King's College, Cambridge, Keynes MS 135; also in Brewster, n.2 above, vol. 2, pp. 95–97.

14. The role we have proposed for Humphrey Newton contrasts sharply with that of assistants in Boyle's laboratory, where they took a very active role in a wide variety of operations. See Steven Shapin, "The Invisible Technician," *American Scientist* 77, no.6 (1989): 554–63.

15. The problems of chemical nomenclature are discussed at length in M. P. Crosland's *Historical Studies in the Language of Chemistry* (London: Heinemann, 1962). Particularly valuable in our context is part 4, chapter 1, "Alchemical Symbols," pp. 227–44.

16. Torbern Bergman would appear to be the first chemist to be acutely aware of the need for using purified reagents in the laboratory. (*Opuscula Physica et Chemica,* 2, Upsala: Holmiae, 1780), p. 85.

17. See, for example, Boyle's *Works* (London: A. Miller, 1744), 1: 230–31, in which Boyle points out that coinage silver always contains some copper and must therefore be used with care as a laboratory material.

18. University Library, Cambridge; MS Add 3973, fol. 11a (Experiment 47).

19. University Library, Cambridge; MS Add 3973, fol. 5 (Experiment 18).

20. University Library, Cambridge; MS Add 3973, fol. 8 (Experiment 28).

21. University Library, Cambridge; MS Add 3973, fol. 7 (Experiment 23).

22. P. E. Spargo and C. A. Pounds, "Newton's 'derangement of the intellect'; New Light on an Old Problem," *Notes and Records of the Royal Society of London* 34 (1979): 11–32.

23. University Library, Cambridge; MS Add 3973, fol. 6 (Experiment 22).

24. University Library, Cambridge; MS Add 3973, fol. 6a (Experiment 19).

25. University Library, Cambridge; MS Add 3973, fol. 8 (Experiment 28).

26. University Library, Cambridge; MS Add 3973, fol. 7a (Experiment 22).

27. University Library, Cambridge; MS Add 3973, fol. 7 (Experiment 23).

28. Fitzwilliam Museum, Cambridge; Newton Notebook (unpaginated).

29. University Library, Cambridge; MS Add 3973, Inside back cover.

30. University Library, Cambridge; MS Add 3975, p. 270.

31. This MS is now in the Joseph Regenstein Library, University of Chicago; MS 1075–73.

32. Newton's sketches of chemical furnaces bear an interesting resemblance to those of the diarist John Evelyn, who attended a course of lectures on chemistry at about the same time. See F. Sherwood Taylor, "The Chemical Studies of John Evelyn," *Annals of Science* 8, no.4 (1952): 285–92.

33. The *Chymicall Dictionary,* by "J.F." (London: Thomas Williams, 1650) defines this piece of apparatus as follows: "*Ovum Philosophicum* is a glasse in the forme of an egge which Philosophers use in their operations."

34. University Library, Cambridge; MS Add 3973, fol. 14a (Experiment 55).

35. See, for example, Evelyn's *Diary,* ed. E. S. de Beer (Oxford: Oxford University Press, 1955), 5: 592, 600.

36. I have recently come across a similar view expressed by J. R. Partington in his 1951 Presidential Address to the British Society for the History of Science:

> We might surmise that, if Newton had applied his powerful mind to chemistry he would have made advances in that science which could have linked up with the epoch-making work of Lavoisier without the necessity of the slow progress and decline of the theory of phlogiston during the eighteenth century. Yet Newton did apply himself to chemistry, and most assiduously, and the results of his labours were practically negligible. Few historians of chemistry find it necessary to mention Newton at all, and in the history of that science he fills a very modest place.

(*Bulletin of the British Society for the History of Science* 1, no.6 [1951]: 132).

Reasoning From Phenomena: Newton's Argument for Universal Gravitation and the Practice of Science

WILLIAM L. HARPER

Introduction

Newton's theory of universal gravitation is one of the most celebrated achievements in the entire history of science. His argument for it, in *Principia* Book 3, is an extraordinarily powerful piece of scientific reasoning. It established, for the first time, a unified theory of terrestrial and celestial motion phenomena. It is widely recognized that the example of Newton's stunning achievement helped shape the transformation of natural philosophy into today's physical science. I hope that the following sketch will make it clear that studying details of Newton's argumentation can still teach us interesting things about the practice of this science.

The two parts of this paper correspond to the two parts Newton distinguished in his famous summary of his experimental philosophy in the General Scholium, *In hac philosophia propositiones deducuntur ex phaenomenis, & redduntur generales per inductionem*.[1] In part 1, I shall concentrate on Newton's practice of inferring forces from phenomena of motion. I shall argue that attention to certain details of what I. B. Cohen has called "Newton's mathematical style" will reveal that interesting equivalence theorems back up these inferences. These theorems indicate that Newton's inferences are far more cogent than they have often been taken to be. Moreover, once one notices the use of such equivalence theorems in these inferences it becomes plausible to see a more pervasive role for them. This allows us to see Newton's celebrated explanations of diverse motion phenomena as examples of a new mathematical style of causal explanation. In such explanations the phenomenon—the effect to be explained—can be con-

144

strued as a measurement of the critical parameter of the cause that explains it.

These relations between Newton's phenomena and the propositions he inferred from them give content to some of his controversial rules of reasoning and other methodological remarks. In the second part I shall say something about how this shows up when we look at details of his actual argumentation in the course of his attempt to generalize from propositions inferred from phenomena to establish universal gravitation. Unification, generalization, and the protection of the results of induction from being undercut by hypotheses are familiar themes in Newton's methodological remarks. I shall argue that Newton still has valuable lessons to teach about each of them. I shall also say something about his claim that he did not affirm that gravity is essential to bodies.

Inferring Propositions from Phenomena

In hac philosophia propositiones
deducuntur ex phaenomenis, . . .

Propositions 1 and 2 of *Principia,* Book 3, are the opening moves of Newton's argument for universal gravitation. They are the most paradigmatic examples of his practice of inferring propositions about forces from phenomena of motion. Here is a recent translation of the relevant passages.[2]

Proposition 1, Theorem 1
The forces by which the circumjovial planets [or satellites of Jupiter] are continually drawn away from their rectilinear motions and are maintained in their respective orbits are directed to the center of Jupiter and are inversely as the squares of the distances of their places from that center.

The first part of the proposition is evident from phen. 1 and from prop. 2 or prop. 3 of Book 1, and the second part from phen. 1 and from corol. 6 to prop. 4 of book 1.

The same is to be understood for the planets that are Saturn's companions [or satellites] by phen. 2.
Proposition 2, Theorem 2
The forces by which the primary planets are continually drawn away from rectilinear motions and are maintained in their respective orbits are directed to the sun and are inversely as the squares of their distances from its center.

The first part of the proposition is evident from phen. 5 and from prop.

2 of Book 1, and the latter part from phen. 4 and from prop. 4 of the same book. But this second part of the proposition is proved with the greatest exactness from the fact that the aphelia are at rest. For the slightest departure from the doubled ratio [or the ratio of the square] would (by bk. 1, prop. 45, corol. 1) necessarily result in a noticeable motion of the apsides in a single revolution and an immense such motion in many revolutions.

In the first parts of Propositions 1 and 2 Newton infers that the forces deflecting the satellites of Jupiter (and Saturn) and the primary planets are directed to the centers of their respective primaries. In the second parts of these propositions he infers that in each system these centripetal forces vary inversely as the squares of their distances from the center.

EQUIVALENCE THEOREMS

When we look at the offered proofs we see three forms of argument. One corresponds to the appeals to Theorems 2 or 3 of Book 1 and is used to establish the first parts of both propositions. A second form of argument corresponds to the appeals to Proposition 4 of Book 1 and is used in the proof of the second part of Proposition 1 and in the first proof of the second part of Proposition 2. The third form corresponds to the appeal to Corollary 1 of Proposition 45, Book 1, and is used in the second proof of the second part of Proposition 2. I shall argue that each of these forms of argument is backed up by theorems that establish a systematic equivalence between a phenomenal quantity and a corresponding theoretical magnitude that is specified in the proposition being inferred from the phenomenon.

The first form of argument infers centripetal forces from orbital motions that satisfy the Keplerian law of areas. Motion satisfies the Keplerian law of areas just in case the orbiting body describes areas proportional to the times of description by radii to the center of the primary. Given Newton's assumptions, Theorems 2 and 3 of Book 1 show that such Keplerian motion implies that the orbiting body is deflected from tangential motion by a force directed to the center about which these equal areas are described in equal times.[3] The relevant cited phenomena are Phenomenon 5 and the first part of Phenomenon 1. These specify respectively that the orbits of the primary planets and of the moons of Jupiter satisfy this law of areas.

The second form is used in the proof of the second part of Proposition 1 and in the first proof of the second part of Proposition

2. The cited result (Corollary 6 of Proposition 4, Book 1) establishes (from Newton's assumptions) the *equivalence* of two conditions: 1) the periodic times of a system of satellites being as the 3/2 power their distances; and, 2) the centripetal forces deflecting them into their orbits being inversely as the squares of their distances from the center. The first condition, of course, is the Keplerian harmonic law. The relevant cited phenomena are the second part of Phenomenon 1 and Phenomenon 5. These specify respectively that the moons of Jupiter and the primary planets are orbital systems satisfying this harmonic law.

The third form of argument appeals to Corollary 1 of Proposition 45, Book 1. This is one of Newton's most interesting theorems about precessing orbits. Given the assumptions, it establishes *systematic equivalences* between rates of orbital precession and the centripetal force law. Precession forward is equivalent to having the centripetal force fall off faster than the inverse square of the distance. A stable elliptical orbit is equivalent to having the centripetal force vary exactly as the inverse square of distance. Precession backward is equivalent to having the centripetal force fall off slower than the inverse square. The result is quite sensitive. Newton pointed out that having the centripetal force vary inversely as the 2 and 4/243 power (only slightly faster than the inverse square) is equivalent to about 3° precession forward for every revolution.[4]

Precession of an orbit may be regarded as a phenomenal magnitude. Relative to its assumptions, the theorem established a systematic equivalence between values of this phenomenal magnitude and corresponding values of a theoretical magnitude specifying the law of the centripetal force. Such equivalences show that (given the assumptions) values of the phenomenal magnitude carry the information that corresponding values of the theoretical magnitude obtain.[5] Under such conditions one may regard the phenomenal magnitude as measuring the corresponding theoretical magnitude. This wonderful example of Newton's use of mathematical equivalence theorems allowed him to treat the stability of an elliptical orbit as though it were a measurement of the inverse-square variation of the centripetal force deflecting a planet into that orbit.

A closer look at the theorems available to back up the first two forms of argument shows that they too are supported by such systematic equivalences. Theorem 1, Book 1, established the converse of the implication established in Theorem 2. Relative to the assumptions, these two theorems together establish an equivalence between the centripetal direction of a deflecting force and having a constant rate at which radii—from the deflected body to that center—sweep out areas. Corollary 1 of Theorem 2 gives the equivalences according to which an increasing rate corresponds to having the deflecting force be off-center in the direction of tangential motion, while a decreasing rate corresponds to a direction opposed to the tangential motion.[6]

The theorem backing up the second form of argument (Corollary 6, Proposition 4, Book 1) is already in the form of an equivalence between having the periodic times as the 3/2 power of the distances and having the centripetal forces inversely as the squares of those distances. Corollary 7 of Proposition 4 establishes a general equivalence between having the periodic times be as any power n of the distances and having the corresponding centripetal forces be inversely as the power $2n - 1$ of the distances.[7]

Clark Glymour has argued that Newton's inferences from phenomena are examples of a form of inference he has called "bootstrap confirmation."[8] For a bootstrap confirmation the inferred proposition must be *deducible* from the phenomenon together with background assumptions, and those background assumptions must be compatible with alternative phenomena together with which they would have contradicted the proposition. The equivalence theorems to which we have been calling attention show that Newton's inferences do satisfy Glymour's bootstrap confirmation. But they also show more. We have seen that Newton's background assumptions support specific corresponding values of the theoretical magnitude over a whole range of alternative values of the phenomenal magnitude with which the assumptions are compatible. As we have noted, this makes the phenomenon into a measurement of the value of the theoretical magnitude specified in the proposition inferred from it.

GENERAL BACKGROUND ASSUMPTIONS

Several sorts of assumptions are involved in these inferences from phenomena, assumptions in the phenomena themselves, and two sorts of assumptions involved in the proofs of the equivalence theorems. These latter include specific idealizations in the models for which the theorems are proved and the general background assumptions for Newton's mathematical approach to natural philosophy. I want to begin with the general background assumptions.

Newton introduced his laws of motion in a separate section before the beginning of Book 1 of the *Principia*.[9] Book 1 is titled *The Motion of Bodies*. Its opening section is devoted to lemmas developing his method of first and last ratios.[10] These lemmas give the version of the calculus Newton used in *Principia*. The laws of motion and the method of first and last ratios together with Euclidean geometry and other familiar mathematical methods form the general background assumptions within which Newton developed all of his mathematical principles of natural philosophy.

The opening sentence of Newton's scholium to the laws suggests that the laws of motion were acceptable to Newton's intended readers and that they were supported by many kinds of experiments. "The principles I have set forth are accepted by mathematicians and confirmed by experiments of many kinds."[11] When we look more closely at the scholium and his discussions of the laws and corollaries we see several different sorts of arguments. Some are thought experiments having the sort of intuitive compulsion to belief that we find in Euclidean geometrical demonstrations.[12] Some of the confirmations by experiment seem to be like the inferences from phenomena we are investigating.[13] In addition there is the broad base of inductive support provided by the many successful applications of these laws in explanations of diverse motion phenomena.

After the *Principia* was assimilated and the Newtonian framework was entrenched into the practice of science, a number of writers, such as Kant and Whewell began to treat these laws of motion as necessary truths.[14] Nevertheless, Einstein's special theory of relativity transformed these laws into new versions incompatable with the old ones.[15] It may be useful to look at the role played by the development of rival conceptions in such radical transformations. One of the background assumptions involved in Newton's discussion of the laws is the parallelogram method of compositions of motions. This classical kinematics was a new conception, pioneered by Galileo, under which lines and figures represented motions and compositions of motions. Built into this conception was the principle of Galilean invariance to which Poincaré provided an alternative—an alternative Einstein showed us we ought to adopt. Kant and many eighteenth- and nineteenth-century physicists such as Hertz regarded Galilean invariance as immune to empirical refutation.[16] Yet, when Einstein showed that an alternative conception could organize the relevant phenomena as well or better than the Galilean conception, most physicists had little difficulty in coming to regard the Galilean invariance principle as empirically refutable.[17]

For some of these physicists the Galilean kinematical conception may have had a status akin to what Hilary Putnam has called "contextually a priori."[18] They regarded Galilean invariance as an a priori constraint on compositions of motions, so long as they were unaware of any rival alternative. But, once an alternative conception that could rival its role in organizing the relevant phenomena became available, Galilean kinematics lost its privileged status.[19]

Perhaps the most revolutionary aspect of Einstein's general theory of relativity was that it dethroned Euclidean geometry from its status as the received account of the geometry of physical space.[20] School children from the time of Plato's *Meno* have learned to appreciate the warranting powers of intuitive demonstrations in Euclidean geometry. Such demonstrations had seemed to secure a priori status to this geometry, even as an account of physical space in the large. Yet Einstein's theory led to an understanding that phenomena (such as the anomalous precession of Mercury's orbit) would count as empirical evidence counting against this geometry. This was a striking example of how formulation of a serious rival conception could change what had been taken to be an a priori commitment into something to be decided by empirical inquiry.

Newton did not seem to follow most of the thinkers of his time in holding geometry to be a priori. His preface to the first edition of the *Principia* suggests that he understood geometry to be as much open to empirical investigation as mechanics itself.[21] We shall see that Einstein's overthrow of Euclidean geometry can be understood as an example of the methodology Newton advocated in the *Principia*.

SPECIAL IDEALIZATIONS IN THE THEOREMS

Newton's elegant proofs of Propositions (also called Theorems) 1 and 2 of Book 1 impress all who look at them. They are a fascinating example of what I. B. Cohen has called Newton's mathematical style.[22] Newton first proves a result for a system that is so radically idealized that it stretches language to call it a model of the phenomenon. In this simple system, however, the result can be proved by geometrical demonstration. In these propositions Newton proved that instantaneous forces acting at discrete equal time intervals to deflect a body about a center will produce equal triangles by radii drawn to the center if (Theorem 1) and only if (Theorem 2) those forces are directed at the center. He then showed that the result continued to hold as he transformed this simple system into a more realistic one. Newton applied his method of first and last ratios to transform these theorems about triangles produced by discrete instantaneous deflecting forces to the corresponding theorems for areas swept out by smooth curves produced by continuously acting forces.[23]

When Newton applied these results in his argument for Proposition 2, Book 3, he referred to Theorem 2, Book 1, but not to Theorem 3. Theorem 2 holds for centers that are unaccelerated. Thus he treated the sun as an approximately unaccelerated center. His argument for Proposition 1, on the other hand, referred to Theorem 3 as well as to Theorem 2. In Theorem 3, Book 1, Newton appealed to Corollary 6 of the laws of motion, to extend the results to motions relative to centers in nonuniform motion.

Corollary 6 of the Laws of Motion
If bodies are moving in any way whatsoever with respect to one another and are urged by equal accelerative forces along parallel lines, they will continue to move with respect to one another in the same way as they would if they were not acted upon by those forces.[24]

Proposition 1 applies the argument to Jupiter's moons. Jupiter and its moons undergo an empirically significant centripetal acceleration toward the sun, but these accelerations are approximately equal and parallel since the orbits of Jupiter's moons are so small compared to Jupiter's orbit.

Newton made similar use of idealization in his arguments from the harmonic law to establish the inverse-square variation of centripetal forces. The main theorems backing up these inferences (Proposition 4, Book 1, Corollaries 6 and 7) are proved for systems with concentric circular orbits. As Newton pointed out in his discussion of Phenomenon 1, Book 3, the idealization of concentric circular orbits did not conflict with the available empirical data about Jupiter's moons.[25] In the case of the primary planets, however, this idealization could only be an approximation since small eccentricities in these orbits had been empirically established. His reference in the proof of Proposition 2, Book 3, was Proposition 4, Book 1, but not to the more specific reference to Corollary 6 of that proposition, which he made in his argument for Proposition 1. This suggests that he may have intended to include Corollary 8 in the reference.

Corollary 8
In cases in which bodies describe similar parts of any figures that are similar and have centers similarly placed in those figures, all the same proportions with respect to the times, velocities, and forces follow from applying the foregoing demonstrations to these cases. And the application is made by substituting the uniform description of areas for uniform

motion, and by using the distances of bodies from the centers for the radii.[26]

This generalization removes the need for the assumption of concentric circular orbits.[27]

Newton's second form of argument for establishing inverse-square variation of centripetal forces also allows the assumption of concentric circular orbits to be weakened. As we have seen, according to Corollary 1, Proposition 45, Book 1, the absence of precession in an elliptical orbit establishes the inverse-square variation of the centripetal force.[28] This precession theorem is more general in other ways.[29] It can be applied to systems (like the lunar orbit) where there is only one satellite so that Kepler's harmonic law does not apply.

Our brief look at Newton's treatment of the theorems backing these inferences from phenomena has revealed progressions from very radical idealization toward more realistic systems. But some striking idealizations remain. The orbiting bodies are treated as point masses, and, perhaps most striking to us, all the theorems are one-body idealizations. Even in the harmonic law theorems, which apply to relations among several orbits, the only forces considered are the centripetal forces on the orbiting bodies. There are neither forces of interaction among these bodies, nor any actions by any of them on a central body. These theorems show that Kepler's laws are laws of orbital motion for bodies that can be construed as point masses undergoing isolated inverse-square centripetal forces.

PHENOMENA

In Proposition 1, Book 3, Newton inferred inverse-square centripetal forces on Jupiter's moons from Phenomenon 1—which specifies that these orbits satisfy a Keplerian harmonic law and law of areas. In Proposition 2 he inferred centripetal forces on the primary planets from Phenomenon 5: "The primary planets, by radii drawn to the earth, describe areas that are in no way proportional to the times, but, by radii drawn to the sun, traverse areas proportional to the times."[30] He inferred that these centripetal forces are inverse-square from Phenomenon 4: "The periodic times of the five primary planets and of either the sun about the earth or of the earth about the sun—the fixed stars being at rest—are in the sesquialteral ratio [or as the 3/2 power] of their mean distances from the sun."[31] These phenomena are neutral between Kepler's model where the earth orbits the sun and an alternative Brahean model where the sun orbits the earth while the primary planets orbit the sun. Kepler's model takes the sun as the fixed center for determining true motions. The Brahean model takes the earth as the center relative to which true motions are defined. Each of these models is equally able to fit the empirical data about the motions among themselves of the bodies in the solar system.[32]

At a later stage (Proposition 12, Book 3) Newton used a dynamical argument to establish that Kepler's model was the approximately correct account. He began this argument by explicitly assuming an hypothesis—that the center of the system is at rest.[33] He then appealed to Corollary 4 of the laws of motion to infer that this hypothesized resting center is the common center of gravity of the solar system (Proposition 11, Book 3).

Corollary 4 of the Laws of Motion
The common center of gravity of two or more bodies does not change its state whether of motion or of rest, as a result of the actions of the bodies upon one another; and therefore the common center of gravity of all bodies acting upon one another (excluding external actions and impediments) either is at rest or moves uniformly straight forward.[34]

He went on (Proposition 12) to argue that: "The sun is in continual motion but never recedes far from the common center of gravity of all the planets."[35] He did this by calculating the common center of gravity for various configurations of sun and planets. This calculation appealed to the universality and direct proportion to quantity of matter of gravity (Proposition 7, Book 3) and to the result that gravitating spheres can be treated as point masses (Proposition 8, Book 3) to measure the relative masses of the sun and several planets from the orbits of their satellites.[36]

True motions are those that count as accelerations (more accurately, changes in momentum) to be explained by actual forces. Newton's center-of-mass construction will yield an approximately unaccelerated frame that can be used to determine the true motions among themselves of a system of bodies, so long as there are no significantly differential external forces or impediments. By "significantly differential" I mean that they correspond to accelerations that are not approximately equal and parallel; such forces or impediments would violate the condition of Corollary 6 of the laws. The precession theorem allows stability of orbits (or precessions limited to those that can be accounted for by interaction among the bodies) to count as a mark that the system is sufficiently free from these differential external forces or impediments.

Universal gravitation showed that Kepler's model was far closer to the truth than its Brahean rival, but it also showed that Kepler's model was only an approximation to the true motions. Newton's progression toward more realistic systems went beyond the harmonic law phenomenon. I. B. Cohen has identified Proposition 60, Book 1, as a transformation of Kepler's harmonic law from the one-body idealization (in Proposition 4, Book 1) where the only forces are the centripetal forces on the satellites to a more complex two-body idealization where the satellites are allowed to interact with the central body (though not with each other).[37] Karl Popper had used a version of this transformed law to demonstrate that universal gravitation entails violations of Kepler's harmonic law even if one ignores interactions between satellites.[38] The violations are entailed because universal gravitation requires interactions between the satellite and the primary.[39] The question this has set for philosophers of science is whether an argument like Newton's could ever be successful when success would require undercutting the very premises it starts from.

In order to answer this question we shall look at the role of idealization in what Newton counts as phenomena. Kepler's laws are no mere summaries of the relevant observations. They are *generalizations* that extend and idealize the data. William Whewell construed the transformation of the data into Kepler's laws as a form of induction he called colligation of facts. According to Whewell,

> The colligation of ascertained Facts into general Propositions may be considered as containing three steps, which I shall term *the selection of the idea, the construction of the conception, and the determination of the magnitudes.*[40]

He also tells us

> These three steps correspond to the determination of the *Independent variable,* the *Formula,* and the *Coefficients,* in mathematical investigations, or to the *Argument,* the *Law* and the *Numerical data* in a Table of an Astronomical or other Inequality.[41]

I think a key to understanding the role of idealization in Newton's phenomena is to understand the step in colligation that Whewell calls "constructing the conception."

Let us see how this might help with violations of the harmonic law. Kepler's harmonic law is a higher order colligation in which the data are coefficients of the orbits already colligated from observations of the planets. Where R is the average distance (the major semi-axis of the ellipse) and T is the period of a orbit, the ratio of R^3/T^2 is the same for each of the planets (when the same units are used). Newton offered a summary of the data supporting Kepler's harmonic law in his discussion of Phenomenon 4.

> This ratio, which was found by Kepler, is accepted by everyone. In fact, the periodic times are the same and the dimensions of the orbits are the same, whether the sun revolves around the earth, or the earth about the sun. There is universal agreement among astronomers concerning the measure of the periodic times. But of all astronomers, Kepler and Boulliau have determined the magnitudes of the orbits from observatons with the most diligence, and the mean distances that correspond to the periodic times [that is, as found from the periodic times by computation from the above ratio] do not differ sensibly from the distances that these two astronomers found [from observations], and for the most part lie between their respective values, as may be seen from the following table.[42]

To see how colligation works here, take as arguments the squares of the periodic times and assign as data the cubes of the distances estimated by the astronomers. This will give a graphical representation of Newton's table in which the *law* corresponds to the selection of a straight-line as the best fitting curve. The slope of this line is the constant ratio $R^3/T^2 = k$. It is fixed by our choice of units; a convenient choice is to use astronomical units [the mean earth-sun distance] and sidereal years [the period of the earth's orbit, which Newton has at 365.2565 days]. This will make the constant k equal 1.0. The selection of $R^3/T^2 = k$ as the law or formula for this best fitting curve is the step Whewell would call "constructing the conception."

Newton knew that universal gravitation required corrections to the harmonic law. In Proposition 15, Book 3 he even provided a method of computing what the correction should be, due to the interaction between each planet and the sun, but this did not prevent him from using the harmonic law as a premise from which to infer inverse-square variation of the centripetal forces on the planets. Moreover, he did so in the first edition of 1687 even though he had been speculating since 1684 that there might be observable violations of the harmonic law produced by interactions of Jupiter and Saturn.[43] By 1694 he had information that he took to establish sensible violations of the law of areas due to mutual perturbations of Jupiter and Saturn at their conjunction of 1683.[44] Yet in the second edition of 1713 he continued to use both the law of areas and the harmonic law as premises in his argument. Indeed, it was in the second edition that he promoted these laws from the rank of hypotheses to that of phenomena.[45] Newton never did regard any theoretical or even observable violations of these laws due to interactions among bodies in the system as a problem for his inferences to inverse-square centripetal forces on the planets.

We can now make our puzzled philosopher's question more specific by asking how (and if) this could have been reasonable. I think when we see the question in this specific context the answer is clear. Newton's practice was reasonable. This was because the formula for a best-fitting curve for such a perturbed orbit is properly conceived as a formula for a composition of motions one of which is the Keplerian orbit that fits the law and the other of which is the perturbation produced by the interaction. According to such a conception the Keplerian phenomenon is there to be found in the data.[46] It is, however, transformed from a claim about the total motion to a claim about that component of the total motion caused by the inverse-square centripetal force.

Newton's use of inferences from phenomena is part of a process of theory construction that is, in some ways, more like an information feedback process than like the arguments—either deductive or inductive—from fixed premises with which contemporary philosophers usually deal. I. Bernard Cohen has suggested to me that G. E. L. Owen's interesting discussion of Aristotle's *tithenai phainomena* can illuminate the role played by Newton's phenomena.[47] According to Owen Aristotle's *tithenai phainomena* are the common conceptions on the subject accepted by all, or most, or by the wise.[48] They are not rock-hard data, but starting points that can be corrected as the investigation proceeds. Paul K. Feyerabend has also mentioned Aristotle's *tithenai phainomena* in connection with Newton's phenomena.[49] But he seems to have been mainly interested in contrasting what he took to be the clear-cut, easily established agreement characteristic of the conception of observable knowledge in Aristotle's philosophy with the theory-ladenness and vulnerability to correction of Newton's phenomena. Though he refers to Owen's paper, he does not say anything about Owen's interesting discussion about how Aristotle's starting points are subject to correction as investigation proceeds.

The special problem for Newton's argument is, however, that he maintained the forces that he inferred from the original phenomena as constraints on his developing theory, even after he corrected those phenomena from which he had inferred them. He inferred inverse-square forces directed toward the sun on each planet. He then used the theory to infer other forces to compose with these, e.g., the equal and opposite inverse-square forces toward each planet acting on the sun. These perturbing forces required correcting the original phenomena, but he did not construe these corrections as undercutting his commitment to the original component forces. This is the procedure that I claim to be reasonable. I claim that as long as the corrections were such as to maintain the original phenomena as true *component motions,* his inferences to the *component forces* they measure could continue legitimately to constrain the theory he was constructing.

What makes this answer work is that universal gravitation provided a principled way of decomposing motions into components. The component motions are specified by accelerations toward other bodies.[50] The Newtonian methodology for specifying such component motions was so powerful that sometimes perturbations remaining after accounting for the components toward known bodies were then used to locate new bodies. So it was that in 1846 Neptune was discovered after Adams and Leverrier estimated where to look for an unknown planet if that planet were to account for deviations in Uranus's orbit, for which the known planets could not account.

THE COGENCY OF NEWTON'S INFERENCES

On the view I am advocating, Newton's phenomena are idealizations that correspond to motions produced by forces limited to those specified in the equivalence theorems supporting Newton's inferences from them. This may make it appear that the phenomenon is an ad hoc idealization constructed to fit the proposition being inferred from it. Feyerabend, for example, has made just this criticism. He suggested that such inferences are logically vacuous, even if they are effective pieces of rhetoric.[51] I have pointed out that a phenomenon is constructed by settling upon a conception of what is to count as a best-fitting curve for the data. This point, also, may lead some to suggest that the phenomenon is more our own creation than something fixed by nature. What these suggestions emphasizing "free creation" and "mere rhetoric" overlook is the powerful constraining role played by the data in what could count as a phenomenon colligated from it.

The fit of Kepler's harmonic law to the data was extremely good, independent of Newton's theory.[52] I have said that plotting R^3 values against T^2 values leads to a clear selection of the straight line of slope $R^3/T^2 = K$ as the best-fitting curve. One quite simple indication of how closely clustered about this line these points are is how closely clustered the R^3/T^2 values are about K, which is the constant slope of the line. Where R is in astronomical units and time is in sidereal years, this constant $K = 1.0$. The closeness of the clustering of the ten values (Kepler's and Boulliau's values for each of the five planets) is indicated by the mean 1.00077 and standard deviation .0059. These indicate that the constant line $R^3/T^2 = 1$ fits this data very closely indeed.

In Proposition 15, Book 3, Newton told us how to use the theory of gravitation to correct the harmonic law.

Proposition 15, Problem 1

To find the principle diameters of the [planetary] orbits. These diameters are to be taken in the inverse sesquialteral proportion [or as the 2/3 power] of the periodic times by bk. 1, prop. 15; and then each one is to be increased in the ratio of the sum of the masses of the sun and each revolving planet to the first of two mean proportionals between that sum and the sun, by bk. 1, prop. 60.[53]

I carried out this calculation for Jupiter and Saturn using Newton's estimates of the masses of these planets and that of the sun (Proposition 8, Corollary 1).[54] In the case of Jupiter the corrected value is slightly closer (.575 standard deviations as against .689 standard deviations) to the mean of the measurements from the R calculated from the period using the harmonic law.[55] In the case of Saturn the corrected value is farther away from the mean of the measurements than the harmonic-law value. This may be due to an error in Kepler's low R measurement—the corrected value is closer to Boulliau's value. In each case it would take better data than that of Newton's table to make the corrections empirically significant.

Supose Newton had data at hand that was precise enough to make these corrections fit significantly better than the harmonic-law estimates. The new estimates are explicitly constructed as corrections to the harmonic-law estimates. This shows that the conception of this new specification as a perturbation of Keplerian values is not an ad hoc device to save the Keplerian phenomenon. Moreover, if we wanted to recover the R's corresponding to the unperturbed centripetal force from this data we could apply the correction in reverse.[56] That the centripetal component force is inverse-square would be shown by the close fit of a straight line to the R^3/T^2 values computed for these recovered R values. So, this property of the centripetal component force can be measured from the data even in the presence of perturbations.

I have just defended the harmonic-law argument for the inverse-square variation of the centripetal forces on the planets. One might wonder why, if this argument is as good as I say it is, did Newton go on to offer the precession argument as well. On the interpretation I am proposing there is a *special advantage to inferences to a proposition from alternative phenomena*. Each such inference is a measurement of the value of the relevant magnitude specified in the proposition. An inference to this same proposition from another phenomeneon is *an independently agreeing measurement of this same magnitude*. The precession theorem makes the relative stability of the planetary orbits into a measure of the inverse-square variation of the centripetal forces holding the planets in their orbits. The two arguments together establish the inverse-square variation of the centripetal forces as a common cause of the harmonic law and the stability of the orbits. This sort of unification and its epistemic virtues will be my next theme.

Generalizing Propositions by Induction

. . . & redduntur generales per inductionem.

UNIFICATION

As we have seen, Newton's harmonic-law argument for Proposition 2 showed that the agreement in the ratios R^3/T^2 for different planets was a measure of the inverse-square variation of the centripetal forces on those planets. This was the beginning, but not the end, of Newton's account of the dynamical significance of Kepler's harmonic law. From Proposition 5 Book 3 he could conclude that these various forces are all *accelerative measures,* at their various distances, *of the sun's gravity;* therefore, he could conclude that the constant ratios R^3/T^2 were each a measure of the absolute quantity of centripetal force of this common center.[57] From Proposition 7, he could conclude that this absolute quantity of the sun's gravity is directly proportional to its quantity of matter. These propositions establish equivalences according to which *each of these ratios R^3/T^2 is a separate measurement of the mass of the sun.*

The orbit for Mars is a certain ellipse specified as the best-fitting curve for the astronomers' data about the motions of that planet.

The mean distance R (the greatest semi-axis) and the period T are parameters of this ellipse that are measured from the data. These measurements fix the ratio R^3/T^2 for Mars and this ratio may be regarded as a phenomenon colligated from the data measuring R and T. The empirical base for Kepler's harmonic law was the remarkable agreement among these various ratios.[58] Newton's unification explained this remarkable agreement by showing that these ratios were all effects that measured the same common cause.

A special virtue of this unification was that it brought the evidential force of the empirical agreement to bear on estimates of the mass of the sun. When we get such an agreeing measurement for a magnitude from a new phenomenon the data colligated by that phenomenon is added to the data already supporting the estimate. This provides two kinds of additional support.[59] First, closely agreeing values in a larger data set will, typically, support higher confidence levels for estimates than would be supported by equally close agreement in smaller subsets of that data set. Second, closely agreeing values in the larger data set will, typically, make the estimates more robust with respect to perturbation by new data. This is because the estimates from the new data will be evaluated by goodness of fit with respect to the whole combined data set.

In our everyday reasoning and in our scientific practice was put great weight on finding agreement among measurements of a magnitude by means of different phenomena. The foregoing remarks show that this practice is backed up by elementary facts about data analysis. When a new phenomenon is found to be an agreeing measure of a magnitude we are already using to explain other phenomena, which are measures of it that also agree, we have a successful unification. Our magnitude is now seen to be a common cause for the new phenomenon as well as for the old ones.[60] Such unification increases support for our supposition by adding the data from which the new phenomenon is colligated to the set used to estimate our magnitude. There are two conditions required for such a unification to be successful. There is an empirical part: the agreement in measurements. Without this the new phenomenon may erode rather than add to the support for our supposition. The second condition is that we be able to generate the relevant equivalence theorems from our background assumptions. Without this we might not be able to legitimately combine the data sets at all.

Newton's first two rules of reasoning appear to be general policies to guide our reasoning from effects to their causes. One important role they play is to endorse the sort of unification we have been exploring.

Rule 1   No more causes of natural things should be admitted than are both true and sufficient to explain their phenomena.[61]

Rule 2   Therefore, the causes assigned to natural things of the same kind must be, so far as possible, the same.[62]

We can read these rules, together, as telling us to opt for common causes wherever we can succeed in finding them. This seems to be exactly the role they play in their most central application in the *Principia*. This is their application to support Newton's inference in Proposition 4, Book 3, to identify the inverse-square centripetal force on the moon (argued for in Proposition 3) with terrestrial gravity.

Newton's famous moon test showed that on the assumption that the form of the centripetal force on the moon was inverse square, its centripetal acceleration would equal the acceleration of gravity at the surface of the earth. Here is his appeal to Rules 1 and 2 to argue from this agreement to the identification of the centripetal force on the moon with terrestrial gravity.

And therefore that force by which the moon is kept in its orbit, in descending from the moon's orbit to the surface of the earth, comes out equal to the force of gravity here on earth and so (by rule 1 and rule 2) is that very force which we generally call gravity.[63]

We have two phenomena: the centripetal acceleration of the moon and the law of a seconds pendulum at the latitude of Paris. Each measures a force producing accelerations at the surface of the earth. These accelerations are equal and equally directed toward the center of the earth. This agreement in the measured values is a higher order phenomenon, like Kepler's harmonic law. Identifying the forces explains the agreement by the claim that each phenomenon measures the very same force. This makes these phenomena into effects of a single common cause.

We can see how each rule applies here. For Rule 1, note that if we do not identify the forces we will have to admit two causes rather than one to explain the two basic phenomena. Moreover, we will either have to leave the higher order phenomenon of the agreement unexplained or introduce yet another cause to explain it. The identification certainly does minimize the number of causes we have to admit. For Rule 2, note that our basic phenomena are of the same kind in so far as each measures a force producing the same centripetal accelerations at the surface of the earth. On the identification we assign the very same cause to each. It also extends the sense in which we understand them as phenomena of the same kind by making them each measure the same force.

When Newton makes the identification of the force on the moon with terrestrial gravity he is imposing a constraint on the theory that he is constructing.[64] After Proposition 4, any equivalences that make a phenomenon into a measurement of gravity also make it into a measurement of the centripetal force on the moon. So, to the extent that balls rolling down inclined planes, projectile motions corrected for air resistance, or rates of fall in a vacuum can be made to measure the acceleration of gravity they will also measure the centripetal force on the moon. Similarly, of course, any measurement of the centripetal force of the moon measures gravity. The identification also transforms the common notion of terrestrial gravity by making it now vary inversely with the square of distance from the center of the earth.

GENERALIZATION

Newton applied Rule 2, on its own, in his argument for Proposition 5. From the case of the moon, he generalized the identification of inverse-square centripetal forces of satellite toward primary with gravitation to the moons of Jupiter and Saturn, and to the primary planets.

Proposition 5, Theorem 5

The circumjovial planets [or satellites of Jupiter] gravitate toward Jupiter, and the circumsaturnian planets [or satellites of Saturn] gravitate toward Saturn, and the circumsolar [or primary] planets gravitate toward the sun, and by the force of their gravity they are always drawn back from rectilinear motions and kept in curvilinear orbits.

For the revolutions of the circumjovial planets about Jupiter, of the circumsaturnian planets about Saturn, and of Mercury and Venus and the other circumsolar planets about the sun are phenomena of the same kind as the revolutions of the moon about the earth, and therefore (by Rule 2) depend on causes of the same kind, especially since it has been proved that the forces on which those revolutions depend are directed toward the centers of Jupiter, Saturn, and the sun, and decrease according to the same ratio and law (in receding from Jupiter, Saturn and the sun) as the force of gravity (in receding from the earth).[65]

The unification here is to explain all these phenomena as effects of instances of the same kind of force—gravitation of a satellite toward its primary. On the unification we can understand each phenomonon as a measurement of general features of this kind of force—

centripetal direction and inverse-square accelerative measures. To the extent we have an inference to a common cause it is to a common kind rather than to the very same instance. This is unification under a theory.

Rule 2 says we are to adopt such unification *as far as possible*. What does it take to be able to count a theory as achieving a unification of this sort? I think our previous discussion of the role of equivalence theorems in Newton's inferences from phenomena can illuminate this question. Newton established a new mathematical ideal to guide our attempts to give causal explanations. According to this ideal *our theory should deliver equivalences that make the phenomenon to be explained into a measurement of the relevant features of the magnitudes which figure in its explanation.* One criterion for successful unification under a theory is the capacity of the theory to deliver such equivalences for the phenomena to be explained.

Our discussion of common-cause unification indicated that the confidence and robustness of our estimates of a magnitude ought, generally, to increase as it receives an agreeing measurement by a phenomenon that had not been known to measure it before. As the network of such cohering explanations generated by a theory becomes more extensive and more unified we will legitimately make our estimates of its magnitudes with more confidence and make them more resistent to perturbation by new data. This suggests that support for the theory, also, ought generally to become stronger as this network of successful measurements of its magnitudes becomes larger and more unified. Query 31 of the second English edition of the *Opticks* indicates that Newton might agree.

> And although the arguing from experiments and observations by Induction be no Demonstration of general conclusions, yet it is the best way of arguing with the Nature of Things admits of, and may be looked upon as so much the stronger, by how much the induction is more general.[66]

I am reading the last clause so that the induction is more general just in case the scope of its base of support in phenomena it explains is more general. On my reading, Newton is endorsing the idea that the argument to a generalization may be looked upon as stronger as it successfully unifies more phenomena under it.

This role for inferences from phenomena gives a structure to the support for a theory that hypothetico-deductivists have not paid sufficient attention to. Just as the confidence and robustness of estimates of the various magnitudes will differ according to the support they receive from specific measurements by phenomena, so too will the various general suppositions of a theory receive differential support from the specific phenomena unified under them. These differences in support among its various suppositions are especially important when a theory needs to be revised. They help fix the relative immunity to revision that guides the modification of what Quine (1970) would call the "web of belief."[67]

Newton's argument for Proposition 6 *Principia,* Book 3, is a prime example of what he called "arguing to general conclusions by induction" in our quotation from Query 31.

Proposition 6, Theorem 6

All bodies gravitate toward each of the planets, and at any given distance from the center of any one planet the weight of any body toward that planet is proportional to the quantity of matter which the body contains.[68]

He cited several phenomena and other propositions as the basis for this induction. First, he cited the direct proportionality of weight of terrestrial bodies to their quantities of matter, which he demonstrated by an ingenious pendulum experiment testing the equality of the proportion for gold, silver, lead, glass, sand, common salt, wood, water, and wheat.[69] Second, he cited the property of gravity toward any planet and the sun to produce equal accelerative measures on unequal bodies.[70] This was supported by generalization from the case of the earth; by the inverse-square accelerative measures of gravitation toward the sun, Jupiter, and Saturn inferred from the respective harmonic law phenomena of their satellites; and by the generalization of this gravity to the planets without moons in Corollaries 1 and 2 of Proposition 5. Another phenomenon was the stability of the orbits of Jupiter's moons—which he argued would be very disturbed if the proportions of their weights toward the sun to their respective quantities of matter were different from the corresponding proportion for the weight of Jupiter toward the sun to its quantity of matter.[71] By this same argument the stability of the moons of Saturn and the relative stability of the orbit of our moon counted as additional phenomena.[72] Finally, he also cited and argued for the proposition that the weights of the parts of a planet toward any other planet are to one another as the quantity of matter in the individual parts.[73]

It is clear that Newton *made considerable effort* to measure the variation of the motive force of gravity to a planet on bodies at equal distances by as many different sorts of phenomena as he could and that in each case the motive forces on the bodies were directly proportional to the quantities of matter in them.[74]

In Corollary 2 of Proposition 6 Newton appealed to an explicit Rule of reasoning to bolster this argument by induction.

All bodies universally that are on or near the earth are heavy [or gravitate] toward the earth, and the weights of all bodies that are equally distant from the center of the earth are as the quantities of matter in them. This is a quality of all bodies on which experiments can be performed and therefore by Rule 3 is to be affirmed of all bodies universally.[75]

Here is Rule 3.

Those qualities of bodies that cannot be intended and remitted [that is, qualities that cannot be increased and diminished] and that belong to all bodies on which experiments can be made should be taken as qualities of all bodies universally.[76]

The condition that it is qualities that cannot be intended or remitted to which this rule applies limits its application to qualities with invariant measures. The quality to be affirmed of all bodies in Corollary 2 seems to be gravitation toward the earth. The invariant measure is that the weights of all bodies at equal distances from the center of the earth are directly proportional to the quantities of matter in them.

In his rather lengthly discussion of Rule 3 Newton illustrates its application to argue that gravity is *universal*.

Finally, if it is universally established by experiments and astronomical observations that all bodies on or near the earth gravitate [are heavy] toward the earth, and are so in proportion to the quantities of matter in each body, and that the moon gravitates [is heavy] toward the earth in proportion to the quantity of its matter, and that our sea in turn gravitates [is heavy] toward the moon, and that all planets gravitate [are heavy] toward one another, and that there is a similar gravity [heaviness] of comets toward the sun, it will have to be concluded by this third rule that all bodies gravitate mutually toward one another.[77]

Here the whole range of *sorts* of phenomena appealed to in his argument for Proposition are included in the inductive base for the inference. Once again the invariant measure is the direct proportionality to quantity of matter.

"YET I AM BY NO MEANS AFFIRMING THAT GRAVITY IS ESSENTIAL TO BODIES"

Newton's discussion of Rule 3 concludes in a somewhat puzzling way. Here are these concluding remarks. They follow immediately upon the passage we have just been quoting.

> Indeed, the argument from phenomena will be even stronger for universal gravity than for the impenetrability of bodies, for which, of course, we have not a single experiment, and not even an observation, in the case of the heavenly bodies. Yet I am by no means affirming that gravity is essential to bodies. By inherent force I mean only the force of inertia. This is immutable. Gravity is diminished as bodies recede from the earth.[78]

Here Newton on the one hand shows that gravity satisfies Rule 3 as well as or better than impenetrability; while, on the other hand, he refrains from affirming it as essential on the grounds that it does not satisfy Rule 3 because it is diminished with distance. As Koyré pointed out,

> And as to the statement of the third edition of the *Principia,* it certainly was beside the point: it was not *gravitas* as weight *(pondus)* that was the question but *gravitas* as an attracting power, the *pondus* of which was only an effect. Thus it could—and according to Newton himself did—remain constant, all changes in *weight* notwithstanding.[79]

On Newton's theory the gravity of a body—as the absolute measure of the centripetal force it generates—is fixed by its quantity of matter. This is just as its inertia is; therefore, according to Rule 3, it ought to have as good a claim as inertia to be considered an essential quality. Moreover, the key features of the gravitational attractions that this body produces on other bodies—directly proportional to their quantities of matter and inversely proportional to the square of distance from the center of gravity of the first body—are also invariant in just the way required by Rule 3. J. E. McGuire has cited draft passages where Newton explicitly claims that the direct proportionality of the force of gravity to quantity of matter cannot be intended or remitted.[80]

As we shall see, Newton had reason to refrain from using Rule 3 to argue that gravity was essential. But, an examination of the text of Rule 3 reveals that it only claims that qualities satisfying its criterion are *universal,* not that they are essential. Moreover, the arguments we have been examining from the explanation of Rule 3 and in its application in Corollary 2, Proposition 6, only purport to show universality and not essentiality. So, why did Newton formulate his disclaimer in such a way as to suggest that gravity does not satisfy the criterion of not admitting intention or remission of degree?

McGuire argues that Newton *did* construe the criterion of not admitting intention or remission of degree as a criterion of *essential* and not merely of *universal* qualities of matter. He suggests that this essentialist use of the criterion is deeply imbedded in Newton's own conception of matter, whose essential features, he thought, had to be able to be revealed to our senses in experiments. According to McGuire, Newton appealed to and transformed to his own use an essentialist tradition going back to Aristotle, as he wrestled with the philosophical difficulties of developing this conception. If this is correct, then it gives a reason why Newton was unable to cut away the essentialist implications of the criterion when he used it in Rule 3 as part of his criterion for picking out universal qualities of matter.

Newton's generalizations by induction that we have been examining are remarkably different from the inductions Popper and Carnap disputed. They are not examples where the probability gradually increases that an arbitrary body will gravitate toward planets as support for this indefinite singular claim grows as new phenomena are added to the inductive base. Nor are they examples of the somewhat more controversial inductions, where it is the probability of the universal generalization itself that is increased, as its support by instances grows. Instead, what we have here is something much more daring—a jump from the inductive base to outright acceptance of the universal generalization as true, or, to use a qualification we shall see in Rule 4, very nearly true.

Relatively mundane examples of this sort of inference abound in science as we know it. When we measure the specific gravity of a metal alloy we jump from the inductive base in samples we have measured to infer the same value for all samples of that alloy. Similarly, when we measure the charge on an electron we jump from the inductive base of those we have measured to attribute the same charge to all elections universally. I call such inferences "natural-kind inferences"[81]. They have not been studied nearly enough by philosophers of science, but some things about them seem clear. These inferences are not backed up by just their inductive bases alone, they also depend on accepted theories. The theories specify certain qualities as ones that have constant fixed magnitudes for objects of the kind in question, so that a correct estimate of such a magnitude for any object of that kind may be taken to measure the value of that magnitude for all objects of that kind universally. In this respect these background theories act like natural-kind conceptions and the magnitudes they specify for such inferences are treated as essential properties of the kind.

If Newton's generalizations by induction are to be counted as examples of this legitimate scientific practice of making natural-kind inferences, then his argument to establish that gravity is universal to bodies can succeed only if a theory that makes gravity essential to bodies is also established. We have been discussing one aspect of the old idea of essential qualities that, I have suggested, is built into the legitimate scientific practice of making natural-kind inferences. There are other aspects of this idea that are not built into this practice and that, I think, gave Newton good reason for this reluctance to affirm that gravity was essential to bodies.

One of these aspects of essentialism can be illustrated by the comparison Newton makes with inertia. This aspect is the idea that to recognize something as essential is to recognize that any call for causal explanation of it is illegitimate. Thus, it was a fundamental insight of the new dynamics that maintaining the state of motion did not need a causal explanation—that, rather, it was changes in this state that needed to be explained. Newton certainly wanted the enterprise of looking for a deeper causal explanation of gravity to count as a legitimate scientific project. He was, perhaps, as sensitive to the oddness of the idea of action-at-a-distance as Huygens and the other mechanical philosophers and at least as eager to provide a mechanical explanation of gravity. His genius was not that he did not share their metaphysical qualms, but rather that he did not let failure to give a mechanical explanation of gravity undercut his unified theory of terrestial and celestial motion.[82]

Our next topic will be Newton's efforts to protect his generalizations by induction from being undercut by hypotheses, including metaphysical ones such as the claim that all causes are mechanical. Before we turn to it, however, we note in passing that the resistance to change of a theory entrenched as a natural-kind conception is supported by the absence of a serious rival conception. In the absence of a rival to universal gravitation that could provide an alternative unified conception of motion phenomena, scientists could legitimately treat the Michelson-Morley experiment and the anomalous precession of Mercury (to use two of the famous examples discussed by Kuhn [1962] in his seminal treatment of scientific revolution) as puzzles or problems to be worked on rather than as empirical evidence against the theory.[83]

PROTECTION FROM HYPOTHESES

Whewell offered a criticism of Rule 2 that can set the stage for the discussion of Newton's efforts to protect his inductions from being undercut by hypotheses. According to Whewell, the problem of deciding when effects are of the same kind renders the application of Rule 2 vacuous.

Are the motions of the planets of the same kind with the motion of a body moving freely in a curvilinear path, or do they not rather resemble the motion of a floating body swept round by a whirling current? The Newtonian and Cartesian answered this question differently. How then can we apply this Rule with any advantage?[84]

It is only when it appears that comets pass through this plenum in all directions with no impediment, and that no possible form and motion of its whirlpools can explain the forces and motions which are observed in the solar system, that he [the Cartesian] is compelled to allow the Newtonian classification of events of the same kind.[85]

The first passage indicates that an application of Rule 2 can be undercut by an alternative hypothesis about what count as effects of the same kind. The second passage suggests that the difficulties raised in Book 2 of *Principia* for the vortex theory prevented it from undercutting Newton's appeal to Rule 2, when he argued (in Proposition 5, Book 1) for the identification of the orbital forces on the planets with gravitation toward the sun.

Newton did open his General Scholium (at the end of Book 3) with a recapitulation of difficulties that he had argued (in Book 2) beset the vortical hypothesis[86]. His famous "hypotheses non fingo" passage toward the end of that scholium, however, suggests that such hypotheses can just be ignored.

But hitherto I have not been able to discover the cause of those properties of gravity from phenomena, and I frame (feign[87]) no hypotheses; for whatever is not deduced from the phenomena is to be called an hypothesis; and hypotheses whether metaphysical or physical, whether of occult qualities or mechanical, have no place in experimental philosophy. In this philosophy particular propositions are inferred from the phenomena, and afterwards rendered general by induction. Thus it was that the impenetrability, the mobility, and the impulsive force of bodies, and the laws of motion and of gravitation were discovered. And to us it is enough that gravity does really exist, and act according to the laws which we have explained, and abundantly serves to account for all the motions of the celestial bodies, and of our sea.[88]

According to this passage the reason that the alternative conception of the vortex theory was a mere hypothesis was because it was not deduced from phenomena (or a result of generalization by inductions from propositions inferred from phenomena). What this suggests to me is that the vortical alternative conception was to be ignored unless it could deliver equivalence theorems to make a sufficiently rich base of orbital phenomena into measurements of vortical magnitudes.

In Rule 4 Newton explicitly codified this protection of inferences from phenomena and their generalizations by inductions from undercutting by hypotheses.

> Rule 4   In experimental philosophy, propositions gathered from phenomena by induction should be considered either exactly or very nearly true notwithstanding any contrary hypotheses, until yet other phenomena make such propositions either more exact or liable to exceptions.
>
> This rule should be followed so that arguments based on inductions may not be nullified by hypotheses.[89]

Some such protection is needed even to get off the ground. Even the most minimal generalization of data into a phenomenon can be undercut if the set of data points themselves is allowed to count as a rival candidate for best-fitting conception.

Newton raised the stakes required to get into the game as a serious competitor. According to his new rule merely qualitative analogies, such as the vortex theory Descartes had provided were not enough to make an alternative into a rival to be taken seriously.[90] In this new game an alternative conception could undercut a theory that had been generated by the methods of the experimental philosophy only if it could provide its own equivalence theorems to rival those of the original theory.

Bas van Fraassen has suggested that Rule 4 is an excessively conservative principle, and that it was limited in its descriptive application to the superseded scientific theories of a bygone age.[91] One role of Newton's standard for what is sufficient to count as a serious rival was that alternative theories that could meet the standard could end up overturning even the most cherished of accepted conceptions. This happened, right at the beginning, when Newton did not let his failure to find a mechanical cause of gravity interfere with his acceptance of the theory.[92] In the absence of such an account or some other mechanical theory that could meet the standard, the demand for a mechanical explanation was not allowed to carry any weight; it would not undercut his inferences from phenomena or their generalizations to the theory of universal gravitation.[93] So it was, I believe, in 1905 and 1916 that even such deeply entrenched conceptions as classical kinematics and Euclidean geometry had to give way when relativity theory was developed into a more successful unification of electromagnetic and motion phenomena.[94] And, so it may be, even today, when quantum mechanics is almost unchallenged because of its enormous empirical success, even though it is shot through with theoretical puzzles. I. Bernard Cohen has quoted Murray Gell-Mann:

All of modern physics is governed by that magnificent and thoroughly confusing discipline called quantum mechanics, invented more than fifty years ago. It has survived all tests and there is no reason to believe that there is any flaw in it. We suppose that it is exactly correct. Nobody understands it, but we all know how to use it and how to apply it to problems; and so we have learned to live with the fact that nobody can understand it.[95]

Cohen suggests a strong analogy between our situation with quantum mechanics now and the situation with gravity when natural philosophers began to learn to live with action-at-a-distance.[96]

Notes

This paper has benefited from advice and criticism by many who have read earlier drafts or heard me talk on this material. Among these, I. B. Cohen, Howard Stein, and Curtis Wilson deserve special mention for pointing out errors in detail or in conception.
Corlis Swain read and made valuable stylistic comments on a draft I had already revised from the version I sent the editors in September 1989. In January 1990 I received from Paul Theerman an astonishingly extensive and astute set of proposals for stylistic revisions of the earlier draft. The present paper has benefited greatly from these comments and revisions.
    1. Isaac Newton, *Philosophiae Naturalis Principia Mathematica*, the third edition (1726) with variant readings assembled and edited by Alexandre Koyré and I. Bernard Cohen (Cambridge: Harvard University Press, 1972), p. 746 (hereafter Koyré and Cohen). Translated "In this philosophy particular propositions are inferred from the phenomena, and afterwards rendered general by induction" in Isaac Newton, *Mathematical Principles of Natural Philosophy and System of the World*, trans. Florian Cajori (Berkeley: University of California Press, 1962), p. 547 (hereafter Cajori).
    2. Isaac Newton, *Mathematical Principles of Natural Philosophy*, trans. I. Bernard Cohen and Anne Whitman (Cambridge: Harvard University Press, 1987), pp. 496–97 (hereafter Cohen and Whitman).
    3. Cohen and Whitman, pp. 65–68.
    4. This is approximately the precession of our moon. It made his inference to

inverse-square variation of the centipetal force on the moon less straightforward than the inferences we are looking at.

5. I believe that attention to Newton's inferences from phenomena can illuminate discussions by philosophers (e.g., Dretske, Lewis, and Stalnaker) about implications for knowledge and inquiry of the idea of information carried by systematic causal dependencies. Newton's reasoning suggests that theory can specify standard conditions relative to which the information carrying equivalences hold, even if these standard conditions would fail if some of the alternative values of the phenomenal magnitude were actually to obtain.

In his interesting discussion of scientific observation, Dudley Shapere makes background theory very important. His view is strongly supported by the role background theory plays in generating the equivalences in virtue of which the phenomenon can carry the information that the proposition inferred from it is true.

F. Dretske, *Knowledge and the Flow of Information* (Cambridge: MIT Press, 1980); D. K. Lewis, "Veridical Hallucination and Prosthetic Vision," *Australasian Journal of Philosophy* 58 (1980): 239–49; R. Stalnaker, *Inquiry* (Cambridge: MIT Press, 1984).

6. This corollary was first published in the second edition (Koyré and Cohen, p. 93).

7. This corollary, also, was first published in the second edition (Koyré and Cohen, p. 99).

8. Clark Glymour, *Theory and Evidence* (Princeton: Princeton University Press, 1980), pp. 203–25. This form of inference has received considerable attention from philosophers of science, e.g., the essays in John Earman, ed., *Testing Scientific Theories* (Minneapolis: University of Minnesota Press, 1983).

9. Cajori, pp. 13–28; Koyré and Cohen, pp. 54–72.

10. Cajori, pp. 29–39; Koyré and Cohen, pp. 73–88.

11. Cohen and Whitman, p. 35.

12. Perhaps one such example is the thought experiment defending the application of Law 3 to attractions (Cohen and Whitman, p. 92). Another might be the argument for Corollary 2 of the laws (Cohen and Whitman, pp. 25–26).

13. For example, the argument in the Scholium (Cohen and Whitman, pp. 35–36) from Galileo's experiments may be construed as an inference from phenomena where the background assumptions are Law 1, Euclidean geometry, and what we may call Galilean kinematics. The argument in the Scholium (Cohen and Whitman, p. 42) for applying Law 3 to attractions from floating lodestone experiments may be an inference from phenomena where the background assumptions include Law 2 as well as the others we have just mentioned.

14. Immanuel Kant, *Metaphysische Anfangsgrunde der Naturwissennchaft*, originally published in Riga. Available in vol. 4 of *Kants Werke Akademie Textausgabe* (Berlin: Walter de Gruyter & Co., 1968); trans. James Ellington as *Metaphysical Foundations of Natural Science* (New York: Bobbs-Merrill Company, 1970). William Whewell, "On the Nature of the Truth of the Laws of Motion," *Transactions of the Cambridge Philosophical Society* 5 (1834): 149–72.

15. Albert Einstein, "Zur Elektrodynamik bewegeter Körper," *Annalen der Physik* 17 (1905): 891–921. Trans. W. Perett and G. B. Jeffrey as "On the Electrodynamics of Moving Bodies," in *The Principle of Relativity* (1923; reprint ed. New York: Dover Publications, 1952), pp. 35–66.

16. Kant, *Metaphysische Anfangsgrunde;* Heinrich Hertz, *Die Prinzipien der Mechanik, in neuem Zusammenhange dargestellt*, ed. Philipp Lenard (Leipzig: 1894). Trans. D. E. Jones and J. T. Walley as *The Principles of Mechanics, Presented in a New Form* (1899; reprint ed., New York: Dover, 1956).

17. There was considerable initial resistance to the special theory of relativity, including a rejection in 1907 by the physics faculty at Bern of Einstein's submission of his 1905 relativity paper as his *Habilitationsschrift*. Gerald Holton, "Einstein's Scientific Program: Formative Years," in *Some Strangeness in the Proportion*, ed. Harry Woolf (Reading Mass.: Addison-Wesley, 1990), p. 58. One factor in some of the resistance was the claim by Kaufmann to have found experimental support for Abraham's electron theory against the theories of both Einstein and Lorentz (Holton, pp. 57–58). Planck's logical analysis of difficulties in Kaufmann's experimental interpretation helped promote Einstein's theory. (See E. Zahar, *Einstein's Revolution: A Study of Heuristic* [La Salle, Illinois: Open Court Publishing Company, 1989], pp. 201–26, for an excellent account of Kaufmann's experiment and Planck's analysis.) By 1911 or so Einstein's theory was fairly widely accepted among Germany physicists. Christa Jungnickel and Russell McCormmach, *Intellectual Mastery of Nature: Theoretical Physics from Ohm to Einstein*, 2 vols. (Chicago: University of Chicago Press, 1986), 2:247–48.

The transition from classical kinematics to the kinematics of special relativity has been cited by Feyerabend as a prime example of incommensurability. I believe that a close investigation of this transition would give little support to those who, like Richard Rorty, would suggest that such incommensurability renders the comparative evaluation of rival theories as little open to rational critique as political debate.

P. K. Feyerabend, "Against Method: Outline of an Anarchistic Theory of Knowledge," in *Analysis of Theories and Methods of Physics and Psychology*, Minnesota Studies in the Philosophy of Science, vol. 4, ed. Michael Radner and Stephen Winoker (Minneapolis: University of Minnesota Press, 1970), pp. 17–130, see pp. 81–89; Richard Rorty, *Philosophy and the Mirror of Nature* (Princeton: Princeton University Press, 1979), pp. 328–31.

18. Hilary Putnam, *Realism and Reason* (New York: Cambridge University Press, 1983), p. 95.

19. In "Conceptual Change, Incommensurability and Special Relativity Kinematics," I developed a conceptual change model that would capture this sort of contextually a priori conceptual commitment. It was applied to the transition from classical kinematics to the kinematics of special relativity. In the model a Hertzian physicist who initially treated Galilean invariance as a conceptual commitment transformed his conceptual commitments to those of the kinematics of special relativity. This transition displayed most of the radical characteristics Feyerabend and Kuhn had suggested were the basis for their ideas about incommensurability. For example the change affected even observation—like claims about relative lengths. In the model, however, my agent was endowed with the capacity to carry out hypothetical reasoning to evaluate the alternative kinematical proposal. He was then able to use empirical evidence to decide between it and his original commitments once he had evaluated it as a serious alternative candidate. I think anyone who appreciates the facility with which human beings can reason about nonstandard mathematical systems (once they are made aware of them) will have little doubt that real physicists are, also, capable of such rational, even if radical, revisions of their conceptual commitments.

W. L. Harper, "Conceptual Change, Incommensurability and Special Relativity Kinematics," *Acta Philosophica Fennica* 30, no. 4 (1979): 431–61. T. S. Kuhn, *The Structure of Scientific Revolutions* (Chicago: University of Chicago Press, 1962), pp. 111–59.

20. Albert Einstein, "Die Grundlage der allgemeinen Relativitätstheorie," *Annalen der Physik* 49 (1916): 769–822. Trans. W. Perett and G. B. Jeffrey as "The

Foundation of the General Theory of Relativity," in *The Principle of Relativity* (1923; reprint ed., New York: Dover Publications, 1952), pp. 109–64.

21. Another way to look at this is to treat mechanics as an arbitrary axiomatic system, as free from constraint by actual phenomena as an abstract axiomatic treatment of Euclidean geometry. This was the tradition of rational mechanics. See I. Bernard Cohen, *The Newtonian Revolution: With Illustrations of the Transformation of Scientific Ideas* (Cambridge: Cambridge University Press, 1980), pp. 94–99 for a discussion of Newton's reaction to critics who wanted to limit the *Principia* to this abstract mathematical discipline.

22. Cohen, *The Newtonian Revolution,* pp. 52–64.

23. There is some controversy about whether Newton's specific formulation of the calculus in his lemmas on first and last ratios can legitimately support this generalization to continuously acting forces. E. J. Aiton, "Polygons and Parabolas: Some Problems Concerning the Dynamics of Planetary Orbits," *Centaurus* 31 (1989): 207–21; see pp. 207–10.

24. Cohen and Whitman, p. 35.

25. Cohen and Whitman, p. 489.

26. Cohen and Whitman, p. 71.

27. This suggests that a number of philosophers (e.g., Glymour, *Theory and Evidence,* p. 208, and myself [in talks]) have made too much of the assumption of concentric circular orbits in the proof of Proposition 4. Jon Dorling was the first one to call my attention to the significance of Newton's generalization of Proposition 4 to elliptical orbits. He suggested that the inference for elliptical orbits is an easy extension of Proposition 15, Book 1. Howard Stein pointed out to me that the variations in eccentricities in the orbits of the planets (from .007 for Venus to .206 for Mercury) make the application of Corollary 8 of Proposition 4 somewhat problematic.

28. This precession theorem (Proposition 45, Book 1) applies to motions of apsides in orbits that differ very little from circles. Ric Arthur has reconstructed Newton's interesting proof of this theorem that includes a step that is essentially taking a limit as the ellipse approaches a circle and Ram Valluri has calculated the effects introduced by explicitly taking into account various small eccentricities. The three of us hope to present some of this material in a joint paper on Newton's proof of and uses of Propositions 44 and 45, Book 1.

29. The precession theorem (Proposition 45, Book 1) is like the harmonic law theorem, a one-body idealization in which the only forces considered are centripetal forces on the orbiting body. But Newton used it as a tool to account for precession caused by the action of a third body on an orbiting body. He handled third-body interference on an orbit corresponding to an inverse-square centripetal force by adding a second centripetal component (e.g., the earth-moon axial component of the action of the sun on the moon). Composition of this new centripetal component results in a new total centripetal force that will no longer be inverse-square. This allowed the basic one-body idealization to be applied to compute the corresponding precession. Newton established a numerical formula to give the precession as a function of the foreign force expressed as a fraction of the basic inverse-square centripetal force (Corollary 2, Proposition 45, Book 1).

When Newton applied this method to the lunar precession he found that using the correct value of this fraction for the action of the sun on the moon only accounted for about one half of the lunar precession. This was the failure that made his lunar theory ultimately disappointing. D. T. Whiteside, "Newton's Lunar Theory: From High Hope to Disenchantment," *Vistas in Astronomy* 19 (1975):

317–328; C. B. Waff, "Isaac Newton, The Motion of the Lunar Apogee, and the Establishment of the Inverse Square Law," *Vistas in Astronomy* 20 (1976): 99–103.

In the late 1740s Clairaut finally developed a model that could take into account transverse as well as centripetal components of the action of the sun on the moon. He found that the contribution of this component was to augment the precession contributed by the centripetal component enough to account for the empirically observed precession. This augmentation of the precession produced by the transverse component of the action of a third body makes relative absence of orbital precession an even better indication of relative absence of perturbing forces than Newton knew. C. B. Waff, "Universal Gravitation and the Motion of the Moon's Apogee: The Establishment and Reception of Newton's Inverse-square Law, 1687–1749" (Ph.D. diss., Johns Hopkins University, 1976), especially pp. 201–15.

30. Cohen and Whitman, p. 494.

31. Cohen and Whitman, p. 493.

32. If one construes the Brahean orbits as ellipses this model results from the transformation of Kepler's model corresponding to the change of reference frame from the sun to the earth. Though these models are exactly equivalent if the only relative motions to be accounted for are those of the bodies in the solar system among themselves, they are not equivalent if motions of the planets relative to the stars are included. The absence of observed stellar parallax counted as evidence in favor of Brahe rather than Kepler, since on Kepler's model, the large axis of the earth's orbit (2 A.U.'s) provided a base for parallax. Brahe had used this very point to argue against the Copernican hypothesis. A. Van Helden, *Measuring the Universe* (Chicago: University of Chicago Press, 1986), pp. 66–67.

Brahe's own system used eccentric circles rather than ellipses. Perhaps this is an additional clue to the puzzle over why Newton did not build elliptical orbits into what he counted as the phenomena from which to infer inverse-square centripetal forces. Glymour, *Theory and Evidence;* C. A. Wilson, "From Kepler's Laws, So-called, to Universal Gravitation: Empirical Factors," *Archive for History of Exact Sciences* 6 (1970): 89–170.

33. This is one of the very few places in the third edition of the *Principia* where there is an explicit appeal to a hypothesis.

34. Cohen and Whitman, p. 31.

35. Cohen and Whitman, p. 518.

36. See Howard Stein, "Newtonian Space-Time," *Texas Quarterly* 10, no. 3 (1967): 174–200, especially pp. 178–84 for an assessment of this argument drawing upon an analysis of the constraints on space-time structure generated by Newton's laws of motion. M. Friedman ("The Metaphysical Foundations of Newtonian Science," in *Kant's Philosophy of Physical Science*, ed. R. E. Butts [Dordrecht: Reidel, 1986], pp. 31–57) has offered an interesting account of the important role this argument of Newton's played in Kant's *Metaphysical Foundations of Natural Science* of 1786.

37. Cohen, *The Newtonian Revolution*, p. 224.

38. Karl Popper, *Objective Knowledge, An Evolutionary Approach* (London: Oxford University Press, 1972), pp. 200–201. Cohen (*The Newtonian Revolution*, p. 224) pointed out that if one uses Newton's data for the relative masses of Jupiter and the sun (the worst case) the difference between harmonic law and this transformed version is about one part in a thousand. See nn. 52 and 54 below for the details of such calculations for Jupiter and Saturn.

39. Violations of the law of areas, on the other hand, require interactions between satellites. Newton's two-body generalization (Corollary 3, Proposition 58,

Book 1) of his areas-law theorems (Propositions 1–3, Book 1) agrees with these one-body results. This is so, because in the two-body system equal areas will be swept out by radii drawn from one body to the other if and only if equal areas are swept out by both radii drawn to their common center of gravity (Corollary 3, Proposition 58, Book 1). Interactions between satellites, however, will violate the law of areas (see n. 44). Universal gravitation requires such interactions so it is strictly incompatible with the area law (as well as the harmonic law) for any system with multiple satellites.

40. William Whewell, *Novum Organon Renovatum,* 3d ed. with additions (London: John W. Parker and Son, 1858), p. 187. Facsimile reprint (Ann Arbor, Mich.: University Microfilms, 1971).

41. Whewell, *Novum Organon Renovatum,* Aphorism 35, p. 186.

42. Cohen and Whitman, p. 493.

43. Newton's table for Phenomenon 4 has Kepler's observed value for Saturn's distance at 9.51 A.U., while Boulliau's value is 9.54198 A.U. The distance he computed from the period by the harmonic law was 9.54006 A.U., which agrees very well with Boulliau's number but less well with Kepler's. In a letter to the Astronomer Royal dated 30 Dec. 1684 (Newton to Flamsteed, *Correspondence,* 2:407–8) Newton speculated that Kepler's low number might be due to the action of Jupiter on Saturn at their conjunction. Flamsteed's reply (Flamsteed to Newton, 5 Jan. 1685, *Correspondence,* 2:408–11) was ambiguous and Newton continued to press for more information about Saturn and Jupiter off and on until at least 1691 (Newton to Flamsteed, 10 Aug. 1691, *Correspondence,* 3:164).

The Correspondence of Isaac Newton, ed. H. W. Turnbull, vols. 1–3 (Cambridge: Cambridge University Press, 1959, 1960, and 1961).

44. In May of 1694 David Gregory visited Newton at Cambridge. His memoranda of their discussions include references to mutual perturbations of Jupiter and Saturn. One is the following:

> The mutual interactions of Saturn and Jupiter were made clear at their very recent conjunction. For before their conjunction Jupiter was speeded up and Saturn slowed down, while after their conjunction Jupiter was slowed down and Saturn speeded up. Hence corrections of the orbits of Saturn and Jupiter by Halley and Flamsteed, which were afterwards found to be useless and had to be referred to their mutual action. (*Correspondence,* 3:318)

Another of these memoranda contains this remark:

> Flamsteed's observations when compared with those of Tycho and Longomontanus prove the mutual attraction of Jupiter and Saturn in their most recent past conjunction in 1683. (*Correspondence,* 3:337)

This remark suggests that Newton may have been able to make the case from some of the information Flamsteed eventually provided. Perhaps some of this was in the paper he refers to as included in his long-delayed answer to Newton's letter of 10 Aug. 1691 (Flamsteed to Newton, 24 Feb. 1692, *Correspondence,* 3:199–205. See especially 201, n. 9). Howard Stein ("Newtonian Space-Time," pp. 180–81) has discussed the first reference from Gregory's memoranda and has suggested that it would be quite useful to know when and in what form Newton got the information that Saturn and Jupiter sensibly disturb each other's orbits.

45. Paraphrasing Alexandre Koyré, *Newtonian Studies* (Cambridge: Harvard University Press; London: Chapman and Hall, 1965), p. 263. Koyré and Cohen, pp. 561–63.

46. My treatment here [and in "Consilience and Natural Kind Reasoning," *An Intimate Relation, Studies in the History and Philosophy of Science*, ed. James Brown and Jürgen Mittelstrass (Dordrecht: Klewer Academic Publishers, 1989), pp. 119, 131–34] has benefited from Malcolm Forster, "Unification, Explanation and the Composition of Causes in Newtonian Mechanics," *Studies in History and Philosophy of Science* 19 (1988): 55–100. Forster has very useful things to say about Whewellian techniques for finding such component motions in the data. He also mobilizes these ideas in an explicit and cogent defense of inferences to Newtonian component forces against a recent skeptical argument of Nancy Cartwright, *How the Laws of Physics Lie* (Oxford: Clarendon Press, 1983), pp. 54–72.

47. I. B. Cohen, personal communication, November 1987.

48. G. E. L. Owen, "Tithenai ta Phainomena," in *Aristotle*, ed. J. M. E. Moravesik (Garden City, N.Y.: Doubleday & Company, Inc., 1967), pp. 170–73. There is controversy over the extent to which each of Kepler's laws was accepted before Newton's *Principia* was published (e.g., J. L. Russell, "Kepler's Laws of Planetary Motion: 1609–1666," *The British Journal for the History of Science* 2 [1964]: 1–24; Wilson, "From Kepler's Laws;" C. A. Wilson, "Newton and Some Philosophers on Kepler's Laws," *Journal of the History of Ideas* 35 [1974]: 221–58; Cohen, *The Newtonian Revolution*, pp. 224–25; B. S. Baigrie, "Kepler's Laws of Planetary Motion before and after Newton's *Principia:* An Essay on the Transformation of Scientific Problems," *Studies in History and Philosophy of Science* 18, no. 2 [1987]: 177–208; and many others). We have seen that Newton's phenomena include the law of areas and the harmonic law, and that they are compatible with some version of a Brahean model. Probably, there was some appropriate group of experts for which we can reasonably suppose the harmonic law would count as presumed common knowledge. With the law of areas the case is less clear. (See especially Wilson, "From Kepler's Laws.")

49. P. K. Feyerabend, "Classical Empiricism," in *The Methodological Heritage of Newton*, ed. R. E. Butts and J. W. Davis (Toronto: University of Toronto Press, 1970), p. 151.

50. Such specification requires a sufficiently unaccelerated frame with respect to which these accelerations can be determined; however, as we have seen (n. 29) the precession theorem allowed Newton to use the relative stability of the orbits of the major planets to count as a mark that the center of mass of the solar system is sufficiently unaccelerated to be used in specifying the mutual accelerations among the bodies in it.

51. Feyerabend, "Classical Empiricism," pp. 101–65, especially p. 155.

52. This is an example where the perturbations are small relative to the effect from which the feature of the component force is inferred. Forster, ("Unification, Explanation, and the Composition of Causes," pp. 170–72) argues that the method can work even if the effect to be accounted for is small in relation to the perturbations and random fluctuations.

53. Cohen and Whitman, p. 523.

54. In Proposition 8, Book 1, Newton used the R^3/T^2 values to estimate the masses of the sun, Jupiter, Saturn, and the earth to be respectively

$$1, \quad 1/1067, \quad 1/3021, \quad \text{and} \quad 1/169282.$$

His estimates for Jupiter and Saturn agree fairly well with the true values, which are about 1/1048 and 1/3497, but his value for the mass of the earth is more than twice as large as it should be (about 1/332949).

According to Proposition 60, Book 1, the corrected distance R' (where the sun and planet orbit their common center of gravity) is to the harmonic law distance R (where the sun is fixed) as S + P is to the first of two mean proportionals between S + P and S (where S is the mass of the sun and P is the mass of the planet). This makes (see Cajori, n. 24, p. 651)

$$R'/R = (S + P) / ((S + P)^2 \ S)^{1/3}$$

We compute the corrections for Jupiter using as uncorrected R values those computed from the harmonic law and the periods in Newton's table for Phenomenon 4 (Cajori, p. 404). For Jupiter R = 5.20096 A.U. and the correction factor is

$$(1068) / \sqrt[3]{(1068)^2 \ (1067)} \ = \ 1.00031$$

so that the corrected distance R' = 5.2026 A.U.

55. For Jupiter the mean of Kepler's (5.1965 A.U.) and Bouillau's (5.2252 A.U.) empirical estimates of R is 5.21085 and the trivial standard deviation is .01435 A.U. so that the uncorrected harmonic law distance R (5.20096 A.U.) is .00989 A.U. or .689 standard deviations out. The corrected R' (5.2026 A.U.) is .00825 A.U. or .575 standard deviation out. With only two measurements, these standard deviations are not to be taken too seriously. We can be sure, however, that the improvement from the correction is not empirically significant in these data.

For Saturn the corrected distance R' = 9.54097 A.U., while the uncorrected R = 9.54006 and the mean of the empirical estimates is 9.52601. Here the correction actually makes things worse.

Newton did not estimate the masses for planets without moons, but these corrections would surely have been too small to measure from his data. The earth is the most massive of the smaller planets and even with Newton's incorrect estimate of this mass (more than twice as big as it should be) the correction factor is only 1.00000197.

56. In this two-body idealization, finding the unperturbed motion by factoring out the perturbation is particularly straightforward. In more complicated cases, solving for the true motions with all the perturbations is done by the approximation methods of perturbation theory. See J. Kovalevsky, *Introduction to Celestial Mechanics* (New York: Springer Verlag, 1963), pp. 44–68. Here too, however, we can factor out given perturbations to recover what the component motions without those perturbations would be.

57. In Definitions 6 and 7, Newton distinguished the *absolute quantity* and the *accelerative quantity* of centripetal force.

> Def. 6 The absolute quantity of centripetal forces is the measure of this force that is greater or less in proportion to the efficacy of the cause propagating it from the center through the surrounding regions. (Cohen and Whitman, p. 6.)

This is what the constant slope R^3/T^2 in our graph for Kepler's harmonic law measures for the centripetal force of the sun.

Def. 7   The accelerative quantity of centripetal force is the measure of this force that is proportional to the velocity which it generates in a given time. (Cohen and Whitman, p. 6.)

This is the measure of centripetal force toward the sun that varies inversely as the square of distance from it.

That these *accelerations* are all inversely as the square of distance, even though the masses of the various planets are so very different is testimony to a very special property of this centripetal force. Newton also defined the *motive* quantity of centripetal force.

Def. 8   The motive quantity of centripetal force is the measure of this force that is proportional to the motion which it generates in a given time. (Cohen and Whitman, p. 7.)

His example is *weight*, which we know is the product of mass and the acceleration of gravity. The special property of the centripetal force to the sun is that its motive quantity on any planet must be proportional to that planet's quantity of matter, otherwise the accelerations produced would not satisfy the inverse-square proportion of distance. This is part of the data for the identification of this force as gravitation to the primary in Proposition 5. (Howard Stein was the first to call my attention to the importance of this point.) It is also part of Newton's evidence for the direct proportionality of the *motive* force of gravitation on any body to the quantity of matter of that body in Proposition 6.

58. This agreement among these separately fixed ratios is an example of what Whewell (*Novum Organon Renovatum,* pp. 87–90) would call the "consilience of inductions." See W. L. Harper, "Consilience and Natural Kind Reasoning in Newton's Argument for Universal Gravitation," pp. 125–27.

59. These points support Michael Friedman's conjecture that each new phenomenon unified adds confirmation to a common explanatory hypothesis. See Michael Friedman, "Theoretical Explanation," in *Reduction, Time, and Reality,* ed. R. A. Healy (Cambridge: Cambridge University Press, 1981); and Michael Friedman, *Foundation of Space-Time Theory (Relativistic Physics and Philosophy of Science)* (Princeton: Princeton University Press, 1983), pp. 320–39.

60. Forster ("Unification, Explanation, and the Composition of Causes," pp. 73–80, 97–100) called my attention to the importance of common-cause unification to Newtonian science. For more on my own views and some differences with Forster see "Consilience and Natural Kind Reasoning," pp. 129–31 and 140–47.

61. Cohen and Whitman, p. 485.

62. Cohen and Whitman, pp. 485–86.

63. Cohen and Whitman, p. 500.

64. This may help to illuminate the condition in Rule 1 that we only admit causes that are true. To count the identification as corresponding to a true cause is to be committed to make all the systematic constraints it imposes as the theory is developed. This is one of the differences between a unified theory such as Newton's and an approach such as Descartes's method of hypotheses, which would not require various explanatory hypotheses to cohere together. (See n. 90 for more on this "vera causa" condition.)

65. Cohen and Whitman, p. 502.

66. Isaac Newton, *Opticks,* second Dover edition, p. 404.

67. W. V. Quine, "Natural Kinds," in *Essays in Honor of Carl G. Hempel,* ed.

N. Rescher (Dordrecht: D. Reidel, 1970), pp. 5–23. This structure given to support by inferences from phenomena also illuminates experimental science. Some have recently called attention to progressions in experimental physics that have shown a fascinating autonomy in maintaining their integrity over quite radical changes in higher-level theory. For example, see E. Hiebert, "The Role of Experiment and Theory in the Development of Nuclear Physics in the Early 1930's," in *Theory and Experiment: Recent Insights and New Perspectives on Their Relation,* ed. D. Batens and J. P. van Bendegem (Dordrecht: Reidel, 1988), pp. 55–76. I believe that the relative stability generated for lower-level theory fragments by inferences from phenomena is fundamental to understanding this autonomy.

Experimental science has recently been getting considerable attention from historians and philosophers of science, for example A. Franklin, *The Neglect of Experiment* (Cambridge: Cambridge University Press, 1986); D. Galison, *How Experiments End* (Chicago: University of Chicago Press, 1987); D. Batens and J. P. van Bendegen, eds., *Theory and Experiment: Recent Insights and New Perspectives on Their Relation* (Dordrecht: Reidel, 1988); I. Hacking. "On the Stability of Laboratory Science," *The Journal of Philosophy* 85 (1988): 507–14; and I. Hacking, "Philosophies of Experiments," in *P. S. A. 1988,* vol. 2, ed. A. Fine and J. Leplin (East Lansing, Mich.: The Philosophy of Science Association, 1989).

68. Cohen and Whitman, p. 503.
69. Cohen and Whitman, p. 504.
70. Cohen and Whitman, pp. 504–5.
71. Cohen and Whitman, pp. 500–505.
72. Cohen and Whitman, pp. 500–501.
73. Cohen and Whitman, p. 507.
74. This is the first step of Newton's explicit argument to establish what we would call an identity of gravitational with inertial mass. What Newton calls quantity of matter is the magnitude appearing in the laws of motion. Newton does not just assume that this concept of "inertial" mass is identical with "gravitational" mass. He argues explicitly for the identification from phenomena.
75. Cohen and Whitman, pp. 507–9.
76. Cohen and Whitman, p. 486.
77. Cohen and Whitman, pp. 487–88.
78. Cohen and Whitman, p. 488.
79. Koyré, *Newtonian Studies,* p. 162.
80. J. E. McGuire, "The Origin of Newton's Doctrine of Essential Qualities," *Centaurus* 12 (1968): 233–60.
81. Harper, "Consilience and Natural Kind Reasoning," pp. 144–45.
82. For example, see Cohen, *The Newtonian Revolution,* p. 254.
83. Kuhn, *The Structure of Scientific Revolutions.* Many of the features of radical scientific change that Kuhn has called attention to in his seminal account of scientific revolutions (1962) may be understood as legitimate and reasonable features of scientific practice if accepted scientific theories are to be treated as natural-kind conceptions. This suggests a somewhat less pessimistic view of the implications of those episodes than that of many who have seized upon Kuhn's and Feyerabend's discussion of incommensurability in an attempt to undercut the idea that science produces any very significant warrant for knowledge.
84. Whewell, *The Philosophy of Discovery* (London: J. W. Parker and Son, 1860), pp. 193–94.
85. Ibid., p. 288.
86. Cajori, p. 453.
87. Koyré (*Newtonian Studies,* pp. 325–36) argued that "feign" is a more

accurate translation of "fingo" in Newton's famous phrase "Hypotheses non fingo." See also Cohen (1966) and (1978), pp. 240–45. Koyré (*Newtonian Studies,* pp. 344–35) shows how his translation points up the explicit contrast with Descartes's famous declarations in Part 3 of his *Principia Philosophiae* that he wishes the hypotheses he uses to explain phenomena to be taken as *mere fictions,* some of which it is certain are false.

88. Cajori, p. 547.

89. Cohen and Whitman, p. 488.

90. E. J. Aiton, *The Vortex Theory of Planetary Motions* (London: Macdonald and New York: American Elsevier, 1972), pp. 30–60. Here we see again how the contrast with Descartes illuminates the "vera causa" condition on Rule 1. According to Rule 4 we are to regard as true or very nearly true those propositions gathered from phenomena by induction. Mere hypotheses—that is, propositions not appropriately backed up by inferences from phenomena—do not count as true causes.

91. B. C. van Fraassen, "Empiricism in Philosophy of Science," in *Images of Science,* ed. P. M. Churchland and C. A. Hooker (Chicago: University of Chicago Press, 1985), pp. 265–66.

92. Cohen, *The Newtonian Revolution,* pp. 81, 109–17.

93. As Cohen (*The Newtonian Revolution,* p. 113) has pointed out, the *Hypotheses non fingo* passage was not designed to set a boundary that would rule out-of-court all investigations into the causes of gravity. The very passage itself follows immediately upon a list of properties of gravity that have been established by inferences from phenomena. The purpose of this list is to summarize the data that an investigation of the cause of gravity would have to account for, not argue that such investigations could not in principle have a legitimate place in natural philosophy.

94. Recently Jon Dorling ("Einstein's Methodology of Discovery was Newtonian Deduction from the Phenomena," forthcoming [1987] in a volume being edited by Jarrett Leplin) has argued that Einstein's generation of the Lorentz transformations in his 1905 special relativity paper was a Newtonian deduction from phenomena. According to Dorling (pp. 4–5), the phenomenon was the constancy—including the independence of frame of reference of the velocity of light, and the background assumptions included "the principle of special relativity, i.e., the physical equivalence of all inertial systems, and certain additional more innocuous theoretical constraints such as the linearity of the appropriate coordinate transformations." Dorling (pp. 9–12) also argued that Einstein's field equations in general relativity were fixed (up to the gravitation constant) by a Newtonian deduction from a falling-elevator phenomenon generalized from the Eötvös experiment. The background included "further non-controversial experimental facts, and theoretical requirements which consisted of those theoretical parts of the previously successful theories which seemed still sufficiently plausible." Dorling (see additional listings in references) has over the years pointed out quite a number of examples of the Newtonian inferences from phenomena in scientific developments much later than Newton's own time.

95. Cohen, *The Newtonian Revolution,* p. 147.

96. I have been arguing that equivalence theorems connecting phenomena with theoretical magnitudes are the heart of what is required to meet Newton's standard. Quantum mechanics is certainly rich with such theorems. They connect the magnitudes of the quantum mechanical state of a system to statistical distributions that count as the quantum mechanical phenomena. The connection between these statistical distributions and the quantum mechanical magnitude is just as "deduc-

tive" as the connection between harmonic-law phenomena and inverse-square variation of centripetal forces. What makes quantum mechanics "statistical" is that its phenomena are statistical rather than simple universal generalizations. What makes for the impressive empirical success is the wonderfully good fit of these quantum mechanical distributions to the empirical statistics generated from experiments. See Max Jammer, *The Conceptual Development of Quantum Mechanics* (New York: McGraw-Hill, 1966), pp. 362–65; R. Feynman, *The Character of Physical Law* (Cambridge: MIT Press, 1965), pp. 127–48; and M. Redhead, *Incompleteness, Nonlocality, and Realism* (Oxford: Clarendon Press, 1989), pp. 16–42.

Algebraic vs. Geometric Techniques in Newton's Determination of Planetary Orbits

MICHAEL S. MAHONEY

Newton's *Principia* is now three centuries old. Yet it is only within the last three decades that we have begun to discern through the aura of his eighteenth-century apotheosis the true shape of his genius and the precise nature of his accomplishment. If our progress has demythologized Newton, it has not diminished his genius or his accomplishment. On the contrary, making him human and placing him among his scientific colleagues—showing him, as R. S. Westfall has, "never at rest"—has only enhanced his stature.[1] But it is the stature of a mortal man who learned from others, who discovered and invented some things for himself, and who had his limits. He did not know everything, he did not do everything, he was not capable of everything. One must approach the adulatory claims about him, both those made by his friends and those he himself put about, with the same critical attitude that one takes toward any historical figure who lived, worked, and created at a definite time in a definite place.

To carry out this more critical evaluation it has to some extent simply sufficed to have Newton's works as actually published in his lifetime and shortly thereafter. When set against the writings of eighteenth-century rational mechanists like Euler and the Bernoullis, the *Principia* itself revealed the limitations of Newton's momentum mechanics and the host of problems it left unsolved.[2] Enough of his mathematics existed in print to reveal the essential differences between his method of fluxions and Leibniz's differential calculus and thereby early in this century to right the biased judgment of plagiarism handed down against the latter by a Royal Society jury handpicked by Newton.[3] The third edition of the *Opticks* with its full complement of queries offered more than a hint of Newton's interest in chemistry and in a "subtle fluid" that would serve as the "hypothesis" about gravity he had refused to "make up" in the General Scholium added to the second edition of the *Principia* in 1713.

Yet, the full-scale reevaluation that has characterized books and articles about Newton over the past three decades has only become possible with the publication of the mass of manuscript material left unpublished at his death. The great editions of his correspondence and his mathematical and mechanical writings enable us to sharpen our understanding of his technical achievement and to lay to rest longstanding legends that hitherto could claim possible grounding in some as yet undiscovered papers. Moreover, by revealing the course of Newton's thought from initial insight to developed form, his unpublished papers show us the structure of his ideas and thereby yield a glimpse into what did and what did not lie behind the finished product.

Take, for example, the mathematics of the *Principia.* Tradition (with some help from Newton himself) long told us on the one hand that Newton set forth his mechanics in strict accordance with the canons of classical Greek geometry and on the other that he worked it out in the algebraic system of fluxions before translating it into the ancient mode. The reasons given for such mathematical behavior varied from Newton's expressed distrust of algebra to his desire to shut off possible criticism from any but a handful of fellow mathematicians. None of this bore close scrutiny. As D. T. White-side convincingly argued in the 1969 Gibson Lecture at Glasgow, "The Mathematical Principles Underlying Newton's *Principia Mathematica,*"[4] the geometry of that great work is classical neither in style nor in content, and there exists not a shred of manuscript evidence that Newton ever developed the propositions of the *Principia* in any mathematical form other than the geometry of the published versions.[5] Volume 6 of the *Mathematical Papers* reinforces this last point, and it seems implausible to hold out for yet more papers; the available evidence sets a twenty-year pattern of research that gives no hint of another version in a different mode.[6]

Understanding the mathematics of the *Principia,* then, means understanding its geometry, not merely as a form of expression but as a conceptual framework. Languages carry their own, peculiar categories and habits of thought. That is why translation requires more than a dictionary and a grammar. If Newton did not translate his thinking into geometry from a system of fluxions he knew, it seems unpromising to seek the patterns of his thought by translating his geometry into a symbolic calculus he did not know. At the very least, doing so begs the question of translation itself.[7]

Yet, having insisted on the geometrical character of the *Principia,* Whiteside went on in his lecture to explicate its mathematics in terms of modern calculus, on the premise that the geometry, characterized by the use of "limit-increments of variable line segments," constitutes an equivalent system of infinitesimal calculus that can be expressed in familiar algebraic terms by a relatively simple and straightforward translation of symbols and operations.[8] Thus he referred to the "all-but-trivial analytical redefinition [of Newton's theorem on central forces] by Pierre Varignon, Johann Bernoulli, and Jakob Hermann," and he criticized Pierre Costabel for maintaining that "one would be wrong in thinking that Newton knew how to formulate and resolve problems through the integration of differential equations."[9] Differential equations abound in Whiteside's article, derived from the original geometry by the imposition of polar coordinates r, ϕ and the use of Taylor series expansions to order $d\phi^3$. Despite his admission that the introduction of polar coordinates is "without published Newtonian precedent," and despite the fact that one finds few series expansions, Taylor or otherwise, in the *Principia,*[10] Whiteside presented its mathematics as algebraic, infinitesimal analysis in all but form.

But taking the elements of such a translation from various places in Newton's work, using them to change propositions of the *Principia* into a form in which Newton never expressed them, and on that basis attributing to Newton the conceptual structure and the technical potential underlying the transformed system seems risky at best. Close studies of Fermat, Galileo, Descartes, Huygens, and other figures of Newton's era suggest caution in combining elements in that way. Fermat could find the tangent to a curve and the area under it, but he never discerned the inverse relation of the two problems.[11] Galileo could use inertial motion to derive the parabolic trajectory of a cannon ball at the same time that he asserted the natural conservation of motion in a circle about the center of force. Descartes knew the implications of relativity of motion and yet violated the principle in his laws of impact. There are many other, similar examples.

What Newton could or could not have done by way of an algebraic mechanics is less important historically than what he did. He wrote the *Principia* as a geometrical treatise. He did so for several reasons. First, he was primarily concerned with a geometrical object, namely the orbit or trajectory of a moving body, and with the geometrical elements that one must measure to determine that object uniquely. Second, he found the treatment and investigation of conic sections by recently developed affine and quasi-projective techniques better suited than algebraic methods to his immediate purposes.[12] Third, the method of first and last ratios—the "limit-increment" method—had its demonstrative roots in the classical Greek method of exhaustion and lent itself best to geometrical expression. Fourth and, for the moment, finally, with few exceptions the standard style of his day for mechanical investigations was geometry, not algebra; Huygens's *Horologium oscillatorium* was the model.[13]

But while Newton worked always in geometrical terms, they were not the same geometrical terms throughout. The style of analysis Newton used in the early sections of Book 1 to investigate centripetal motion on conic sections differs from that applied later to the general problem of motion under central forces. The first exploits the classic properties of the conic sections, adding to the classical repertoire Newton's classically based method of first and last ratios. Its application features a continuing effort to locate, and often to relocate, dynamic parameters as elements of the geometrical configuration, which is basically a drawing of the orbit itself. The second style employs Newton's method of geometrical fluxions, based on the contemporary technique of transmutation of areas,[14] and moves from the relocation of parameters to the juxtaposition of the orbital configuration with graphs (or functional curves) of those parameters, and mapping corresponding elements of the two configurations on one another by means of their fluxions.

The two styles do not equally well fit their subjects. Both reveal the difficulties of trying to represent time and motion in the same two-dimensional framework as space. But, while the first leads to a constructive solution of the question of planetary motion, the second hides rather than reveals such important structural information as the number and nature of the constants involved or the dimensionality of the various elements of the analysis.

Moreover, as a reduction of the mechanics of the orbit to the mathematics of curves, Newton's account leaves open the question of how those curves are to be constructed and measured. Without Newton's method of fluxions, then unpublished and not widely known, the propositions of Sections 7 and 8 show that problems in the dynamics of central forces can be solved in principle, but not how to solve them in practice.[15] That is what Continental mathematicians set out to provide, using algebraic means they thought better suited to that goal. A close reading of several examples from the *Principia* illustrates how its geometrical form conditioned its mathematical content and indicates the ways in which the analytical redefinition undertaken by Continental mathematicians in the period 1690–1725 reflected a fundamental shift in the substance as well as the form of mathematics and mechanics.

Book 1 of the *Principia* bears the title "On the motion of bodies"; with rare exceptions that motion results from the action of central forces in a nonresisting medium.[16] In the first ten sections, Newton restricted his analysis to point masses moving about centers of force that are either at rest or moving inertially. The shift to point masses interacting according to the Third Law occurs in Section 11 and is made relatively simple by the form of the preceding analysis; succeeding sections turn to other subjects. From the outset, Newton concentrated on the orbit or trajectory of motion. Keeping before his and the reader's eyes the path of the body's motion, he located within the geometrical configuration structural elements proportionally representative of the dynamical, nongeometrical parameters determining that motion. The task was made all the more difficult, of course, by the fact that in most cases those parameters are continuously variable.

Proposition 1, Theorem 1, establishes the pattern of Newton's analyses. To demonstrate that a body under the action of a centripetal force moves in a plane and by the radius drawn to it sweeps out areas proportional to the time, Newton broke up into finite parts

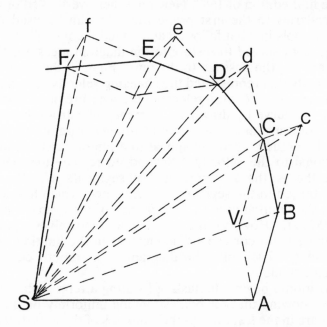

Figure 1

the two parameters he could not get into the diagram. During equal small intervals of time, the body moves along rectilinear segments at uniform velocities proportional to those segments; the velocity changes instantaneously at the end of each interval by the action of a centripetal impulse equal to the total force acting over the interval.[17] Beginning at A, then, the body moves uniformly to B, where an impulse shifts it to path BC, and so on. It is an easy matter to show that triangles SAB, SBc, and SBC are equal; by a similar argument, $\Delta SBC = \Delta SCD = \Delta SDE = \ldots$. To move to the case of a continuously acting force, Newton then decreased the size of the intervals of time, simultaneously increasing their numbers and hence also the number of impulses. None of these changes shows up directly in the diagram; rather, rendering them graphically would require a series of diagrams in which the length of the segments decreases and their number increases until in the limiting case the body moves along a smooth curve passing through A, B, C. . . . The justification for moving to the limit is found in the third of the eleven lemmas that constitute Section 1, "On the method of first and last ratios," for the limiting curve can be caught between two sequences of inscribed and circumscribed triangles that respectively form figures differing from one another by less than any assigned value.

In the first edition of 1687, Newton either overlooked or omitted two corollaries to the first proposition, though he used them as analytical tools in what followed and added them explicitly in later editions. Both served to restore to the diagram measures that had disappeared in the passage to the limit. First, in the limiting case, the segments that represented the changing velocity vanished into a continuous curve. Newton relocated the measure of velocity at any point of the curve by drawing a perpendicular from the center of force to the tangent to the curve at the given point; that perpendicular, he argued, again by appeal to a limiting case, is inversely proportional to the velocity.[18] Second, to replace the measure of impulse, that is, the Δv represented by segments cC, dD, eE, etc., Newton turned to the segments BV, EZ, etc., which he took to be equal in the limiting case to the versines of the evanescent angles BSA, DSF, etc. However, the versines are themselves relations for which there is no corresponding element in the limiting case. Velocity had got back into the diagram of a smooth curve; force remained outside it.

Proposition 4 begins the task of locating a representation of the force by showing that for bodies moving uniformly on circles the versines are in the last ratio as the "squares of the arcs described in the same times, applied to the radii," i.e., *acceleration* ∝ *versine(arc)* ∝ *arc²/R*. Again, the major work of the theorem has already been accomplished in the introductory lemmas, here Lemma 11. Since the motion is uniform, the velocity is as the arc traversed, whence the measure of centrifugal force immediately follows. To move from uniform motion on circles to varying motion on other curves, in particular conic sections, Newton established in Proposition 6 a general framework for the analysis of forces and with it a general expression that, when interpreted for a particular curve, yields a measure of the acceleration.

> If body *P* revolving about center *S* describes any curved line *APQ*, and the straight line *ZPR* is tangent to that curve at a point *P*, and to the tangent from some other point *Q* of the curve *QR* is drawn parallel to the length *SP*, and the perpendicular *QT* is let fall on the length *SP*, I say that the centripetal force is inversely as the solid $(SP^2 \cdot QT^2)/QR$, provided that one takes as the quantity of this solid its final value when the points *P* and *Q* come together.

Newton later made Proposition 6 more mathematical in form by relating *QR* to the versine of arc *PQ*, but in substance it rested on a dynamical principle. *QR* is the change of velocity caused by the central force acting over the nascent arc *PQ* and pulling the body back from its inertial path *PR*. By the Second Law that change is proportional to the magnitude of the force, and by Lemma 10 it is proportional to the square of the time during which the force acts. Hence, it is proportional to their product conjointly.[19] The time is nothing other than the area of triangle *PSQ*, which is proportional to the product *SP·QT*. Everything is in the diagram where Newton and the reader can see it, though not everything is as yet part of the orbit itself. That comes next.

By Proposition 6 Newton had arrived at the general expression for centripetal force that would guide his analysis of orbital motion on conic sections. He would evaluate the final ratio of $QR/(SP^2 \cdot QT^2)$ as *QT* and *QR* evanesced with the coincidence of *Q*

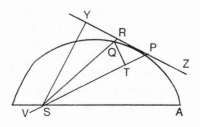

Figure 2

and P. Or rather, he would seek to transform that expression into one involving the parameters peculiar to the curve to which it was being applied. In quick succession to close Section 2, he considered motion: (P7) on a circle, where the center of force is eccentric; (P8) on a semicircle with an infinitely distant center of force; (P9) on an Archimedean spiral, toward the center; and (P10) on an ellipse, toward the center. Section 3 then opened with the curves of main concern to him, namely the conic sections with the center of force located at a focus. In each case (P11–13), the final ratio of QT^2/QR reduces to the latus rectum of the particular conic section, with the result that the centripetal force varies inversely as the square of the distance SP from the center of force.

With Proposition 14 Newton left his general expression to concentrate on the special case of the inverse square and to seek the various elements of the conic sections that could serve as measures of the body's velocity at points on them. Here he seems at first to have had no special mathematical secrets. Propositions 14, 15, and 16, the latter with its long list of corollaries, simply follow from his intimate knowledge of the conic sections.

Yet, his mathematics does have one feature of interest to the development of analytic mechanics. By using ratios and proportions, Newton could avoid questions of the precise constants involved in the relations he was unwrapping. But by excluding constants, Newton also excluded parameters. That is, his analysis did not point the way from specific results to general considerations by considering the constants of an expression and what occurs if they are varied. Seeing the relations between the constant coefficients of an equation and its roots had constituted the main achievement of the new symbolic algebra, opening the way toward the study of the structure of equations.[20] Varignon and the other Continental readers of the *Principia* saw in Leibniz's calculus a vehicle for bringing that analytical power to bear on mechanical relations as well. In their eyes, Newton had failed to do that, despite the seemingly analytical style of his geometry. Whatever his treatment of central forces revealed about the structure of the orbits, it gave little insight into the structure of mechanics.

Orbits, of course, and in particular conic-section orbits were Newton's main concern. Proposition 17 set out the general procedure by which, given the absolute measure of an inverse-square force, the center of force, and the initial position, velocity, and direction of the moving body, one can determine the orbit. Fig. 3, which is adapted from Newton's original, exemplifies the line of analysis: line PR, point P, and point S (the center) are given; by the reflective properties of the conic sections, one has line *PH;* from that and *SK   PH,* one has *KP;* from the velocity at *P* ($\propto 1/SR$) and *SP,* by Proposition 14 one has the latus rectum and, from *SP, KP,* and the latus rectum, the value of *(SP + PH)/SP,* from which, given *SP, PH* follows, locating *H,* the other focus of the conic section. *SP + PH* is the major axis, and *(SP + PH)·latus rectum* is the square of the minor axis. To know what kind of conic section one is dealing with, it suffices by Proposition 14 to consider the relation

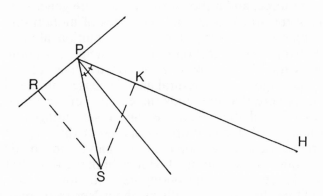

Figure 3

$$L = 2(SP + KP) \implies \begin{array}{l} < \quad :\text{ellipse} \\ = \quad :\text{parabola} \\ > \quad :\text{hyperbola} \end{array}$$

Proposition 17 completes, then, the theoretical solution of Newton's immediate problem, that is, the motion of bodies under an inverse-square force, which (as Newton forewarned the reader in Corollary 6 of Proposition 4) observational data and Kepler's third law show to obtain in the solar system. But those same observational data do not readily lend themselves to direct construction of planetary orbits, nor had Newton so far made clear how the data might be used to verify the theoretical analysis. Sections 4 and 5 attend to the first matter, Sections 6 and 7 to the second.

Sections 4 and 5, comprising Propositions 18–29 and Lemmas 15–27, treat the construction of conic sections given varying combinations of foci, points, and tangent lines, in particular the construction of an orbit through five points. It is here that the *Principia* looks most like a geometrical treatise in the classical mode, even though, as Whiteside pointed out, the geometry dates from the seventeenth century (e.g., de la Hire) and not from the third century B.C. Nonetheless, it is geometry, and some of it rests on projective methods. As such, it is not a cover for algebraic techniques; the algebraic, analytic geometry of the time was not up to the problems being solved in these sections, and an algebraic mechanics would have avoided them.

Section 7 makes a transition both in substance and in style. In substance, it completes the study of inverse-square forces by solving the problem of motion along a straight line toward or away from the center of force, and it then turns to the more general problem of determining the trajectories and the times of motion on them for central forces of any sort, starting with motion along a radius; Section 8 extends the question to orbits. In style, Section 7 moves from the by-now-familiar pattern of constructive solutions to an altogether new mode of structural analysis based on the method of fluxions. It is here that one finds the essence of Newton's analytic mechanics with all the advantages and disadvantages of its geometrical form, and it was Propositions 39–41 that constituted the starting point for the development of analytic mechanics in the algebraic mode on the Continent. Hence both the transition and the new methods are worth considering here, the former for what the contrast within the *Principia* tells us about Newton's mathematical style, the latter for discerning what went into the "translation" of Newton's methods into the calculus.

Newton treated rectilinear motion under an inverse-square force as a degenerate case of orbital motion, taking the limit as each of the conic sections narrows down to its axis. At the limit, however, the geometrical parameters of motion located in the elements of the orbits disappeared with them. Again, as earlier, Newton had to find auxiliary constructions that would not disappear in the limiting case and in which the parameters could be relocated, and, as earlier, that need set the path of his analysis. So too did his search for constructive solutions, which forced him to treat the three conic sections separately.[21] It will suffice here to consider the elliptical case.

If (P32) the body at A did not fall perpendicularly, it would follow a conic section, say, the ellipse *ARPB* with the center of force at the lower focus, and the time of motion to point *P* would be proportional to the focal sector *ASP*. If *ADB* is a semicircle on the major axis *AB* of the ellipse, it follows that the area of sector *ASD* is also proportional to that time of motion, since for any ordinate *CPD*, the ratio *CP:CD* is constant and therefore, by a combination of Euclid's

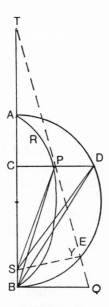

Figure 4

Elements and Cavalieri's principle, so to is the ratio of the sectors *ASP* and *ASD*. Hence, the use of the ellipse's focal sectors to measure time can be transferred to those of the circle.[22] That is the case for all ellipses, however narrow. Hence, Newton argued, it is true in the case of the narrowest ellipse, namely the degenerate ellipse that coincides with its major axis. As the ellipse evanesces, however, the focus *S* moves to coincide with vertex *B,* which becomes the center of force in the limiting case; as a result, the measure of time lies in the peripheral sector *ADB* of the circle. To find the position of the body on *AB* at a given time, determine the point *D* corresponding to that time and drop the perpendicular *DC;* the body will be at *C.*

So far Newton had simply reduced the problem, not solved it, since he had given no means for finding point *D.* For that he needed a measure of the falling body's velocity at any point *C.* Proposition 33 establishes that that velocity is to the velocity of the body revolving uniformly on a circle of radius *BC* under the same force as the square root of the ratio of *AC* to the semidiameter *AB*/2; the derivation rests largely on the corollaries to Proposition 16. From there Newton moved in Proposition 35 to yet a third circle, the center of which lies at *B* and the radius of which is half the latus rectum of the equant figure. The uniform motion of point *K* on that circle corresponds, he showed, to the uniform change of the sector *ASD,* and with that he had a constructible measure of the time of fall over *AC* (Proposition 36).

In this last proposition, Newton turned from indivisibles to fluxions as his means of proving two areas equal. The heart of the demonstration is the argument that the *particulae KSk* and *SDd* of the two areas are equal in the last instance for any point *C,* that is that the two sectors are increasing by the same amounts for the same *lineola Cc* of the path. That method of comparing curvilinear areas is the foundation of the general analysis of central-force motion to which Newton then turned in Propositions 39–41.

Reversing the order of his analysis of inverse-square motion, Newton began his general investigation with rectilinear motion. The configuration of Proposition 39 shows by contrast with those of Propositions 32 and 35 how different the two treatments will be. The proposition sets out the general problem: given any sort of centripetal force and assuming the quadrature of curved figures, and given a body rising or falling along a radius of force, find both the velocity of the body at any point on the radius and the time taken to reach that point. The diagram begins with the path *AC,* with *C* the center of force and *A* the origin of motion. On that diagram Newton then drew curve *BG* as a graph of the force at each point on the path. The velocity at any point *E,* he asserted, is as the square root of the area *ABFE* under that graph. Moreover, if one

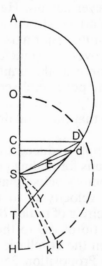

Figure 5

constructs *EM* inversely proportional to that velocity, then the area under the curve *ALM* will be as the time of motion from *A* to *E*.

New to Newton's previous line of analysis is the curve *BFG* and with it the assertion that the area under it varies as the square of the velocity. No such curve had figured in Propositions 32–37, where it would have been defined by the relation $EG \cdot EC^2 = const$. New also is the argument by which he proved the assertion; it is important to hear his own words:

In line *AE* take a very small line [*linea quam minima*] *DE* of given length, and let *DLF* be the position of line *EMG* when the body passed through *D;* and if that be the centripetal force and the square root of area *ABGE* be as the velocity of the falling body, that area will be in the duplicate ratio of the velocity. That is, if for the velocities at *D* and *E* we write *V* and *V + I*, area *ABFD* will be as V^2, area *ABGE* as $V^2 + 2VI + I^2$, and, by difference, area *DFGE* as $2VI + I^2$; hence *DFGE/DE* will be as $2I(V + I/2)/DE$. That is, if we take the first ratios of the nascents, length *DF* will be as the quantity $(2I \cdot V)/DE$, and therefore also as the half $(I \cdot V)/DE$ of this quantity. Moreover, the time in which the falling body describes the little line *DE* is directly as that line and inversely as the velocity *V,* and the force is directly as the increment *I* of the velocity and inversely as the time, and therefore, if the first ratios of nascents be taken, as $(I \cdot V)/DE$, that is, as the length *DF* or *EG.* Hence a force proportional to this *DF* causes the body to fall at a speed that is as the square root of the area.

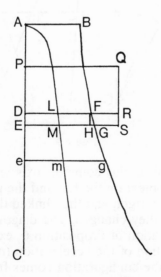

Figure 6

The use of fluxions here—that is, showing that the curve increases at a certain rate—clearly accomplishes little more than its task of demonstrating the assertion in question. Newton took no advantage of the opportunity to use them as the vehicle of a general treatment of mechanics, nor did he adapt his language to them. Here, as throughout the *Principia,* he retained the geometrical language of ratios and proportions, and only by hindsight could one discern in the passage just quoted anything akin to equations of motion.[23]

The pattern holds as well as Proposition 41, where Newton finally set out the general problem later to be called the "inverse central-force problem":

> Assuming any sort of centripetal force, and granting the quadrature of curvilinear figures, required are both the trajectories in which the bodies move and the times of motions in the trajectories found.

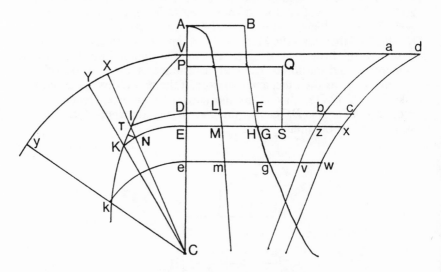

Figure 7

Here Newton joined at the common axis *AC* a drawing of the physical geometric object on the left and the graphs of the parameters of motion on the right, and then linked them by means of the fluxions[24] by which they changed. The diagram of the orbit is related to the configuration of Proposition 6, except that it includes no direct representation of the acceleration (nor did Newton seek one). The center of the configuration comes from the propositions immediately preceding. The two rightmost curves emerge by the process, familiar from the earliest propositions, of relocation of parameters.

The proposition is quite intricate, but its flow can be described briefly. Over the very short interval of time represented by sector *ICK,* the increment *IK* of the trajectory can be taken as proportional to the velocity at *I* which, by Proposition 40, is the velocity of rectilinear fall at *D* and hence is measured by the square root of the area *ABFD.* For the given interval of time, $\Delta\ ICK\ =\ IC\ \cdot\ KN/2$ is fixed. Hence $KN\ =\ Z\ =\ Q(const)/IC(radius)$; and, by Proposition 40, if $\sqrt{ABFD}\ /Z=IK/KN$ at any instant, the relation holds for all instants. From the squared relation $ABFD/Z^2=IK^2/KN^2$, it follows that $IN/KN\ =\ \sqrt{(ABFD\ -\ Z^2)}/Z,$ whence the sector *ICK* can be expressed as $Q\cdot IN/\ \sqrt{(ABFD\ -\ Z^2)}$, where Q is the constant of proportionality for the time.

This expression is the basis for the construction of two auxiliary curves. Aiming at a construction of the orbit, Newton next moved the measure of time to a circle *VXY* of fixed radius *VC* about *C.* The sectors *NCK* (which Newton is taking as equal to *ICK*) and *XCY* are to each other as $(YC/KC)^2$, and hence sector $XCY\ =\ (Q\cdot IN\cdot CX^2)A/\sqrt[2]{(ABFD\ -\ Z^2)}$. Constructing *Db* such that *Db·IN* equal to the sector of the trajectory, and *Dc* such that *DC·IN* is equal to that of the circle, Newton had two new graphs of motion *abzv* and *dcxw,* the fluxions of which, *DbzE* and *DcxE,* are equal respectively to the fluxions of the sectors *VCI* of the trajectory and *VCX* of the equant circle and measure the flow of time over them. Given, then, the time of motion from *V,* the equivalent area *VabD* will locate point *D* and hence the distance $IC\ =\ VC$ of the body from the center, while the equivalent area *VdcD* will yield sector *VCX,* which is directly constructible, and by means of it the angle *VCX* (i.e., *VCI*) by which point *I* is uniquely located.

The general result of Proposition 41 for a given curve of forces, then, is a means of constructing a trajectory and an equant. The demonstration is, one might say today, an "existence proof"; its effectiveness as a guide to action in any particular instance depends on mathematical methods of quadrature and reduction not belonging to it. Its generality is deceptive, for it is in fact no more general than the mathematics available for applying it. That is true, moreover, in whatever direction one applies it, whether to find the trajectory given the forces or to find the forces given the trajectory. Perhaps that is why Newton's analysis of the forces producing motion on conic sections and of inverse-square forces in the first seven sections of Book 1 has little or nothing to do with the general proposition derived in Section 8. That analysis took special advantage of Kepler's law and of the properties of conic sections, and Newton shaped this method accordingly.

Perhaps that is also why, when mathematicians on the Continent read the *Principia,* they focused their attention squarely on Section 8 and developed precisely the general theory Newton had only sketched out. Unlike Newton, who was interested in orbits and who had a clear astronomical goal in sight, the Europeans were interested in mathematics—indeed a particular kind of mathematics—and its power to probe questions of natural philosophy. They began their reading of Newton's great work by translating it into their own mathematical terms, reshaping it in the process. A brief example may suggest in conclusion what that transformation involved and where it led.

Pierre Varignon's recasting of the *Principia* into analytical terms began with a memoir, "Manière générale de déterminer les forces, les vitesses, les espaces, & les temps, une seule de ces quatre choses étant donnée dans toutes sortes de mouvement rectilignes variés à discrétion."[25] In it, he captured Newton's theorems on rectilinear centripetal motion in two "general rules," from which all else followed by the techniques of ordinary and infinitesimal analysis. Varignon's modification of the configuration of Proposition 39 of the *Principia* reveals both the different form of mathematics he was working with and the different ends to which he was applying it.

All the rectilinear angles in the adjoined figure being right, let *TD, VB, FM, VK, FN, FO* be any six curves, of which the first three express through their common abscissa *AH* the distance traversed by some body moved arbitrarily along *AC.* Moreover, let the time taken to

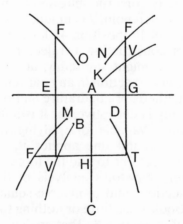

Figure 8

traverse it be expressed by the corresponding ordinate HT of the curve TC, the speed of that body at each point H by the two corresponding ordinates VH and VG of the curves VB and VK. The force toward C at each point H, independent of [the body's] speed (I shall henceforth call it central force owing to its tendency toward point C as center) will be expressed similarly by the corresponding ordinates FH, FG, FE of the curves FM, FN, FO.

The axis AC, with the center of force at C, stemmed from Newton. The six curves, however, were inspired by Leibniz.[26] They represent graphically the various combinations of functional dependency among the parameters of motion: the "curve of times" TD represents time as a function of distance; the "curves of speed" VB and VK, the velocity as functions of distance and time respectively; and the "curves of force" FM, FN, and FO, the force as functions of distance, time, and velocity respectively.

To translate those designations into defining mathematical relations, Varignon turned to algebraic symbolism. At any point H on AC set the distance $AH = x$, the time $HT = AG = t$, the speed $HV = AE = GV = v$, and the central force $HF = EF = GF = y$. "Whence," Varignon concluded from the perspective of the calculus,

one will have dx for the distance traversed as if with a uniform speed *(comme d'une vitesse uniforme)* v at each instant, dv for the increase in speed that occurs there, ddx for the distance traversed by virtue of that increase in speed, and dt for that instant.

The first of the two general rules expressed symbolically an assumption that became fundamental to analytic mechanics: over infinitesimal distances ds traversed in infinitesimal times dt a motion may be considered as uniform, whatever its actual nature in the finite realm. That is, by Galileo's law of uniform motion, $ds = vdt$. Expressed in another form, since "speed consists only of a ratio of the distance traversed by a uniform motion to the time taken to traverse it," $v = dx/dt$. Then, by the rules of differentiation,[27] $dv = ddx/dt$.

The second rule took account of the change of speed and of the increment of distance that results from it.

Moreover, since the distances traversed by a body moved by a constant and continually applied force, such as one ordinarily thinks of weight, are in the compound ratio of that force and of the squares of the times taken to traverse them, $ddx = ydt^2$, or $y = ddx/dt^2 = dv/dt$.

The rule appears to have stemmed in the first instance from the *Principia*. The first half of the measure expresses the second law, and the second half translates Lemma 10 into the language of the calculus. Varignon's version of the rule literally brings a new dimension to it, however, by capturing through the second differential *ddx* that the effect is a second-order variation of the motion of a body.

Those two rules, $v = dx/dt$ and $y = dv/dt$, sufficed, Varignon maintained, to give a full account of forced motion along straight lines. For, given any one of the six curves set out above, one can use the rules to carry out the transformations necessary to produce the other five. That central proposition reduces the mechanics in question to a matter of mathematics, and for the remainder of the memoir Varignon pursued an essentially mathematical point. In essence, the solution of the differential equation $v(x) = dx/dt$ yields the curve *DT* determined by $HT = t(x)$, and, if $VH = v(x)$, then $v'(x)dx = dv = ydt$ will produce $y(x,t)$, which can take two forms, depending on how the curve *DT* is expressed. Either $FG = y(x(t),t)$ or $FH = y(x,t(x))$. The other curves emerge by similar transformations. As Varignon noted at the outset, the general claim rests on

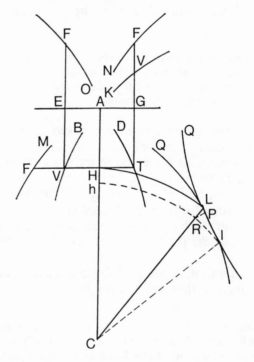

Figure 9

the dual assumption of complete solvability in the two realms of analysis: the resolution of any algebraic equation (i.e., getting $x(t)$ from $t(x)$) and the integration of any differential equation. The limits of the mechanics in question were those of the calculus.[28]

Varignon's second memoir of 1700, "Du mouvement en général par toutes sortes de courbes; & des forces centrales, tant centrifuges que centripètes, nécessaires aux corps qui les décrivent," applied the two general rules of the first memoir to the motion of bodies along curves. The basic mathematical configuration remained the same; to it Varignon added a trajectory QL (which he called the "curve of paths," the *"courbe des chemins"*) referred to the axis AC and to the center of force C via what may be termed "arc coordinates": the point L is located by means of the abscissa AH ($= x$) and the arc-length HL ($= z$). If $AC = a$ and $CH = r$, then L is also located by the radial distance $r = a - x$, in which case, $dx = -dr$. The coordinates point to Varignon's source: the diagram came from Propositions 40 and 41 of the *Principia,* though only in part; Varignon needed only the trajectory on the left of Newton's configuration, not the auxiliary curves on the right.[29]

As in the memoir on rectilinear motion, Varignon set out two "general rules" by means of which, given any two of the seven curves of his configuration, he could construct the remaining five. The first rule followed straightforwardly from the first memoir by replacing the axial interval dx by the infinitesimal increment ds of the distance QL ($= s$) along the trajectory. That is, by Rule 1, $v = ds/dt$. The crux of the memoir lies in Varignon's derivation of the second rule by analysis of the change of motion caused by a central force acting obliquely on the moving body over the infinitesimal distance Ll ($= ds$). Proposition 40 of the *Principia* provided the necessary hint, which Varignon developed in a manner peculiar to the infinitesimal analysis he was using.

Following Newton, or perhaps rather joining Newton in following Galileo, Varignon superimposed on the infinitesimal curvilinear triangle LRl the rectilinear configuration of an inclined plane and resolved the "absolute force" acting on the body at L (along the radius LR) into its normal and tangential components PR and LP. But then Varignon went his own way. Concerned here with acceleration along the arc Ll, rather than with the normal component of the centripetal force represented by PR, he focused his attention on LP. By the law of the inclined plane,

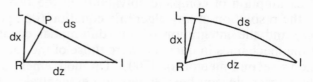

Figure 10

absolute force:motive component $= LR:LP = Ll:LR = ds/dx$.

If, that is, the absolute force is y, then the motive component is ydx/ds. That component acts as a constant force over the interval, producing an acceleration measured by the relation set out in the first memoir; that is, by Rule 2, $dds = (ydx/ds)dt^2$, or $y = (ds/dx)(dds/dt^2) = vdv/dx$.

The particular form of the general rules lent itself well to the expression of the central forces determining a given trajectory. In Varignon's coordinate system, $ds^2 = dr^2 + dz^2$, which may be rewritten in the form $ds^2/r^2dz^2 = [(dr/dz)^2 + 1]/r^2$. By Kepler's hypothesis, $r^2dz^2 = dt^2$, and, since dt is constant, differentiation of the left-hand side of the previous equation yields $2dsdds/dt^2$. Dividing that result by $-2dr$ produces the value y of the central force. Assuming that dz can be expressed in terms of r and dr, one can rewrite the right-hand side of the equation in the form $[1 + (dz/dr)^2]/[dz/dr]^2r^2$ to get an expression containing only r, dr, and constants. Differentiating that expression and dividing the result by $-2dr$ will therefore give the value of y for the curve $z = z(r)$. Conversely, an expression for y in terms of r will constitute a differential equation, the solution of which yields a trajectory $z = z(r)$.

Varignon's analysis of the motion of bodies under central forces effectively tied the development of mechanics to that of the calculus. Given the wholly algorithmic nature of differentiation, finding the forces had become a matter of straightforward computation. By contrast, finding the trajectory depended on the resources of the integral calculus and the theory of differential equations. As those resources developed over the eighteenth century, the realm of mechanics grew with them, extending well beyond the orbits of planets into dimensions that geometry could not contain.

Notes

1. Richard S. Westfall, *Never at Rest. A Biography of Isaac Newton* (Cambridge: Cambridge University Press, 1980).

2. C. A. Truesdell, "A Program toward Rediscovering the Rational Mechanics of the Age of Reason," *Archive for History of Exact Sciences* 1 (1960–62): 3–36.

3. J. E. Hofmann documented Leibniz's independence in *Die Entwicklungsgeschichte der Leibnizchen Mathematik während des Aufenthalts in Paris, 1672–76* (Munich, 1949; Engl. trans. *Leibniz in Paris 1672–76: His Growth to Mathematical Maturity,* Ithaca: Cornell University Press, 1974), and two recent publications have exposed even more clearly the inanity of the dispute: A. R. Hall's *Philosophers at War* (Cambridge: Cambridge University Press, 1980) and vol. 8 of D. T. Whiteside's edition of *The Mathematical Papers of Isaac Newton,* 8 vols. (Cambridge: Cambridge University Press, 1966–83); cf. my essay review of the latter, "On Differential Calculuses," *Isis* 75 (1984): 366–72.

4. Published both separately and, with fuller notes, in the *Journal for the History of Astronomy* 1 (1970): 116–38.

5. "The published state of the *Principia*—one in which the geometrical limit-increment of a variable line segment plays a fundamental rôle—is *exactly* that in which it is written." Whiteside, "Mathematical Principles," p. 119.

6. D. T. Whiteside, ed., *The Mathematical Papers of Isaac Newton,* vol. 6 (1684–1691).

7. It also obscures the question of why Newton shifted from the geometrical to the analytical mode on the few occasions when he did, for example, to account for indefinite parameters, to analyze solvability, or to establish computational techniques.

8. "Note that we have done nothing more than cast Newton's geometrical measure [of centripetal force] into an equivalent analytical form *without* adding any new logical precision or structural redefinition to it." Whiteside, "Mathematical Principles," p. 123.

9. Whiteside, "Mathematical Principles," p. 126 (re: Varignon et al.) and p. 130 (re: Costabel). While quoting (p. 120) Clifford Truesdell to the effect that "the *Principia* is 'a book defense with the theory and application of the infinitesimal calculus,' " Whiteside declared himself (p. 137, n. 53) "hesitant to fall uncritically in with" Truesdell's even broader claim in the same source that "Except for certain simple if important special problems, Newton gives no evidence of being able to set up a differential equations of motion for mechanical systems." (Truesdell, "A Program toward Rediscovering," p. 9.)

10. There are some, e.g., 1:40—"Required are the motions of the apsides of orbits which are most close to circles," or 2:10—"Let the uniform force of gravity tend directly to the plane of the horizon, and let the resistance be as the density of the medium and the square of the velocity conjointly; required are both the density of the medium at each point, which causes the body to move in a given curved line, and the velocity of the body at the same points." But on the whole, Newton uses the technique of infinite series quite sparingly.

11. See M. S. Mahoney, *The Mathematical Career of Pierre de Fermat* (Princeton: Princeton University Press, 1973), pp. 278–79.

12. On these new geometrical techniques, see D. T. Whiteside, "Patterns of Mathematical Thought in the Later 17th Century," *Archive for History of Exact Sciences* 1 (1961–62): 271–89. For their role in Barrow's mathematics, see in addition M.S. Mahoney, "Barrow's Mathematics: Between Ancients and Moderns," in *Before Newton: The Life and Times of Isaac Barrow,* ed. Mordecai Feingold (Cambridge: Cambridge University Press, 1990). Leibniz was only beginning to think about how to extend algebra to questions of coincidence and position to form what he called "analysis situs."

13. As a careful analysis of Huygens's work shows and as will become clear for

Newton below, that model had its limitations, requiring in particular that the mathematician continually refocus his diagram to accommodate the several dimensions in which the quantities and relations under investigation fall. See M. S. Mahoney, "Diagrams and Dynamics: Mathematical Perspectives on Edgerton's Thesis," in *Science and the Arts in the Renaissance*, ed. J. W. Shirley and F. D. Hoeniger (Washington, D.C.: The Folger Shakespeare Library, 1985), pp. 198–220, at pp. 210–13; "Infinitesimals and Transcendent Relations: The Mathematics of Motion in the Late Seventeenth Century," in *Reappraisals of the Scientific Revolution*, ed. D. C. Lindberg and R. S. Westman (Cambridge: Cambridge University Press, 1990), ch. 12, especially pp. 476–85.

14. On the transmutation of areas, see *inter alia* Mahoney, "Barrow's Mathematics," and *Fermat*, ch. 5.

15. As Euler put it, "Newton's Mathematical Principles of [Natural] Philosophy, by which the science of motion has gained its greatest increases, is written in a style not much unlike [the synthetic geometrical style of the ancients]. But what obtains for all writings that are composed without analysis holds most of all for mechanics: even if the reader be convinced of the truth of the things set forth, nevertheless he cannot attain to sufficiently clear and distinct knowledge of them; so that, if the same questions be the slightest bit changed, he may hardly be able to resolve them on his own, unless he himself look to analysis and evolve the same propositions by the analytic method." *Mechanica, sive motus scientia analytice exposita* (St. Petersburg, 1736), Pref., p. [iv].

16. All references to the *Principia* are based on the first edition (London, 1687), reprinted in facsimile by Dawson, London, n.d.

17. That, by the way, is why the Second Law is expressed in terms of an impulse rather than of a continuously applied force.

18. The argument is tricky and rests on the generalization of a constant relationship. For any segmented figure, the velocities are as the segments, which in turn are reciprocally as the perpendiculars from the center of force to them, since the products of the segments and the corresponding perpendiculars measure the triangles that are proportional to the times of motion over the segments and hence equal. What holds for any figure of finite segments, however small, holds for the figure constituted of segments at the moment of evanescence, when they become the point bases of triangles and their extensions become the tangents to the curve. At heart, the principle is none other than Leibniz's Principle of Continuity.

19. Lemma 10. Spatia, quae corpus urgente quacunque vi regulari describit, sunt ipso motus initio in duplicata ratione temporum. Corollary 2. Errores qutem qui viribus proportionalibus similiter applicatis generantur, sunt ut vires & quadrata temporum conjunctim.

20. See M. S. Mahoney, "Infinitesimals," and "The Beginnings of Algebraic Thought in the Seventeenth Century," in *Descartes: Mathematics, Physics, and Philosophy* (Totowa, N.J. and Brighton, Sussex: Barnes & Noble and Harvester Press, 1980), ch. 5.

21. The initial conditions of motion determine which of the three serves as the basis for the limiting case.

22. The measure of time for hyperbolic orbit is transferred to an equilateral hyperbola, that for a parabola to an equilateral parabola. Newton termed the constant curves "equants."

23. In a letter to Jakob Hermann published in *Mémoires de l'Académie Royale des Sciences* (1710), Johann Bernoulli termed Newton's demonstration of Proposition 40 "trop embarrassée" and emphasized that he was replacing it with an analytical version.

24. Note that these are geometrical, rather than analytical fluxions. Newton is comparing elements of curves, not algebraic quantities.

25. *Mémoires de l'Académie Royale des Sciences* (1700), pp. 22–27.

26. G. W. Leibniz, "Considérations sur la différence qu'il y a entre l'analyse ordinaire et le nouveau calcul des transcendantes," *Journal des sçavans* (1694), repr. in C. I. Gerhardt, ed., *Leibnizens mathematische Schriften*, 7 vols. in 5 (Berlin and Halle, 1849–63), 5:306–8.

27. Varignon takes dt as constant, i.e., $ddt = 0$. Doing so in Leibnizian calculus is equivalent in the modern form to choosing t as the independent variable, i.e., as the variable with respect to which one is differentiating. The Leibnizian version allows a greater flexibility in analyzing differential relations, as Bernoulli had shown in his lectures on the method of integrals and as Varignon shows in his next memoirs, to be discussed below. On the mathematical point, cf. H. J. M. Bos, "Differentials, Higher-Order Differentials and the Derivative in the Leibnizian Calculus," *Archive for History of Exact Sciences* 14 (1974): 1–90.

28. Two examples would suffice, he thought, to illustrate the power of his method of analysis. The first stemmed from Galileo. On the assumption that $v = \sqrt{x} = dx/dt$, it follows directly that $dt = dx/\sqrt{x}$, or $t = 2\sqrt{x}$, or $x = t^2/4$. That, in shape at least, is the law of falling bodies. Newton provided the basis of the second example, presented as a "Remarque" or scholium. From $dt = dx/v = dv/y$ it follows immediately that $ydx = vdv$, or $v^2/2 = \int ydx$ which, solved for v, is the first part of Proposition 39 of the *Principia*. Rewritten in the form $dt = dx/\sqrt{(2\int ydx)}$, it leads to the second part by way of the inverse curve $VH = 1/\sqrt{(2\int ydx)}$.

29. One of which was incorrect in any event. Indeed, Newton may have reborrowed Varignon's version for later editions of the *Principia*.

Comets & Idols: Newton's Cosmology and Political Theology

SIMON SCHAFFER

Newton as Idol

Men are apt to vary, dispute and run into partings about deductions. All the old Heresies lay in deductions; the true faith was in the text.

—Newton

The author of the *Principia* was, almost too obviously, an idol of the eighteenth, the "Newtonian," century. Traditional eighteenth-century historiography, exemplified by Carl Becker or Ernst Cassirer, saw the clear and unproblematic interpretation in "the unreserved admiration and veneration of the eighteenth century for Newton"; indeed "the whole eighteenth century understood and esteemed Newton's achievement." Pope's famous couplet expressed "the veneration which Newton enjoys in the thinking of the Enlightenment." Materialist irreligion, which claimed Newton's mantle, for example, was dismissed as "an isolated phenomenon of no characteristic significance" since it misinterpreted the true legacy.[1] No doubt the lack of ambiguity in the meaning Newton intended and his followers espoused played a crucial role in the making of this tradition—it did so, too, in the interpretative strategies used by historians of Newtonianism. However, it is now some time since Newton's statue stood alone in this pantheon: historians have charted the survival and vitality of non-Newtonian and anti-Newtonian strands in the intellectual culture of the long eighteenth century.[2]

Some of the most interesting interpretative issues arise in the fracturing rather than the multiplication of this idol: varieties of Newtonianism have been spotted in the eighteenth-century development of Newton's program. Thus I. Bernard Cohen was concerned to document the vigor of an experimental Newtonianism derived from the *Opticks* because it had previously "been obscured by the glamour of the other," the tradition of the *Principia*.[3] More complex taxonomies emerged after Kuhn's erection of this distinction between experimental and mathematical sciences, and after Schofield's work on what he called the "mechanist" and "materialist" accounts of classical British natural philosophy.[4] Finally, the luxuriance of anti-Newtonianisms in the eighteenth century has recently attracted its own chroniclers—M. C. Jacob examined the career of freethinking deists, Larry Stewart has written of Tory nonjurors, Chris Wilde documented the vigorous British Hutchinsonian challenge to Newtonian orthodoxy, and John McEvoy and J. E. McGuire have given a detailed reconstruction of Joseph Priestley's place in the tradition of "rational dissent."[5]

Yet just as authors writing on "Newtonianism" have appealed to the true and original intention of Newton's texts, so, commonly, alleged anti-Newtonians often claimed at the time that they were giving Newton's intentions in the proper manner. For John Toland, Newton was "the greatest man in the world" and all of his greatest achievements and utterances were "capable of receiving an interpretation favourable to my opinion."[6] The freemason Andrew Ramsay appealed to "the very words of Sir Isaac Newton" in defending neo-Platonist spiritism by sensitively editing texts from the *Opticks*.[7] Behmenist sectaries and Camisard enthusiasts claimed Newton as their follower: a disciple of Boehme, the Bristol cleric Richard Symes, wrote that if Newton had lived to see the effects of electrical machines fully developed, then "it would have enabled him to demonstrate to the sight his attractive and repulsive powers, and saved him the labour of writing so much about it."[8] Famous cases of such interpretative ingenuity include both the optical debates around the work of Thomas Young and the chemical debates at the time of John Dalton. But the trouble was endemic. In 1741 Christian Wolff attacked Voltaire's construction of "Newtonianism" in these terms:

> Nothing can be more absurd than Voltaire's desire to make Newton into a metaphysician. Newton only expressed imaginary notions of things about which metaphysics must establish real concepts. The attempt to establish a parallel between Newton and Leibniz is simply fantastic. Where reasons fail (Voltaire) seeks to win approval by Newton's authority. I am sorry for Newton's sake that his memory is darkened by those who have no comprehension of what he really did.[9]

This hermeneutic chaos is not evoked merely to suggest that the historiographic garden is full of weeds and needs a new, simplified botanic order. A more general problem is at work here: highly esteemed scientific texts function like idols of interpretation. They are taken as stable, pre-given, clear, and precise in ways utterly absent in literary texts. The goal of interpretation then becomes "consistency," to be achieved by constituting a reliable, consensually agreed, and unambiguously authored and dated archive from which to save the inherent and uniform intention. Vitally important disputes among Newtonian scholars, such as those over the dating and meaning of *De gravitatione,* the *Clavis,* "On Motion in Ellipses," and *De aere et aethere,* are testimonies to this need. In these cases, historians have argued over the relation of the manuscript with Newton's authorship, over the possible implications of redating and reattribution for the picture of a smoothly continuous or consistent pattern in his intellectual development, and, last, over the place such texts should be allowed to occupy in the canon of major, reliable testimonies of his true views.[10]

Coherence is grounded in a variety of ways. The relative authority of manuscript and published texts, admirably summarized for the Darwin archive by Martin Rudwick in his analysis of the "continuum of relative privacy," is but one of these. This trouble hinges on the difficulty of deciding whether the most private or the most public utterances of an author are those most to be credited. Both views have been sustained in the secondary literature—some have used the privacy of Newton's alchemical and theological papers to argue for the deep significance that these works had for his program, while others have credited only those texts which he did publish and the immediately connected drafts of these works. Other troubles center on authorized and pirated texts; on the distinction between "literal" and "metaphorical" senses of Newtonian obiter dicta; or, last, on the underlying mathematical structure of a passage of reasoning.[11]

Each of these variations has peculiar relevance to the problem of Newtonian hermeneutics. Newton's celebrated correspondence with Richard Bentley has received ample glossing by historians. The usage of terms such as *hypothesis* or *idea* in these passages might be held to depend on the audience for which Newton was writing, the appropriate public or private status these letters should be granted, and the precise literal or figurative sense of terms such as *not Material.* Thus the Halls suggest that Koyré read the sentences in these letters on immaterial mediation of attractive force by supposing that Newton was "much too intelligent not to perceive that the mechanical hypotheses of forces lead to infinite regress: therefore he must on the contrary have believed that forces are non-mechanical, quasi-spiritual." Yet such reading is easily challenged, as in the interpretation of the same passages offered by Gerd Buchdahl, for whom they represent allusions to the problem of conceivable material interaction.[12]

The attempt to save a consistent account of the *Principia*'s doctrine of gravity and its causation is always troubled in this way. According to Ernan McMullin, Newton's utterances on gravity are characterized by "conceptual incoherence." Newton's followers and critics were cleared of the charge of misunderstanding Newton, because of these ambiguities. McMullin suggests that "what prevented Newton's position from being open to an immediate charge of inconsistency was, of course, the ambiguity of the term gravity as he used it." McMullin rejected Koyré's charge that Cotes "misunderstood" the doctrine of essential qualities. Apparently Cotes was rightly "puzzled" in making gravity "essential" on Newton's say-so. Koyré drew a hard line between Newton's physical metaphors and his mathematical literalism; McMullin married these, so giving him resources to explain, if not explain away, the disciples' interpretative practices and predicaments.[13]

Both the difficulties of piracy and those of hidden structure in published texts infect Newtonian hermeneutics. Figala and Westfall have both charted the tortuous procedures that Newton followed in releasing (and then denying) his chronological writings.[14] The same troubles are notoriously characteristic of the war waged, often under the cloak of anonymity, against Leibniz and his allies. This was, after all, a war fought under the leadership of two experienced interpreters. The review of the *Commercium epistolicum* was only one of several such cases. Newton drafted prefaces for papers that appeared under the name of Desaguliers. He may well have aided protagonists such as Clarke in similar ways. Further, the dispute with Leibniz was famously devoted to arguments about the meaning of terms: Clarke and Leibniz argued over the proper use of dictionaries, while the Newtonians accused Leibniz of "subterfuge, by departing from the received sense of words."[15] Such maneuvers inevitably prompt the common historiographic move that searches for metaphor, for figures of speech, for allusion and complexity beneath the surface or to one side of Newton's explicit utterances. We have noted that Koyré often produced brilliant readings of Newton's texts on gravity and attraction by supposing that physical language posed a metaphor for underlying mathematical realities.

During the calculus dispute, indeed, the whole of the *Principia* was presented by its author and his allies as a reworked expression of an original text composed in analytic language. This picture of a "hidden" analytic core was a significant move in the priority dispute. Newton wrote soon after 1716 that "I was writing for scientists (ad philosophos) steeped in geometry and putting down geometrically demonstrated bases for natural philosophy," while noting that "the analytical method through which we found these propositions shines out everywhere."[16] Yet, as Whiteside has pointed out, "there are no extant autograph manuscripts which could conceivably buttress the conjecture that he first worked the proofs in that book by fluxions before remoulding them in traditional geometrical form."[17] It needs to be emphasized that these tactics of Newton were interested maneuvers in the local setting of violent dispute. The moves show that retrospective identification of essential authorial intention is a powerful tactic in the erection of an intellectual authority.

Because of this power, it is scarcely surprising, though it is revealing, that seventeenth- and eighteenth-century natural philosophers engage in just the essential tactics of interpretation that historians have used. Both historians and their subjects stipulate an inner essence in which lies the true meaning (or irreducible ambiguity) of a text, so as then to argue that different readings must be due to "misunderstanding," a category that needs no further analysis. Those who misunderstand can then be charged with incompetence, if the interpreter is charitable, or with base interest, if the interpreter is polemical. No doubt this, in turn, licenses a common division of labor among historians. It is insinuated that those actors who follow the reading achieved by the historian are acting disinterestedly, merely following the letter of the law; other actors will need to be analyzed by appeal to local biases. So historians of ideas can concentrate on their chosen "tradition," leaving sociologists to chart the history of error. This division is unsatisfactory on many grounds—the one of relevance here is that it is a distinction that relies upon an asymmetry between true and false readings, thus attributing causal efficacy over actors to the essential sense of a text.[18]

There is a wealth of evidence to document the claim that eighteenth-century readers of the *Principia* behaved like essentialist interpreters. It was a commonplace, as Derham put it in 1733, that the *Principia* was written "designedly abstruse . . . yet so as to be understood by able Mathematicians." We have seen how this claim worked to good effect in the dispute with the Leibnizians.[19] In the debate between James Stewart and Lord Kames over the laws of motion and inertia, Stewart defended Newton from the charge that it imputed activity to matter. Stewart linked Newton's intentions with the consistency of his thought and the prevalence of interested misunderstanding: "it seems to have been far from Sir Isaac's intention to ascribe activity to matter in any shape; tho' his meaning has sometimes been mistaken. To do so would be a manifest contradiction to the primary laws of motion, delivered by himself in the beginning of the *Principia.*"[20] We find the same interpretative moves in Priestley's selective use of texts from the *Principia,* extracted with the help of the natural philosophers John Michell and Thomas Melvill; and also in Monboddo's *Antient Metaphysics:*

> I think that it is evident that Sir Isaac has used an improper, as well as an unnecessary expression, when he said that the motion is carried on by a vis insita, which certainly leads us to believe that it is by some power inherent in the body . . . [But] as he constantly distinguishes it from the vis impressa by which the body is set in motion, I think it is plain that he believed the one power not to be intrinsic, or belonging to the nature of Body.[21]

For Monboddo, as for the more devout adherents of Newtonian idolatry, improper expressions in the text were referred to the author's (inaccessible) states of belief and the surrounding circumstances, the "opinion of the times and the prejudices of men."

Essentialist readings of Newton are prompted by his subterfuges in establishing and concealing his authorship in order to build up his authority. They are also elicited by his own manifold efforts to stipulate the proper usage of the statements he released. In a scholium to *De motu corporum* of winter 1684–85 he noted:

> the aim of explaining all these things at length is, that the reader may be freed from certain vulgar prejudices and imbued with the distinct principles of mechanics. . . . But ordinary people who fail to abstract thought from sensible appearances always speak of relative quantities, so much so that it would be absurd for wise men or even prophets to speak to them otherwise. Hence both the sacred writings and the writings of the theologians are always to be understood in terms of relative quantities, and he who would on this account bandy words with philosophers concerning the absolute motions of natural things would be labouring under a gross misapprehension.[22]

This text opens up a fresh perspective on the problem of hermeneutics, to be addressed in the second half of this paper. For here Newton explicitly connects together the need of the secular author to enforce proper reading and to destroy the idols of prejudice with the problem of prophetic language and its interpretation. The reflection is significant: in his theological manuscripts he repeatedly argued that "the world loves to be deceived, they will not understand, they never consider equally, but are wholly led by prejudice, interest, the praise of men and authority of the Church they live in." This was the obverse side of essentialism: essentialism suggested that texts did carry with them proper, intentional readings, and it also pointed to social structure and corruption as the sources of any other reading.[23]

To lay the ground for an examination of Newton's interpretative work, it is necessary briefly to indicate the character of the texts at which historical attention has been directed, and with which Newton sought to make his own authority. Koyré usefully points to the multiplication of readings of Newton's authored words. In answering Cotes on gravity, as Koyré puts it, "Newton's answer to Cotes is thus simply (a) a statement that Cotes misunderstood him and (b) a suggestion that he re-read, or re-study, the relevant texts of the *Principia*."[24] This proliferation of apparent "misunderstanding" was here answered with literalism. The problem is then vitiated in the obiter dicta collected by such as John Conduitt for projected biographies of the great man. One text in particular has been of considerable significance here, and plays an important role in the second section of this paper. It contains records of a conversation of March 1725 between Newton and Conduitt on cosmology, and has been used by historians such as D. C. Kubrin and D. Castillejo to license a convincing rendition of Newton's speculations on comets, planets, stars, and active matter.[25] Newton was apparently asked "why he would not publish his conjectures as conjectures." Recall the views of Newtonian commentators of the period, such as Desaguliers and Coste, that Newton's queries were mere disguises for statements of fact.[26] Newton reportedly answered Conduitt in 1725 that "I do not deal in conjectures." When Conduitt pointed out extracts from the *Principia* that might provide ground for establishing such a conjectural cosmology, and "asked him, why he would not own as freely what he thought of the Sun as what he thought of the fixed stars—he said that concerned us more, and laughing added he had said enough for people to know his meaning."

It is obvious that records, such as those kept by Conduitt of Newton's supposed cosmological views, will always provide complex problems of hermeneutics. Several resources are available to the historian in approaching these problems.[27] We may recognize the indevicality of meaning, noting the linguistic setting in which terms of the Newtonian lexicon are most typically used. This was the force of Newton's comments in *De motu corporum* cited above.

We may make stipulations about explicit or implicit authorial intention, conscious of the extraordinary flexibility of such imputation both now and in the historical period under examination. From the early 1670s, Newton was an Arian: utterance of the creed could and was often read as the cunning subterfuge of such heretics. Last, we may ground our imputations by pointing to implication of use. Historians have indicated Newton's providentialism through the use of his statements by providentialists. Tories saw Newtonianism as a mask for deism, pointing at men they held to be deists who cited Newton. A Tory reader of the General Scholium in 1713 referred to a group of incautious Arians to decode its true sense: Newton "seems to me to lay open his heart and mind, and tell the world what cause he espouses at this day, viz., the very same which Dr. Clarke and Mr. Whiston have publicly asserted."[28] In settings marked by the work that makes authority in controversy, meaning is radically underdetermined by the text—interpretation is an accomplishment and this accomplishment is the historian's explanandum.

A last question must be addressed before engaging in a detailed examination of Newton's practice in making this authority. Given that the historian can set out to explain the accomplished interpretation, there are many reasons to suppose that goals of interpreters of these texts included social, political, or religious factors. In the settings in which Newton made interpretations of the sacred texts and made the texts his interpreters began to treat as sacred, there is much evidence that these kinds of goals were crucial. This is closely connected with the prevalence of essentialism for, as we have seen, essentialist interpretation refers to the putatively visible, the open, the public text. Public texts, as Gadamer and his colleagues suggest, are just those given power, and the act of publication is to that extent a political act. For these texts, "sacred hermeneutics" becomes the appropriate practice. As Mazzeo has argued, these were techniques designed to "save the text."[29] Venerated texts such as Daniel, Revelation, or, ultimately, the *Principia*, must have their seeming faults, ambiguities, and even contradictions effaced. To carry authority, the work of the interpreter must then render invisible his or her own secular authorship.[30] This was Newton's task—he read the prophecies, connected them with a cosmological goal, sought to efface his own authorship of such readings, and in so doing made his authority. He also had to provide an account that explained other readings of sacred texts and compelled proper readings of his own. His success, both in the cosmological and theological realms, was painstakingly assessed by

contemporaries and by historians alike. It is no coincidence that
they have used similar tactics, even if they do not all hold Conduitt's
view that Newton's "virtues proved him a Saint and his discoveries
might well pass for miracles."[31]

Idolatry and Cometography

It is now out of fashion for Kings to be Priests and Prophets.
—(Tory pamphlet on the Exclusion Crisis, 1680)

In the first section of this paper, it was suggested that Newton's
work as interpreter could be recovered by paying attention to two
features of his practice. How did Newton's great works become
idols? How did Newton understand idolatry? First, we need to
understand the way he made his own authority and the context in
which that authority was made; second, we must examine the way
he explained the variation in interpretation of sacred texts and of his
own texts. It emerged that a very large range of ideological and
cultural factors might be relevant in an answer to the former question. It was also proposed that Newton gave an historical account of
the "corrupt" interpretations of Scripture that he found and an
essentialist account of how his own writings should be read. In this
section, by contrast, I engage in a survey of one area of Newton's
work that seemed to command widespread authority (though not
consensus) among his interpreters—that of cometography, presented publicly for the first time in the magisterial final propositions
of the *Principia*'s third book, then expounded at ever-increasing
length in subsequent editions of the book and in influential essays
by disciples such as Edmond Halley, William Whiston, and David
Gregory. It is argued that Newton's early—if not premature—commitment to the view that comets move in interplanetary space in
elliptical orbits was intimately connected with his account of idolatry. He identified idolatry with a false method of interpretation of
sacred texts. This falsehood bred false philosophy. True philosophy,
with true cometography as its weapon, could undermine the philosophical corruption that idolatry had spawned. This suggests part
of an answer to the question of the goals that Newton pursued in
making his authority as cosmologist. It allows a case-study in his
work as interpreter and theorist of interpretation. It leaves open, as
a topic for further research, the relation between this goal and
readings of Newton's cosmological texts by his followers and critics.

Throughout Newton's lifetime, comets were a central concern for a wide range of terrestrial intellectuals: priests, astrologers, journalists, astronomers, and natural philosophers. The frequent transits of major comets during that period repeatedly prompted explosive pamphlet wars and *pièces d'occasion* on the topic of cometography. Two issues mattered: first, were comets permanent or transient occupants of the skies? Second, what was their significance for the inhabitants of the earth? Both were very traditional topics of scholastic natural philosophy; both were questions whose sense and implication changed radically in the seventeenth century. The sources Newton read when spurred to initiate his astronomical researches in 1664 agreed that permanent objects must move in closed orbits, while transients would move in open or rectilinear paths.[32] By 1679, Newton had convinced himself of the validity of this conventional wisdom. In the next five years Newton was presented with fresh opportunities to contemplate cometography and to rethink this view. On the one hand, correspondents such as Hooke, Flamsteed, and Thomas Burnet engaged him in debate about the system of the world, stimulated at least partly by the comets of 1677, 1680, and 1682. On the other, the political and theological crisis that wracked Britain during just these years elicited streams of texts on political astrology and polemical philosophy that made use of the signs of heaven and retheorized the heavenly mandate of the monarchy. Both themes left their marks on Newton's complete reorientation of his cosmology, a reorientation that led directly to the composition of his masterpiece of 1687.

The work of Hooke and Flamsteed was extremely important in this process. In his Cutlerian Lectures, which made much of the 1677 comet and its aftermath, Hooke lectured his London audience on the views that comets were transient projectiles wasting into interplanetary aether; that they might remain in our system for "many ages"; and that, given the multiplicity of equivalent *geometrical* models which could save cometary phenomena, a properly natural philosophical analysis of their behavior was essential.[33] Newton also came to share this demand. Flamsteed, as Astronomer Royal, worked hard at Greenwich to contest astrological uses of comets by arguing that they were periodic. He guessed that that of 1677 would return in 1689. He also responded actively to the furore about the comet of 1680 which coincided with the political crisis of the Popish Plot and the envisaged succession of James, Duke of York, as an avowedly Catholic monarch. The young Edmond Halley, then travelling in France, also commented on this astrological and religious connection.[34] In January 1681, Newton was asked by

Thomas Burnet for advice on his recently composed *Sacred Theory of the Earth,* an exercise in sacred physics which aimed to make sense of the "moral and philosophical history of the world" in terms relevant to contemporary Protestant interests. Burnet also mentioned the comet of that winter, which had attracted so much ferment.[35]

One comet had been visible in November 1680, while another appeared from behind the Sun in December. A question strenuously argued between Halley, Flamsteed, and Newton the following spring was whether these comets were one and the same object. Flamsteed, keen to show the natural philosophical grasp of the "magnetic cosmology" and to argue that comets returned, insisted that they were but one body. Newton, following an orthodox view, held that comets were transient, so could not incline at the sun through as great an angle as Flamsteed's hypothesis would imply.[36] The important point about Newton's exchanges with Hooke, Burnet, and Flamsteed is that at this stage, in 1679–81, Newton did not hold that comets return, nor did he claim that they were moved by an attractive force in the sun. On the contrary, as Whiteside and Ruffner have emphasized, in these texts Newton used a Borellian account of cometary and planetary motion, involving the overbalance of tangential and centrifugal forces acting on celestial bodies.[37] There was only the barest hint here of the work based on a centripetal force and an understanding of Kepler's area law. There was no indication that comets returned. That work only appeared from autumn 1684.

This chronology suggests that Newton's cosmology was reconstructed, particularly in respect to cometography, after 1681 and before the heroic period of drafting for the *Principia* initiated with Halley's celebrated visit to Cambridge in summer 1684. Many factors were in play here: a reconsideration of the phenomenal reality of the interspatial aether, which, by 1684, was no longer allowed to disturb planetary or cometary motion; an acceptance of the meaning of the area law for an analysis of this motion; a detailed mathematical presentation of the "centripetal" force that acted from the sun, and, ultimately, from all matter. This was a term Newton coined in his manuscripts of 1684–85. The same documents contain the first statement by Newton that all comets return. With a combination of the area law and an analysis based on an inverse-square law of attraction "we may know whether the same comet returns time and again."[38] Flamsteed ironized about Newton's belated but decisive concession that the comet of 1680, for example, was but one object: "he would not grant it before see his letter of 1681."[39]

From this moment, in the winter of 1684–85, the puzzle of computing the cometary orbits, which Newton already assumed were elliptical, became one of his principal preoccupations. His assumption of closure was reached with no empirical warrant. He had computed no cometary orbits by September 1685. Ultimately, during 1686, when composing the final book of the *Principia,* he found an incredibly accurate graphical method that allowed him to write the lemmata to Proposition 41 of this book. But his cosmological views about comets were already in place in the first version of this book, a System of the World probably written in autumn 1685. There, as in correspondence with Flamsteed and Halley of this time, he argued that comets all move in ellipses for which parabolas may serve as good approximations. This was specifically applied to the comet of 1680–81, which he held truly moved in an ellipse and would therefore return.[40]

The arguments and techniques contained in the last two propositions of the *Principia* are excellent examples of Newtonian texts susceptible to widely varying readings. Newton, and his devoted editor Halley, committed themselves to much work to stipulate the right way of reading these texts. They insisted that parabolas were useful but inexact approximations for comets' orbits. Later readers glossed these texts as indications that the master held that many comets moved in parabolas. Halley and Newton also worked on further cometary data, developing what may be called a "historic method" for estimating cometary periods—comets with similarly oriented and shaped parabolic approximations, moving in the same direction, could be identified with each other. If such comets recurred at regular intervals, a periodic body had been located and its period could be found. This period would then allow a more exact, elliptical shape to be computed.[41]

This was the burden of Halley's work with Newton in the mid-1690s, work that generated the inexact prediction of the return of the comet of 1682 for the late 1750s. The goal of this work was to reinforce the implication that all comets returned and that returns could be made susceptible to analysis by skilled Newtonian cometographers. The inexactness and conjectural basis of these claims were recognized by their authors but not made fit for public consumption.[42] Ultimately, these claims received influential and very widely distributed publicity in Halley's celebrated *Synopsis of the Astronomy of Comets* completed in March 1705. The tables Halley printed there were used in successive editions of the *Principia* and reprinted in astronomy texts by David Gregory.[43] These became a group of texts most often subjected to the interpretative attention of

Newton's successors, just as, after the return of Comet Halley in 1759, they became the most celebrated idols of the Newtonian creed. Thus Voltaire claimed that "when Sir Isaac Newton invented the Hypothesis of the parabolical motion of the comets . . . he *did not actually believe* that the curvatures of their way were true parabolas. (If they were true parabolas rather than ellipses) we should moreover deprive ourselves of ever seeing them again." By contrast, the great French astronomers Lacaille and Delambre both held that Halley's exposition of the historic method was so "unintelligible" that few astronomers could actually follow it in the eighteenth century. As I have argued elsewhere, the interpretation of the events of 1759 as a decisive confirmation of Newtonian cosmology was itself a remarkable exercise in polemical hermeneutics. Fixing a standard reading of this key text of Newtonian orthodoxy was by no means automatically achieved.[44]

In order to connect the significance that Newton gave to his public utterances on cometary return and his work to construct a new authoritative cometography, it is important to stress the extraordinary cosmological role that Newton gave these objects. As is now well known, thanks to the work of D. C. Kubrin and S. Schechner Genuth, Newton gave comets roles in the preservation, restoration, and increase of activity in matter distributed in space.[45] In the 1690s, Newton told Bentley and Gregory that comets acted as agents in the "Growth of new Systems out of old ones," and that "Comets are destined for a use other than that of the planets."[46] At exactly the same period, in 1693–95, Newton discussed with Halley views that the latter had begun to develop in the late 1680s on the possible cometary cause of the Deluge. Newton incorporated suggestions that comets might increase the mass of the earth and so cause the lunar secular acceleration in subsequent editions of the *Principia.*[47] Already, in 1687, he published his view, which helped itself to alchemical research pursued since the 1670s, that "comets seem to be required, that, from their exhalations and vapours condensed, the wastes of the planetary fluids spent upon vegetation and putrefaction and converted into dry earth may be continually supplied and made up."[48]

These claims that comets made new systems, restored activity, and changed terrestrial mass pointed to the need for an account of the material constitution of comets and their tails. Once again, these were texts much in need of clarification and interpretative work. Because Newton made comets' tails the bearers of celestial activity, he had to explain how they worked in void space. Both Gregory and Henry Pemberton, editor of the third edition of the

Principia, found trouble making sense of this thought. "Since the heavens are void of any matter that can give a sensible resistance to the progressive motion of this vapour," Pemberton asked, "what is that *aura aetherea* which by its motion in ascent can carry this vapour along with it?"[49] Using his alchemical and matter theoretic drafts of the 1670s, reworking the initial Definitions to the *Principia* on transmutation of matter, and adding queries to successive editions of the *Opticks* after 1704, Newton spelled out various ways in which the activity of the cosmos could be sustained by the rare fluids carried by comets. They took on a divine role. He told Gregory of "a very small Aura or particle," while in the *Principia* he printed the statement that "it is chiefly from comets that spirit comes . . . so much required to sustain the life of all things with us."[50]

Comets had become the principal transmitters of life and restorers of vitality in the heavens and on earth. In the third edition of the *Principia,* following conversations with Halley, Conduitt, and Gregory on this theme, Newton also rewrote his view that comets would fall into stars. In 1725–26 he announced that such events would produce novae, and it was in this context that, according to Conduitt, Newton joked about the incendiary effects of such events on the Earth.[51] The full role that had emerged for comets, backed with the authority of the *prisca,* substantiated by Halley's data and with Newton's cosmological scheme, then provided eighteenth-century commentators with much fruitful opportunity for dispute about the text. Halley, Bentley, and Pemberton made successive revisions of the ode that prefaced the *Principia* to give widely different readings of this theme of cometary significance. In his astronomy textbook of 1702, David Gregory insisted that "those things which have been observ'd by all Nations, and in all Ages, to follow the Apparition of Comets, may happen; and it is a thing unworthy of a Philosopher to look upon them as false and ridiculous." Citing Gregory's view, both Maupertuis and Buffon held that the Newtonian cometography rightly restored comets in their full terror; Lambert castigated these workers as mere "authorized prophets."[52]

Lambert's striking phrase was well chosen. Prophecy and cometography were intimately connected in Newton's scheme. Our examination of the development of Newton's reinterpretations of cometary motions and cometary meanings shows two important features: until the mid-1680s Newton held that comets were transient and did not return, while after 1684 he held that they were permanent and moved in ellipses; second, just as he made them return, so he made them objects of enormous significance in the history of vital activity in the cosmos and on earth. Just as the texts on cometography were designed to produce a new consensual technology that would render comets the recognized bearers of divine active power, so this doctrine would correct false readings of ancient understanding of the cosmos. Further, just as Newton's utterances about these divine agents became themselves treated as revelations of truth, so Newton produced a technology for reading scripture that would show the basis of his cosmology in the hidden but recoverable sense of the ancients. Finally, where comets had been used by radicals, sectaries, and the godly as marks of the political and theological crisis enveloping the state, so Newton held that with a restored cometography and cosmology, the philosophy that underpinned that corrupted state and church would be discredited. Authority in cosmology and interpretation was a source and a reward of power.

The connection between cometography and idolatry was outlined in autumn 1685 in the draft for the third book of the *Principia* entitled "The System of the World." It was suppressed in the final version. The original argued that geocentrism and the theory of solid spheres were later Greek corruptions of pristine cosmological truths. In the Vestal temples, Newton argued, the ancients constructed an analogue of the true heliocentric and vacuist system of the world. Such claims also appear in the "classical scholia" intended for the *Principia's* second edition, discussed with Gregory in 1694, revised in draft queries for the Latin *Opticks,* in annotations for the General Scholium printed in 1713, and spelled out in a draft preface for the third edition of the *Principia:*

> The Chaldeans once believed that the planets revolve round the Sun in almost concentric orbits and the comets in very eccentric orbits. And the Pythagoreans introduced this philosophy into Greece. . . . This philosophy was discontinued (it was not propagated to us and gave way to the vulgar opinion of solid spheres). I did not discover this, but endeavoured to restore it to light by the power of demonstration.[53]

Newton's claim about the true history of cosmology deployed conventional Renaissance hermeneutics to powerful effect. Using the work of the mythographers of the sixteenth century to help decode the proper sense of the classics, he found warrant and precedent for his cosmology in a priscine philosophy.[54] This much Newton shared in common with his immediate colleagues, such as Henry More or Ralph Cudworth, and disciples, such as David Gregory. But Newton also connected the claims that his new cometography of the 1680s was good old Chaldean cosmology with a claim that if restored now it would undermine false and corrupted religion and politics. This latter view was first presented in a treatise, baptised "Philosophical origins of gentile theology," initiated in 1683–84, at exactly the same time as the construction of his new cosmology. R. S. Westfall rightly emphasizes the central role of this text, in its various versions that range in date up to the 1710s.[55]

The treatise was the source for the extracts noted above that appear in draft prefaces and scholia for Newton's projected published works in natural philosophy. Above all, the "Philosophical origins" was a text on the history and consequences of interpretation. It made false hermeneutics the root of political and theological disaster. It argued that there was a pristine cosmology represented in the Vestal temples and the Pythagorean harmony of the spheres. Corruption followed when such symbols were misinterpreted. "By the harmony of the spheres is to be understood the harmonic motion of the solar soul by which the planets are driven in their orbits by the Sun."[56] Newton gave a similar account in a text he added anonymously to Gregory's astronomy textbook: the symbol of the seven-stringed harp of Apollo rightly allegorized "the sun in Conjunction with the seven Planets for they made him Leader of that Septenary Chorus and Moderator of Nature and thought that by his Attractive Force he acted upon the Planets . . . in the Harmonical ratio of their Distances."[57] The Gentiles interpreted the harmonies wrongly, because they were literalists and held that the spheres must be real and solid. They misread the allegory:

> taking literally the views of the philosophers that the celestial spheres emitted harmonic sounds when moving against each other, they believed that the heavens were fluid, but that the planets and the fixed stars inhered in solid orbs, and this opinion were first introduced by Eudoxus just before the time of Aristotle.[58]

This kind of literalism was Newton's target in many of his campaigns. For example, he held that the disasters of Trinitarianism began when "those who used the language of one usia and one substance did not only receive the language of three persons but many of them began to take the persons for substances."[59] Taking Pythagorean spheres as bodily substances was just as bad. Above all, once this misreading was in place, a false cometography was developed, for solid spheres were incompatible with cometary return—comets were turned into meteors, and the true doctrine of their motion and meaning was lost.

Having established that a misinterpretation had occurred be-

tween the periods of Pythagoras and Eudoxus, and having stressed
the ill effects of this misreading upon cosmology, since now all the
effects of cometary return were suppressed and the law of gravity
forgotten, Newton went on to give a historical, if not political,
explanation of the interests served by this misinterpretation. His
view was that the pristine cosmology was monotheist and thus gave
a correct reading of the system of the world. But when the corrupted interpreters made spheres solid, they gave each sphere a
soul and made those souls the spirits of dead heroes and monarchs.
The doctrine of transmigration was a manifestation of this great fall.
The best characterization of the fall was idolatry, a term that R. S.
Westfall rightly characterizes as Newton's label for "the fundamental sin."[60] Idolatry was, above all, an error of interpretation. It
misunderstood divine power. "The grand occasion of errors in the
faith has been the turning of the Scriptures from a moral to a
metaphysical sense and this has been done chiefly by men bred up
in the theology of the heathen philosophers [who] have been apt to
strain everything to their own opinions."[61] Transsubstantiation was
a perfect example of idolatry as misreading—"if it be said, This is
my body, meaning a symbol of my body, they take it in a metaphysical sense for transsubstantiation."[62] Idolatrous hermeneutics
was responsible for the attribution of the souls of dead kings to the
celestial spheres, to the belief in demons, to papist worship of
relics, to an attribution of God's power to his creatures. In sermons
of the mid-1680s on Biblical passages on the worship of Baal, in
drafts for the General Scholium, and in the "Philosophical origins"
itself, Newton spelled out his view that God should be worshipped
for the power manifest in his actions—since his actions were visible
both in past history and in nature, correct readings of the providential past (through chronology) and the present (through natural
philosophy) would destroy idolatry.[63]

Newton's analysis of idolatry as corrupt interpretation of symbols gave a clue to his explanation of the interests of the corrupt
interpreters. False powers were attributed to natural objects because corrupt men seized false power. This was a theme first developed in Newton's assault on Athanasian Trinitarianism in the
1670s, when he argued that the history of the early church showed
how interested men could pervert the true sense of scripture. According to Newton, the Athanasians learned how to organize themselves into a persecuting and authoritarian interpretative community that set out to ban any rival reading of the pure texts. The
Trinitarians, he suggested, "found by experience that their opinions
were not to be propagated by disputing and arguing, and therefore

gave out that their adversaries were crafty people and cunning disputants and their own party simple well-meaning men, and therefore imposed this law upon the Monks, that they should not dispute about the Trinity." Having banned rival interpreters, the Athanasians "left the success of their Cause to the working of miracles and spreading of Monkery."[64] Newton's relentless exposure of "notable corruptions of scripture" that sustained the Trinitarian interpretations were but one aspect of his polemical use of hermeneutics.

Controversies with rival readers of Daniel and Revelation were another, and a closely connected aspect of his reconstruction of church history. Here interpretation became the central site at which contests took place. For example, Newton gave a close reading of Cudworth's great *True Intellectual System of the Universe* (1678) to provide himself with resources for the "Philosophical origins." Cudworth copied out a passage from Aristotle's *Metaphysics* 12, which argued that most of ancient religion was a system of fables "for the better Persuasion of the Multitude and for Utility of Humane Life and Political Ends to keep men in obedience to civil laws." But, according to both the Stagyrite and the Cambridge Platonist, the animation of the spheres was not a cunning fiction— they seemed to agree that this doctrine was old and true. The deification of the stars was a doctrine of all the ancients. Newton then copied out Cudworth's view, but added in his notes that "Aristotle took all the heathen theology couched in Fables, Allegories & hieroglyphics devised only for the worship of the Stars." Newton's suggestion was that Cudworth was wrong: the text from *Metaphysics* 12 showed that Aristotle was not a virtuous interpreter of the *prisca* but a cunning reader in his own right. Stellar religion was not an old truth distinct from all the ancient political fables, but the best example of an interested fiction. Newton had to reread Aristotle's interpretation so as to reveal the real purposes of Peripatetic cosmology, then contest the reading of Aristotle given by Cudworth.[65] This network of rereadings was characteristic of the whole burden of the "Philosophical origins." In this text, he connected the corruptions of symbols into substances with the corruption that attributed the souls of dead men to the spheres. "Gentile astrology and Theology were introduced by cunning priests to promote the study of the stars and the growth of the priesthood and at length spread through the world."[66]

Divine right and papist monarchy were exposed here, for their authority relied on this idolatrous and false attribution of power: Newton made the comparison between "the deified Heroes with their Idols among the Gentiles" and "the saints with their relics and images among too many Christians."[67] Such idolatrous worship was explicitly developed by the monarchical party: "so ready was the ambition of Princes to introduce their predecessors into the divine worship of the People to secure to themselves the greater veneration from their subjects as descended from the Gods, erected such a worship and such a priesthood as might awe the blinded and seduced people into such an obedience as they desired."[68] The diagnosis for this disease was clear, Newton argued. The historical explanation of the sources of corruption was also an analysis of a possible cure. The restoration of proper cosmology would cut away at the philosophical underpinnings of this religion of the spheres, and, in turn, at the cult of monarchs and priestcraft.

This reading of Newton's lengthy account of the history of interpretation displays important connections with the making of his cosmology. It suggests one of Newton's goals in propagating that world-view, and also establishes a case for the significance of the period when the cosmology was inaugurated. The achievement of our contemporary Newton scholars, notably R. S. Westfall and D. T. Whiteside, in redating Newton's creative accomplishment in cosmology to the 1680s, provides an important resource here, since it can also be shown that this was the period when Newton began writing his major treatise on political theology. A set of important historical tasks remains for the adequate understanding of the scope and scale of Newton's achievement in this period. There are grounds here for a connection between the political and religious crisis of the late 1670s and early 1680s and Newton's new projects of the time. A comparison with our understanding of John Locke is salutary: after the Revolution, Locke worked very closely with Newton in their joint analysis of the corrupt interpretation of scripture that had bred Trinitarianism. Both men were fascinated by the alchemical aspects of the legacy of Robert Boyle. In Richard Ashcraft's recent exemplary analysis of Locke's political career, it is persuasively argued that Locke's epistemology and political doctrines on irenic religion and toleration were systematically developed in the complex context of dissenting opposition to the putatively papist and authoritarian royal regime.[69]

Newton's political experience was strikingly different, and these differences may help explain the peculiar tenor of his own work on these problems. Both men, however, directly experienced the striking consequences of resistance to royal power in the 1680s—Newton, famously, appearing before Judge Jeffreys to defend his university against the imposition of a Catholic master by royal mandamus. Ashcraft supports the judgments of historians of the Popish Plot, the Exclusion Crisis, and the subsequent Tory reaction in pointing to the intellectual and ideological crisis through which Protestant intellectuals passed in the years between 1679 and 1685. He cites the comment of Newton's cocollegiate Roger North, who observed that in 1679–80 "one might have denied Christ with more content than the Plot."[70] As an Arian antipapist, Newton fitted North's phrase rather precisely. It is also clear that the years of the mid-1680s were those of greatest perceived threat for intellectuals of Newton's persuasion. From summer 1684, royal attacks on dissent mounted to a peak; in spring 1685, following James's accession and Monmouth's abortive revolt, this campaign was sustained, many went into exile, and a general toleration for Catholics was widely expected within months.[71] Newton's drafts for the university's case against the king argued forcefully that papism would be a disaster for the faith: "if the fountains once be dryed up the streams hitherto diffused thence throughout the Nation must soon fall off."[72] His assault on idolatry bolstered the claim that papism was as threatening for the purity of religion as for the truths of philosophy.

The connection between Newton's cosmological and political strategies remains circumstantial, based as it is on a comparison between the theological and political *import* of the manuscripts Newton drafted during the period 1679–85 and the public events of the crisis. The problems of interpretation outlined in the first part of this paper are of obvious relevance: we are faced with a common difficulty of the relationship between an author's private meditations and the historically reclaimable circumstances of the institutions to which he belonged. What is required is further research: this should concentrate upon what remains of evidence for Newton's relations with his most immediate colleagues: the Cambridge Platonists and the moderate Anglicans in London. These relations did clearly involve debate on church history, religious crisis, and the politics of interpretation. Most importantly, arguments from *the behavior of Newton's interpreters,* which have characterized the rasher generalizations about the political and theological meaning of Newton's cosmology, should be eschewed. What is required is a detailed analysis of each aspect of the technical construction and declared goals of that cosmology. Interpretation of sacred and classical texts was, as we have argued, seen to be a weapon of ideological warfare. This kind of interpretation was a central endeavor in Newton's practice. His construction of a new cosmology was at least partly an act of interpretation in this sense.

This thought prompts a related suggestion: the need for an adequate model of the role Newton and his contemporaries gave themselves as interpreters. If the consequences of false readings were as disastrous and palpable as Newton suggested, it was also the case that Newton, alongside most of those Protestants who read the prophets, the Fathers, and the *prisca*, believed that their readings showed why the moment had now come for a true reading. That is, interpretation was both ideological (in that it gave authority in political and religious struggle now) and reflexive (in that it explained why this interpretation should be treated as authoritative). In his reading of Revelation, Newton claimed that "amongst the interpreters of the last age, there is scarce one of note who hath not made some discovery worth knowing, and thence it seems that God is about opening these mysteries." Newton detected a record of some success in the Christ's College tradition inaugurated by Joseph Mede and developed by such as Milton, Burnet, More, and Cudworth. He said that "the success of others" prompted him to begin his own work, but he also said that the Medean tradition was "botched and framed without any due proportion." Here, then was a divine mark of both the possibility and the necessity of interpretative work.[73] For Mede and his successors, including Newton, it was a commonplace that the time for right reading had come. It was also a commonplace that part of this right reading would reflexively show why the moment of interpretation had, at last, arrived. But it was, of course, novel to extend this claim in the direction that Newton's cosmology implied. It is in just this sense that Whiston cited scriptural authority to help mark the appearance of Newton's *Principia* as "an eminent prelude and preparation to those happy times of the restitution of all things; which God has spoken of by the mouth of all his holy prophets since the world began."[74] Newton aimed to destroy idolatry with texts that were immediately treated idolatrously. Authorship and authority were connected in a complex technique of interpretation. This is why the history of idolatry should now be put at the center of our own interpretation of his achievement.

Notes

This paper results from work in collaboration with Stephen Shapin and with Robert Iliffe, to whom I express my deep gratitude. I thank the editors of this volume for their patient help, and acknowledge the kind permission extended by the provost and fellows of King's College Cambridge, the William Andrews Clark Memorial Library, the Library of the Royal Society and the Jewish National

Library to cite manuscripts in their possession.

1. Ernst Cassirer, *The Philosophy of the Enlightenment*, trans. Fritz C. A. Koelln and James P. Pettegrove (1932; trans., Boston: Beacon Press, 1955), pp. 43–44, 55.

2. Henry Guerlac, "Newton's Changing Reputation in the Eighteenth Century," in *Essays and Papers in the History of Modern Science* (Baltimore: Johns Hopkins University Press, 1977), pp. 69–91. Robert E. Schofield, "An Evolutionary Taxonomy of Eighteenth-Century Newtonianisms," *Studies in Eighteenth-Century Culture* 7 (1978): 175–92.

3. I. Bernard Cohen, *Franklin and Newton: An Inquiry into Speculative Newtonian Experimental Science and Franklin's Work in Electricity as an Example Thereof* (Philadelphia: American Philosophical Society, 1956), pp. 179, 198.

4. Thomas S. Kuhn, "Mathematical versus Experimental Traditions in the Development of Physical Science," *Journal of Interdisciplinary History* 7 (1976): 1–31. Robert E. Schofield, *Mechanism and Materialism: British Natural Philosophy in an Age of Reason* (Princeton: Princeton University Press, 1970). See R. W. Home, "Out of a Newtonian Straitjacket," *Studies in the Eighteenth Century: IV, Papers Presented at the Fourth David Nichol Smith Memorial Seminar*, ed. R. F. Brissenden and J. C. Eade (Canberra: Australian National University Press, 1979), pp. 235–49.

5. M. C. Jacob, *The Newtonians and the English Revolution, 1689–1720* (Ithaca: Cornell University Press, 1976) and idem, *The Radical Enlightenment: Pantheists, Freemasons and Republicans* (London: Allen and Unwin, 1981). Larry Stewart, "Samuel Clarke, Newtonianism, and the Factions of Post-Revolutionary England," *Journal of the History of Ideas* 42 (1981): 53–72. C. B. Wilde, "Hutchinsonianism, Natural Philosophy, and Religious Controversy in Eighteenth-Century Britain," *History of Science* 18 (1980): 1–24, and idem, "Matter and Spirit as Natural Symbols in Eighteenth-Century British Natural Philosophy," *British Journal for the History of Science* 15 (1982): 99–131. J. G. McEvoy and J. E. McGuire, "God and Nature: Priestley's Way of Rational Dissent," *Historical Studies in the Physical Sciences* 6 (1975): 325–404.

6. John Toland, *Letters to Seneca* (London: B. Linton, 1704), pp. 182–83.

7. Andrew Michael Ramsay, *The Travels of Cyrus, to which is annex'd a Discourse upon the Theology and Mythology of the Ancients*, 2d ed. (London: T. Woodward, 1730), p. 70n.

8. Richard Symes, *Fire Analysed . . . and the Manner and Method of making Electricity Medicinal and Healing confirmed by a Variety of Cases* (Bristol: Thos. Cocking, 1771), p. 36.

9. Wolff to Manteuffel, 27 January 1741, in Irving I. Polonoff, *Force, Cosmos, Monads and Other Themes of Kant's Early Thought* (Bonn: Bouvier, 1973), p. 76.

10. *Unpublished Scientific Manuscripts of Isaac Newton*, ed. A. Rupert Hall and Marie Boas Hall (Cambridge: Cambridge University Press, 1962), pp. 88–90, 293. Richard S. Westfall, *Force in Newton's Physics: The Science of Dynamics in the Seventeenth Century* (London: MacDonald, 1971), pp. 403, 409–10, 513. B. J. T. Dobbs, "Newton's 'Clavis': New Evidence on Its Dating and Significance," *Ambix* 29 (1982): 198–202. D. T. Whiteside, "Before the *Principia:* The Maturing of Newton's Thoughts on Dynamical Astronomy, 1664–1684," *Journal for the History of Astronomy* 1 (1970): 5–19. See the discussion of these issues in Robert Palter, "Saving Newton's Text: Documents, Readers and the Ways of the World," *Studies in History and Philosophy of Science* 18 (1987): 385–439.

11. Martin J. S. Rudwick, "Charles Darwin in London: The Integration of Public and Private Science," *Isis* 73 (1982): 186–206.

12. Alexandre Koyré, *From the Closed World to the Infinite Universe* (Baltimore: Johns Hopkins University Press, 1957), pp. 178–79. *Unpublished Scientific Manuscripts,* p. 194. Gerd Buchdahl, "Explanation and Gravity," in *Changing Perspectives in the History of Science: Essays in Honour of Joseph Needham,* ed. Mikuláš Teich and Robert Young (London: Heinemann, 1973), pp. 167–203.

13. Ernan McMullin, *Newton on Matter and Activity* (South Bend: University of Notre Dame Press, 1978), pp. 59–69, 71, 73, 121, 143 n. 132.

14. Richard S. Westfall, *Never at Rest: A Biography of Isaac Newton* (Cambridge: Cambridge University Press, 1980), pp. 804–15. Karin Figala, "Pierre des Maizeaux's View of Newton's Character," *Vistas in Astronomy* 22 (1978): 477–82.

15. A. Rupert Hall, *Philosophers at War: The Quarrel between Newton and Leibniz* (Cambridge: Cambridge University Press, 1980).

16. *The Mathematical Papers of Isaac Newton,* ed. D. T. Whiteside, 8 vols. (Cambridge: Cambridge University Press, 1967–81), 8:450–51. I. B. Cohen, *An Introduction to Newton's 'Principia'* (Cambridge: Harvard University Press, 1971), pp. 293–94.

17. D. T. Whiteside, "The Mathematical Principles Underlying Newton's *Principia,*" *Journal for the History of Astronomy* 1 (1970): 116–38, quoted from p. 119.

18. David Bloor, *Knowledge and Social Imagery* (London: Routledge and Kegan Paul, 1976).

19. Derham to Conduitt, 18 July 1733, Cambridge, King's College Library, Keynes MS. 133.10.

20. John Stewart, "Some Remarks on the Laws of Motion and the Inertia of Matter," in *Essays and Observations, Physical and Literary* (Edinburgh: Philosophical Society, 1754), pp. 70–140, quoted from pp. 70–71. I owe this and the following citation to the kindness of Michael Barfoot.

21. James Burnett, Lord Monboddo, *Antient Metaphysics, or, The Science of Universals,* 6 vols. (Edinburgh: J. Balfour, 1779–99); Joseph Priestley, *A Free Discussion of the Doctrines of Materialism and Philosophical Necessity, in a Correspondence between Dr. Price and Dr. Priestley* (London: J. Johnson and T. Cadell, 1778), pp. 9, 26–28, 231, 237.

22. *Mathematical Papers,* 6:192.

23. Jerusalem, Jewish National and University Library, Yahuda MS. 1.1, fol. 5–6.

24. Alexandre Koyré, *Newtonian Studies* (1965; reprint ed., Chicago: University of Chicago Press, 1968), p. 276.

25. David Kubrin, "Newton and the Cyclical Cosmos: Providence and the Mechanical Philosophy," *Journal of the History of Ideas* 28 (1967): 325–46. David Castillejo, *The Expanding Force in Newton's Cosmos, as Shown in his Unpublished Papers* (Madrid: Ediciones de Arte y Bibliofilia), pp. 95–96.

26. Cambridge, King's College Library, Keynes MS. 130.11 Isaac Newton, *Traité d'Optique,* trans. Pierre Coste (Paris: Montalant, 1722), sig. e iii. John Theophilus Desaguliers, *A Course of Experimental Philosophy,* 2 vols. (London: J. Senex, 1734–44), vol. 1, sig. cl.

27. Quentin Skinner, "Motives, Intentions and the Interpretation of Texts," *New Literary History* 3 (1972): 393–408. John Law and Barry Barnes, "Whatever Should be Done with Indexical Expressions?" *Theory and Society* 3 (1976): 223–37. A. R. Cunningham, "Getting the Game Right: Some Plain Words on the Identity and Invention of Science," *Studies in History and Philosophy of Science* 19 (1988): 365–90.

28. Eamonn Duffy, "Whiston's Affair: The Trials of a Primitive Christian 1709–1714," *Journal of Ecclesiastical History* 27 (1976): 129–50. John Edwards, *Some*

Brief Critical Remarks on Dr. Clarke's Last Papers (London: F. Burleigh, 1714), p. 40.

29. Joseph A. Mazzeo, *Varieties of Interpretation* (South Bend: University of Notre Dame Press, 1978), chapter 1.

30. Julian Roberts, "The Politics of Interpretation," *Ideas and Production* 1 (1983): 15–32.

31. Cambridge, King's College Library, Keynes MS. 130.14.

32. J. A. Ruffner, "The Curved and the Straight: Cometary Theory from Kepler to Hevelius," *Journal for the History of Astronomy* 2 (1971): 178–94. C. Doris Hellman, "Kepler and Comets," *Vistas in Astronomy* 18 (1975): 789–96. Henry Guerlac, *Newton on the Continent* (Ithaca: Cornell University Press, 1981), pp. 29–40. The discussion that follows is based on Simon Schaffer, "Newton's Comets and the Transformation of Astrology," in Patrick Curry, ed., *Astrology, Science and Society* (Woodbridge, Suffolk: Boydell, 1987), pp. 219–43.

33. Robert Hooke, "Cometa," in *Lectiones Cutlerianae* (London: John Martyn, 1679). J. A. Bennett, "Hooke and Wren and the System of the World: Some Points towards an Historical Account," *British Journal for the History of Science* 8 (1975): 32–61.

34. Halley to Aubrey, 16 November 1679, in Eugene Fairfield MacPike, *Correspondence and Papers of Edmond Halley* (Oxford: Clarendon Press, 1932) pp. 47–48. Flamsteed to Towneley, 11 May 1677, in London, Royal Society Library, MS. LIX.c.10. Derek Parker, *Familiar to All: William Lilly and Astrology in the Seventeenth Century* (London: Cape, 1975), p. 258.

35. *The Correspondence of Isaac Newton,* ed. H. W. Turnbull (vols. 1–3), J. F. Scott (vol. 4), and A. Rupert Hall and Laura Tilling (vols. 5–7), 7 vols. (Cambridge: Cambridge University Press, 1959–78), 2:329–34. M. C. Jacob and W. A. Lockwood, "Political Millenarianism and Burnet's *Sacred Theory,*" *Science Studies* 2 (1972): 265–79.

36. *Correspondence,* 2:315–17, 319–20, 336–39, 341, 351, 358–62.

37. Whiteside, "Before the *Principia.*" D. T. Whiteside, "The Prehistory of the *Principia,*" *Notes and Records of the Royal Society* 45 (1991): 11–61, on pp. 20-21.

38. *Unpublished Scientific Manuscripts,* pp. 283–85. *Mathematical Papers,* 6:57–61.

39. *Correspondence,* 2:421, n. 4.

40. *Correspondence,* 2:419, 437. Sir Isaac Newton, *Mathematical Principles of Natural Philosophy, and His System of the World,* trans. Andrew Motte, rev. and ed. Florian Cajori, 1 vol. as 2 (1934; reprint ed., Berkeley: University of California Press, 1962), 2:615, 619, 626. (Hereafter: *Principia* [Motte/Cajori].) *Mathematical Papers,* 6:483, 485, 498–504.

41. *Principia* (Motte/Cajori), 2:532, 620.

42. *Correspondence,* 4:165, 171–72, 180–83. MacPike, *Correspondence and Papers of Halley,* p. 238.

43. Edmond Halley, *Synopsis of the Astronomy of Comets* (London: J. Senex, 1705). David Gregory, *The Elements of Physical and Geometrical Astronomy,* 2d ed., 2 vols. (London: D. Midwinter, 1726), 2:881–905. Peter Broughton, "The First Predicted Return of Comet Halley," *Journal for the History of Astronomy* 16 (1985): 123–33.

44. Voltaire, *The Elements of Sir Isaac Newton's Philosophy* (London: Stephen Austen, 1738), p. 329. J. B. J. Delambre, *Histoire de l'astronomie au dix-huitième siècle* (Paris: Bachelier, 1827), pp. 127–28, 466–67, 673. Simon Schaffer, "Authorized Prophets: Comets and Astronomers after 1759," *Studies in Eighteenth-Century Culture* 17 (1987): 45–74, on pp. 50–51.

45. Kubrin, "Newton and the Cyclical Cosmos." Sarah Schechner Genuth, "Comets, Teleology and the Relationship of Chemistry to Cosmology in Newton's Thought," *Annali dell'Istituto e Museo di Storia della Scienza di Firenze* 10 (1985): 31–65.

46. *Correspondence,* 3:234, 253–55, 336. I. B. Cohen, "Galileo, Newton and the Divine Order of the Solar System," in Ernan McMullin, ed., *Galileo: Man of Science* (New York: Basic Books, 1967), pp. 207–31.

47. MacPike, *Correspondence and Papers of Halley,* pp. 80–82. Simon Schaffer, "Halley's Atheism and the End of the World," *Notes and Records of the Royal Society* 32 (1977): 17–40. Isaac Newton, *Philosophiae naturalis principia mathematica,* ed. Alexandre Koyré and I. B. Cohen, 3d. ed., with variant readings (1726: Cambridge: Harvard University Press, 1972), p. 758. (Hereafter: *Principia* [Koyré/Cohen].) Henry Pemberton, *A View of Sir Isaac Newton's Philosophy* (London: S. Palmer, 1727), p. 246.

48. *Principia* (Motte/Cajori), p. 529.

49. *Correspondence,* 3:316 and 7:323–25.

50. Dobbs, "Newton's 'Clavis.'" J. E. McGuire, "Body and Void and Newton's *De mundi systemate,*" *Archive for History of Exact Sciences* 3 (1967): 206–48, on p. 220. Schechner Genuth, "Comets, Teleology and Chemistry." *Principia* (Koyré/Cohen), p. 758.

51. *Correspondence,* 2:167, 358–62. *Principia* (Koyré/Cohen), p. 757.

52. Gregory, *Elements of Astronomy,* p. 716. Pierre Louis Moreau de Maupertuis, "Lettre sur la Comète," in *Oeuvres,* new ed., 4 vols. (Lyon: J. M. Bruyset, 1768), 3:209–56, on p. 240. Georges Louis Leclerc, comte de Buffon, *Histoire naturelle, générale et particulière,* 44 vols. (Paris: De l'imprimerie royale, 1749–1803), vol. 1: *Histoire de théorie de la terre* (1749), p. 127. Johann Heinrich Lambert, *Cosmological Letters,* ed. S. L. Jaki (1761: Edinburgh: Scottish Academic Press, 1976), p. 63. Craig B. Waff, "Comet Halley's First Expected Return: English Public Apprehensions, 1755–58," *Journal for the History of Astronomy* 17 (1986): 10–37.

53. J. E. McGuire and P. N. Rattansi, "Newton and the 'Pipes of Pan,'" *Notes and Records of the Royal Society* 21 (1966): 108–43. Paolo Casini, "Newton: The Classical Scholia," *History of Science* 22 (1984): 1–58. *Principia* (Koyré/Cohen), pp. 761–62, 803–7. *Mathematical Papers,* 8:458–59.

54. Casini, "Classical Scholia," pp. 7, 38. D. P. Walker, *The Ancient Theology: Studies in Christian Platonism from the 15th to the 18th Centuries* (London: Duckworth, 1972). C. B. Schmitt, "Perennial Philosophy from Agostino Steuco to Leibniz," *Journal of the History of Ideas* 27 (1966): 505–32. J. E. McGuire, "Neoplatonism and Active Principles: Newton and the Corpus Hermeticum," in *Hermeticism and the Scientific Revolution,* ed. R. S. Westman and J. E. McGuire (Los Angeles: William Andrews Clark Memorial Library, University of California, 1977), pp. 95–142.

55. Richard S. Westfall, "Isaac Newton's *Theologiae Gentilis origines philosophicae,*" in *The Secular Mind: Transformations of Faith in Modern Europe,* ed. W. Warren Wagar (New York: Holms and Meier, 1982), pp. 15–34; and idem, "Newton's Theological Manuscripts," in *Contemporary Newtonian Research,* ed. Zev Bechler (Dordrecht: Reidel, 1982), pp. 129–43. Frank E. Manuel, *The Religion of Isaac Newton* (Oxford: Clarendon Press, 1974), pp. 53–79.

56. Yahuda MS. 17.3 fol. 1–2.

57. Gregory, *Elements of Astronomy,* p. x.

58. Yahuda MS. 17.3 fol. 12r.

59. Maseo Watanabe and Ichiro Tanaka, "A Newton Manuscript in Japan," *Annals of Science* 41 (1984): 159–64.

60. Westfall, *Never at Rest,* p. 314. See C. M. N. Eire, *The War against the Idols* (Cambridge: Cambridge University Press, 1986). S. Kogan, *The Hieroglyphic King* (Rutherford, N.J.: Fairleigh Dickinson University Press, 1986).

61. Yahuda MS. 15.7 fol. 97r.

62. Ibid.

63. Yahuda MS. 21 fol. 1–2. Manuel, *Religion of Newton,* pp. 21–22.

64. Westfall, *Never at Rest,* p. 345.

65. Ralph Cudworth, *The True Intellectual System of the Universe* (London: Richard Royston, 1678), p. 417. Isaac Newton, "Out of Cudworth," William Andrews Clark Memorial Library, MS p. 3. Aristotle, *Metaphysics* 12.8.19–21. See Danton B. Sailor, "Newton's Debt to Cudworth," *Journal of the History of Ideas* 49 (1981): 511–18. "Out of Cudworth" is printed in James E. Force and Richard H. Popkin, *Essays on the Context, Nature and Influence of Isaac Newton's Theology* (Dordrecht: Kluwer, 1990), pp. 207–13.

66. Yahuda MS. 17.3 fol. 9r.

67. Yahuda MS. 21 fol. 1–2.

68. Frank E. Manuel, *Isaac Newton, Historian* (Cambridge: Belknap Press of Harvard University Press, 1963), p. 116.

69. Richard Ashcraft, *Revolutionary Politics and Locke's* Two Treatises of Government (Princeton: Princeton University Press, 1986).

70. Ashcraft, *Revolutionary Politics,* p. 139.

71. John Miller, *Popery and Politics in England 1660–1688* (Cambridge: Cambridge University Press, 1973), pp. 193–205; J. R. Western, *Monarchy and Revolution: The English State in the 1680s* (London: Blandford Press, 1972), pp. 194–238. See Sarah Hutton, "The Neo-Platonic Roots of Arianism", in L. Szczucki, ed., *Socinianism and Its Role in the Culture of the 16th–18th Centuries* (Warsaw, 1983), pp. 139–46.

72. Westfall, *Never at Rest,* p. 479.

73. David Brewster, *The Memoirs of the Life, Writings and Discoveries of Sir Isaac Newton,* 2 vols. (Edinburgh: T. Constable and Co., 1855), 2:313–59. Manuel, *Religion of Newton,* p. 114. Katharine R. Firth, *The Apocalyptic Tradition in Reformation Britain, 1530–1645* (Oxford: Oxford University Press, 1979), pp. 214–41. Richard H. Popkin, "The Third Force in Seventeenth-Century Thought: Skepticism, Science and Millenarianism," in *The Prism of Science: Studies in the History, Philosophy and Sociology of Science,* ed. E. Ullman-Margalit (Dordrecht: D. Reidel, 1986), pp. 21–50. Paul J. Korshin, "Queuing and Waiting: The Apocalypse in England, 1660–1750," in C. A. Patrides and Joseph Wittreich, eds., *The Apocalypse in English Renaissance Thought and Literature* (Manchester: Manchester University Press, 1984), pp. 240–65.

74. William Whiston, "Reflexions upon Sir Isaac Newton's *Observations upon the Prophecies of Daniel,*" in *Six Dissertations* (London: J. Whiston, 1734), pp. 297–302, and idem, *Memoirs of the Life and Writings of Mr. William Whiston,* 2 vols. (London: J. Whiston, 1749), 1:38.

Ether Madness: Newtonianism, Religion, and Insanity in Eighteenth-Century England

ANITA GUERRINI

The concept of insanity in the first half of the seventeenth century was tinged with the supernatural. Magic, possession, witchcraft, astrology, all shared in its interpretation, along with the conventional medical wisdom that attributed madness to an imbalance of the humors.[1]

By mid-century, however, these views were changing. As the mechanical philosophy began to influence biological as well as physical thinking, the mechanics of nervous function became a particular topic of interest. Thomas Willis, among others, coupled mechanical ideas with the bold experimental techniques he had learned from William Harvey to create a powerful investigative tool. His concept of nervous function, and his extension of these physiological theories to neural pathology, dominated British ideas until well into the eighteenth century.[2]

Willis adopted the model of the hollow nerve—that is, that nerves are tubelike structures filled with a fluid known as "animal spirit."[3] This concept dated back to Erasistratus the Alexandrian, and it had been revived by literal-minded mechanists such as Descartes. Willis's notion of nervous function owed much to Descartes, but he also borrowed nonmechanical chemical ideas stemming from Paracelsus. For Willis, the cause of muscular motion was a chemical reaction within the muscle, and not merely the mechanical motion of the animal spirits. Muscle fibers certainly held animal spirits, said Willis, and this spirit was composed of saline particles. These particles reacted in muscular motion with the "nitrosulphureous" particles of arterial blood. When the action of the nerves caused the niter and the salt to combine, they exploded like gunpowder, causing the muscle to contract.[4]

In his work on the pathology of the brain, published in 1667,

Willis showed that all kinds of convulsions and spasms, including those of epilepsy and hysteria, were the result of these explosions gone somehow wrong. The brain, said Willis, transmitted "hetero-geneous" and "unduly explosive" combinations of salt and niter. He used postmortems to prove—though by negative evidence—that such afflictions as asthma were often of nervous origin, since he found no organic lesions.[5]

Although in the work of Willis we find a completely naturalistic explanation of certain forms of insanity, he nonetheless felt it neces-sary to raise the possibility of supernatural origins. Witchcraft, he thought, provided a possible cause of convulsions; "yet all kind of convulsions, which . . . appear prodigious, ought not presently to be attributed to the inchantments of Witches." In most cases, witches were unjustly accused. But in cases exhibiting unusual contortions, extraordinary strength, or voiding of strange objects, then "without doubt, it may be believed that the devill has, and doth perform, his parts in this Tragedy."[6]

Thus even by Newton's era, supernatural explanations had not disappeared. In the 1660s and 1670s, such men as Sir Thomas Browne, Joseph Glanvill, and Robert Boyle had reaffirmed their beliefs in spirits, witchcraft, and possession, beliefs central to their religious and philosophical principles.[7] By this time, however, witchcraft persecutions had declined from their peak at the turn of the century. Louis XIV's minister Colbert abolished the legal charge of "sorcellerie sabbatique" in 1672, and in 1703—the year Newton became president of the Royal Society—a London black-smith's apprentice named Richard Hathaway was convicted of falsely accusing Sarah Morduck of bewitching him. The last witch trial in England was held in 1712 (they went on in Scotland for most of the century), and although Jane Wenham was convicted, the judge pardoned her.[8] Yet in 1693 New England, Cotton Mather, future fellow of the Royal Society, gravely concluded that several girls who suffered from convulsions and other strange behaviors were indeed possessed. His claim refueled the ongoing controversy in Lancashire over the case of the "Surey Demoniack."[9]

One point of contact between possession and mental illness was religious mania or "enthusiasm." Michael Heyd has usefully de-fined an enthusiast as one who claims "to have direct divine in-spiration."[10] Such a claim obviously constituted a threat to the established order, and "enthusiasm" was a pejorative term. Some of the wilder sects of the revolutionary era of the 1640s and 1650s were obviously enthusiasts; the early Quakers stripped naked and danced under the influence of holy benevolence, causing the more orthodox to attempt to draw a line between such behavior and true revelation.[11] After the Restoration, and especially after the Glorious Revolution of 1688–89, religious enthusiasm was increasingly frowned upon as socially disruptive. Heyd plausibly argues that the general trend in the elite culture of this period toward secular rather than supernatural modes of explanation is at least partially owing to this social reaction against enthusiasm.[12]

Some historians have argued that Newtonian natural philosophy, along with the Whig constitution of 1689 and the mainstream Anglican church, supported a campaign to root out enthusiasm in English society. The result was the lofty calm of the Augustan age and its apparent consensus on matters spiritual, the middle ground being sought between atheism and enthusiasm. "A Man is to be cheated into Passion, but to be reason'd into Truth," commented Dryden in the preface to his poem "Religio Laici."[13] An important aspects of this campaign was the secularization of explanations and treatment for insanity.[14]

Yet this scenario of the onward march of rationalism is not entirely convincing. Recent historiography of the eighteenth century has argued that the cultural role of revealed religion diminished less rapidly and less silently than the label of "Enlightenment" suggests.[15] "Enthusiasm" retreated but did not disappear. In particular, explanations of insanity put forth by self-professed Newtonians are not as imbued with Whiggish rationalism as we might think. Far from replacing the supernatural with the mechanical, these explanations instead often resorted to occult causes of another sort. The man cited by many as the quintessential Newtonian alienist, the Scottish physician George Cheyne, turns out to have been more than a little tinged with enthusiasm.

* * *

In 1701, John Freind, a young Oxford physician, sent a letter written in the elegant Latin for which he was renowned to Hans Sloane, editor of the *Philosophical Transactions* of the Royal Society of London. The letter concerned "the history of some extraordinary convulsions." Although not yet a doctor of medicine, Freind obviously was already practicing medicine. Under the tutelage of David Gregory at Christ Church, he had become acquainted with Newton's ideas. With another of Gregory's students, the Scot John Keill, Freind later helped to formulate the basis of a Newtonian theory of matter, based on the concept of short-range forces between atoms, forces analogous to, but not identical to, gravity. In 1704, Freind became Ashmolean lecturer in chemistry at Oxford, and his lectures, published in 1709, expounded these ideas, which he felt were directly applicable to physiological explanation.[16]

But in 1700, while he was still a student, the village of Blackthorn in Oxfordshire witnessed a bizarre phenomenon: five sisters between the ages of six and fifteen began to act very strangely. They crept around on all fours, barking like dogs, their heads nodding violently, their tongues protruding. After a time, the girls would collapse in an epileptic-like fit. Soon after, three girls and a boy from another family began to exhibit similar behavior. Freind was called in, and mindful of the precepts of Willis, he says, he accepted the case. Freind pointed out to Sloane that the symptoms in the case closely resembled those in a case reported by the German physician Jakob Seidel in the sixteenth century, "apart from the fact that this illness is ascribed [by Seidel] not to natural causes but to enchantment."[17]

After closely examining the children's behavior, and their clinical signs during the fits such as their pulses, Freind reached his conclusions. The mothers of the children strongly believed that the affliction was supernatural in origin, but Freind dismissed this: neither the novel form of the ailment, nor its long duration through most of the previous year, proved to him to be sufficient arguments against a natural—that is, a mechanical—explanation. The fits, he said, were simply another form of epilepsy, like St. Vitus's Dance, evidently caused by the unseasonal rising of that volatile fluid, the animal spirits, which stirred up a commotion between the nerves and the muscles. Freind approvingly referred to the work of Willis and to the Italian iatromechanist Lorenzo Bellini as the sources of his explanation. The younger children, he added, may have imitated the older girls, in whom the fits began.[18]

But Freind did not offer any cure to the families; and the concluding remarks in his letter strike us oddly. He noted that the afflicted were first cousins and that therefore the rest of the village was not in danger. He took his leave of Blackthorn, he wrote, "with pleasure," leaving the recalcitrant inhabitants to their horoscopes and sympathies.[19]

Despite Freind's later reputation as a Newtonian, there is nothing very Newtonian about his analysis in this case, except perhaps his arrogance. It serves as an example of mechanistic explanation in which occult forces are explicitly excluded. At about the same time, however, another young physician, Richard Mead, linked the mechanical theory of nervous function with Newton's concept of action-at-a-distance in his first book, *A Mechanical Account of Poisons,* published in 1702.

Mead was the son of a nonconformist preacher, and at the time his book appeared he carried on an unlicensed medical practice out of his father's house in Stepney, in southeast London—not a fashionable district. He had, however, studied briefly in Leiden under Archibald Pitcairne, an admirer of Newton. In addition, Mead was an astute observer of social trends, as well as being ambitious; he accurately chose Newton as the owner of capacious coattails. For Mead, this paid off handsomely, for within a year of the book's publication, he was a fellow of the Royal Society and physician-in-ordinary at St. Thomas's Hospital; he had quickly left behind the preacher's house in Stepney.[20]

In his preface to *A Mechanical Account of Poisons,* Mead credited Bellini, Pitcairne, and Pitcairne's protégé George Cheyne with the reform of medicine along mechanical and mathematical lines. There is, however, very little mathematics in Mead's book, and it is doubtful that he ever read the *Principia.* He proposed that poisons consist of acid salts that, according to the standard mechanical theory of chemistry, were sharp, pointed crystals. The effect of these little knives on the blood was to tear it up and cause its particles to cohere and disjoin in irregular clusters. This cohesion, said Mead, "is the very same thing with the Attraction of Particles one to another, [which] Mr *Newton* had demonstrated to be the great Principle of Action in the Universe."[21]

In this context, Mead went on to discuss the effects of certain kinds of poisons, including the delirium caused by the bites of tarantulas and mad dogs. The poisonous venom in these cases affected the blood by causing its particles to cohere into large clusters that pressed on the blood vessels, especially those in the brain, which were, said Mead, "more easily distended." This in turn put the animal spirits into motion, causing all manner of false perceptions. Mead's treatment for most poisons, and therefore also for this sort of delusion, would either absorb or neutralize the excess acidity of the poison or else stimulate the circulation to prevent the formulation of coagulated clusters of blood. Among stimulants, he strongly recommended cold baths.[22]

In his next work, Mead linked gravity and madness even more directly. In 1704 appeared his *De imperio solis ac lunae in corpora humana et morbid inde oriundis* [Of the influence of the sun and moon on human bodies and the diseases arising thence]. Mead took the familiar climatological disease theories of the Hippocratics and joined to them Newton's theory of the tides. According to Mead, this theory of the lunar and solar pull on the seas applied also to the sea of air surrounding the earth. Winds, he said, followed regular cyclic patterns related to seasonal changes and the lunar cycle; following Newton's rules of reasoning in the *Principia,* he concluded that such similar effects as winds and tides must have the same cause, the gravitational pull of the sun and moon. The stars and planets, he thought, had similar but weaker effects on terrestrial events—did he here attempt to provide a rationalization for the practice of astrology? Obviously the tidal flux of air affected creatures far more than did the oceanic tides; all creatures used air, and its lesser density meant that its tidal effects were greater. Its overall effects on physiology, Mead thought, would be to quicken or retard the circulation of the blood and therefore the formation and flow of animal spirits.[23]

Mead then focused his attention on the diseases of the moon: on lunacy. He reported several case histories, from both his own and Pitcairne's files, which demonstrated the close connection between the waxing and waning of the moon and the cyclical appearances of "mad Fitts," epilepsy, and hysteria. The falling-sickness, he claimed, had "its Recourse every new and full Moon." He recounted that Pitcairne "cured two young Girls, whom the Moon caused to have Epileptical Fits with ridiculous Motions, like those that are troubled with the Dance of St Vitus," a case resembling the Blackthorn episode, but he failed to specify the therapy employed. "It may be justly thought," Mead continued, "that the Ignorance of Causes is the Reason that in the vast Volumes that are writ in Physick, we have no such Relations as these; but when a strict Inquiry shall be made into them, then Notice will be taken of them, for there are many Examples of such like Sympathys." Here, it seems, Mead resurrects, in Newtonian guise, the "horoscopes and sympathies" discarded by Freind only three years earlier. Mead felt that he was being eminently scientific; we can perhaps compare his reasoning with Newton's own probable use of alchemical ideas in formulating the concept of gravity; here too the insensible causes of one genre were translated into another mode of activity. Newtonian interpretation thus far, then, has gone beyond mere mechanism as a cause of insanity to an emphasis on immaterial forces.[24]

*   *   *

For the first twenty-five years after the publication of the *Principia,* Newtonian matter theory centered on atoms and the short-range attractive forces between them. Newton had stated this theory in a long-unpublished 1692 essay entitled *De natura acidorum*. Newton gave a copy of this essay to Archibald Pitcairne soon after he wrote it, and it circulated widely in manuscript for several years before its publication in 1710.[25] He reiterated this concept in the *Opticks*, particularly the 1706 Latin edition, and subsequent development of this notion by John Keill and John Freind closely followed the lines suggested in Query 24 (31) of that edition. Freind and Keill agreed that interparticulate forces were, like gravity, proportional to mass; and like Newton they felt that these forces were not identical to gravity, but analogous. Keill believed that the $1/r^2$ relationship that prevailed between celestial bodies would be $1/r^3$ or even $1/r^4$ at the atomic level—that is, they would be forces acting at very short ranges, which would diminish to nothing at any appreciable distance. An important philosophical corollary of this theory was the congruity of the microcosm and the macrocosm. God would not have designed an incoherent world; therefore, the analogy between the microcosm and the macrocosm was a valid one.[26]

Newton's concept of matter and of attraction began to change during the first decade of the eighteenth century, although the extent of that change has been much debated. The electrical demonstrations and experiments performed between 1704 and 1713 by Francis Hauksbee at the Royal Society fascinated him. To the second edition of the *Principia*, published in 1713, Newton added his General Scholium, which concluded with an almost casual mention of "a certain most subtle spirit" that was "electric and elastic" and caused a variety of phenomena. He tentatively identified this "spirit" with the animal spirits or nervous fluid. This apparent resurrection of the ether, a concept Newton had abandoned in 1679, did not arouse general comment.[27] In his revision of the *Opticks* in 1717 for a second edition, Newton expanded his comments in the General Scholium into two new queries. Eventually this grew to eight new queries, numbered 17 through 24 (between the original sixteen and the additional seven added in 1706). Newton now proposed that the action of an ether could explain the operation of both short-range forces and gravity.[28]

Was Newton's reintroduction of the ether a true retreat to his position of the 1670s; or, a "sop" thrown to his mechanist critics, as Westfall suggests; or the introduction of a new concept?[29] I believe it was the latter, in agreement with an important 1971 article by P. M. Heimann (now Harman) and J. E. McGuire, "Newtonian Forces and Lockean Powers." Heimann and McGuire conclude that this 1718 ether differs greatly from earlier concepts of the ether, which describe it as a particulate, mechanical fluid. Rather, it is a "force-ether," structured in terms of the strong repulsive forces *between* the ether particles. The particles themselves are vanishingly small, or as Newton says "exceedingly smaller than those of air, or even those of Light," with a force of great intensity in inverse proportion to their smallness and solidity. In fact, by this argument matter very nearly disappears, comment Heimann and McGuire, "leaving space containing interparticulate forces clustering around changing networks of foci."[30] This logically develops the implications of Newton's earlier particulate matter theory, in which he had already de-emphasized solid matter to the extent that Joseph Priestley could later comment that all the matter in the world would fit into a nut-shell.[31]

McGuire and Heimann do not mention the probable connection between this concept of a "network" of forces and Newton's alchemical work. Both Westfall and B. J. T. Dobbs have suggested the centrality of the alchemical image of the "net" to the development of Newton's gravitational theory. In addition, Newton's search for an ontological cause of motion was intimately related to his concept of God and God's action in the universe.[32]

The idea of the ether, of subtle fluids, began to dominate Newtonian thinking in the 1720s. Richard Mead, ever the follower of fashion in natural philosophy, was the first to seize upon this notion and apply it to a physiological context. By the 1720s Mead was fabulously successful; he and John Freind were among the wealthiest physicians in London. For both, the reputation of being on the cutting edge of scientific medicine greatly enhanced their careers in the cut-throat London medical scene. The physician Bernard Mandeville complained, "A Man of Wit and good Parts, that has a little smatt'ring of the *Newtonian* Philosophy, is seldom at a Loss now, to solve almost any *Phaenomena*," in contrast to the older style of medical practitioner, who relied on his bedside manner and diligent observation. Mead's own career was especially illustrative of the changes in fashion, since in 1714 he had inherited the practice and the Bloomsbury house of John Radcliffe, the foremost of the old-style practitioners.[33]

In 1721, Mead, a noted antiquarian and book collector, proposed a sumptuous new edition of William Cowper's 1694 treatise on muscle anatomy, *Myotomia reformata*. The bookseller William Innys reported that "this Book will be the most Beautifull Book (both for Paper and Types) ever printed in England." Mead planned to preface his edition with a "Discourse upon Muscular Motion."[34] By the time the edition appeared in 1724, responsibility for the introduction had been passed to one of Mead's protégés, the young physician Henry Pemberton. An aspiring Newtonian, Pemberton happily took Mead's lead and applied Newton's ideas on the ether to physiological explanation in his lengthy essay. After discussing various proposed mechanical causes of muscular motion, Pemberton concluded that "the Fluid contained in the Nerves is probably no other than part of that subtle, rare, and elastic spirit, [Sir Isaac Newton] concludes to be diffused throughout the Universe."[35]

A year later, another young physician, Nicholas Robinson, joined the Newtonian coterie with a book entitled *A New Theory of Physick and Diseases, founded on the Principles of the Newtonian Philosophy.* Like Pemberton, who had received his medical degree in Leiden, and indeed Mead himself, Robinson—a graduate of the University of Rheims in France—operated outside the institutional framework provided by an Oxbridge education. He dedicated *A New Theory of Physick and Diseases* to Richard Mead, his patron.[36]

In his preface, he set the stage for his analysis in appropriately Newtonian language, from the Queries: "is not the Mechanism of the Body conducted by the same laws that support the Motions of the greater Orbs of the Universe? and are not all the Changes and Variations it suffers in Diseases, to be resolv'd from an Alteration of its Matter and Motion?"[37] Yet he couched his account of physiology in the strictly mechanical terms of solids and fluids, following the Italian iatromechanist Giorgio Baglivi. Robinson mentioned attractive particles, such as those of the blood, but not Newton's ether; he drew his description of the animal spirits from the well-known work of James Keill, published in 1708. Robinson conceded that there was one problem in physiology that his atomistic, mechanico-Newtonian analysis could not solve. This was the influence of the mind on the body: "who, from the Texture of the Parts, can account for the Juncture of Matter and Thought?"[38]

He attempted to explain this relationship four years later, in 1729, in his *New System of the Spleen, Vapours and Hypochondriack Melancholy,* dedicated to Hans Sloane, president of both the Royal Society and the Royal College of Physicians of London. In the interim, no doubt with Mead's assistance, Robinson had been admitted as a licentiate to the College of Physicians, and had moved from the crowded alleyways around St. Paul's to a more fashionable neighborhood to the west, not far from Mead's and Sloane's own premises in Bloomsbury. We cannot tell whether this new-found success owed more to Robinson's scientific pretensions than to his patent remedies for the stone.[39] With the *New System of the Spleen* he entered the territory of the most fashionable physicians, the treatment of mental disorders.

Robinson admitted at the outset of his book that the mind-body problem was not capable of "strict proofs," but he believed that by means of probability, one could "lead the Mind, by gradual and easy steps from self-evident Theorems, into the most abstruse Mysteries of Nature." As in his earlier work, Robinson's physiological model was mechanical. He again found that the seat of health resided in the "fibres" or solids, which were themselves composed of attracting particles. The differing power of attraction of these component particles made the fibers more or less "springy" or "elastick."[40]

With regard to mental states, Robinson's theories are characterized by what Porter calls an "aggressive somatism." Robinson found elasticity to be in general a desirable quality, although overly springy fibers led to a bilious, passionate temperament. These overly springy types were also vivacious and witty, but at times this vivacity edged toward madness or "phrensy" when the elasticity became overwound. Robinson added that the most brilliant and witty were also the most erratic; "we frequently observe that those, that are remarkable for a Mastery in the superior Sciences, are generally most strongly impell'd toward Wine, Women, and Musick."[41] We can be fairly certain he did not have Newton in mind here. At the other end of the spectrum, the least springy fibers contributed to a melancholic temperament, in which the individual "resolve[s] long upon the same ideas." He attributed various other mental afflictions to "a clog of matter in the brain" that prevented the free flow of animal spirits through the tubular nerves. Here so far was a typical mechanical rationalization of mental disorder.[42]

Robinson then asked his readers to assume the existence of an immaterial soul that animated the "organiz'd Machines" of the body. This soul he defined as a "self-moving principle" chained to "little Spots of Earth, human Machines." Were it not so chained, it "would naturally rise, and take its Seat in the proper *Ubi* of this immense Theatre." The soul, like gravity, like God, did not change; only the body did. The analogy to gravity seems obvious. Because the soul was unchanging, insanity had no connection with it; it was not a spiritual affliction, but a somatic one.[43]

Robinson did not rest content with this spiritual action-at-a-distance, however. As Newton may have reintroduced an ether to satisfy the demand of his critics for something to fill the gap between force and matter, so Robinson filled the gap between soul and body with something he called the "animal aether." He defined this as "matter resolved into the last degree of fineness," and he equated this matter—which was also of varying springiness—with the passions, for which, he claimed, a strictly mechanical system could not account. To Robinson, the emotions somehow set this ether in motion, which then caused the animal spirits to move through the nerves. Wine, he added, duplicated this impulse to the spirits. This material "animal aether" did not really solve the problem of the gap between matter and force, body and mind; it simply backed it up another step. It most resembles Newton's ether of the 1675 "Hypothesis of Light," described as "the aetherial Animal Spirit in a man [which] may be a mediator between the common aether & the muscular juices to make them mix more freely."[44] But this, as we have seen, was not the same as the ether of 1717–18. It seems clear that Robinson—like most people—did not quite understand what Newton intended when he reintroduced the ether, assuming, of course, that our understanding of his intention is correct. Robinson's ether left the material causes of madness intact, and relegated the soul, like the God of the deists, to a vague spiritual hinterland.[45]

*　*　*

A superficial look at George Cheyne's *The English Malady; or, A Treatise of Nervous Diseases of all Kinds*, published in 1733, would lead one to assume that Cheyne reached very similar conclusions to Robinson. Cheyne commenced, "the Human Body is a Machin of an infinite Number and Variety of different Channels and Pipes, filled with various and different Liquids and Fluids." Nervous distempers, ranging from yawning to apoplexy, were, he concluded, "but one continued Disorder, or the several Steps and Degrees of it, arising from a relaxation or Weakness, and the Want of a sufficient Force and Elasticity in the Solids in general, and the *Nerves* in particular." We may suspect a hidden agenda, however, when Cheyne begins to rail against the excesses of urban life, and of civilization in general, as the cause of the increased frequency of nervous disorders in his time.[46]

Cheyne was no stranger to the dangers and delights of London life, and he included an account of his own case in *The English Malady* as a lesson in both morality and physic. Born near Aberdeen in 1671, Cheyne attended Marischal College in that city. By the late 1690s he was in Edinburgh and Pitcairne's protégé. Cheyne received his M.D. degree by purchase from Aberdeen in 1701 and soon after set out for London to make his fortune, with Pitcairne's blessing. He quickly became a member of the circle around Newton and published several medical treatises in the mechanistic, quasi-Newtonian mold of Pitcairne. He also, as he said in his autobiographical account, moved into the fast lane of London social life:

"I found the *Bottle-Companions,* the *younger Gentry,* and *Free-Livers,* to be the most easy of *Access,* and most quickly susceptible of *Friendship* and *Acquaintance.*" But the young Scot lacked the constitution for prolonged debauchery, and around 1705 he became seriously ill. This illness had important consequences for both his intellectual and physical life.[47]

As a result of this illness, Cheyne abandoned both the good life and the rather facile Newtonianism of his early years in London. He moved to the spa town of Bath, at that time beginning to become prominent as a fashionable resort, and practiced medicine, administering especially, he said, to "low [i.e., melancholic] and nervous *Cases,*" a mainstay of Bath medical practice. His own occasional lapses into physical overindulgence continued into the 1720s, and these lapses were truly on a Hogarthian scale. Cheyne considered moderation to be the consumption of "not above a Quart, or three Pints at most" of wine with his dinner, and not surprisingly, his weight ballooned to 32 stone, or nearly 450 pounds. He finally recovered his health with a moderate regimen and a mostly vegetarian diet, which he continued until his death in 1743.[48]

At the time he wrote *The English Malady,* Cheyne had lived in Bath for twenty-five years, and certainly seems to conform to his usual image of a successful and fashionable physician. He was the author of several medical and scientific treatises, a fellow of the Royal Society, a friend of Mead and Sloane. His patients included members of the nobility and London's political elite.[49] Moreover, in the early pages of *The English Malady,* Cheyne appears to be a rational mechanist much like Robinson. But when we turn to his account of nervous function, the discussion shifts to a different ground.

In a work of 1732, the Irish physician Bryan Robinson (no relation to Nicholas) identified the animal spirits as Newton's ether. Cheyne decisively rejected this identification. No *material* fluid, he wrote, can act in such a way as to cause muscular motion. Any material fluid would be too dense. In addition, the structure of the nerves themselves prohibited such a hydraulic arrangement, for Cheyne denied the common notion of the hollow nerve. Blockages of the tubular nerves therefore could not cause nervous disorders, as Nicholas Robinson had suggested. Cheyne thought that nerves were solid, and that impulses or vibrations were transmitted along them by the action of some subtle fluid that he did not identify. He used an analogy to music: the brain acts as the conductor, and the music flows along the nerves. In Query 24 of the *Opticks,* Newton had similarly described nerves as "solid and uniform, that the vibrating Motion of the Ætherial Medium may be propagated along them from one end to the other uniformly."[50]

However, Cheyne continued, such actions in nature as cohesion and elasticity may be caused by an "infinitely subtil, elastick Fluid, or Spirit," a suggestion "made not improbable by the late *sagacious* and *learned* Sir Isaac Newton."[51] Was Cheyne contradicting himself? He noted that Newton's reference to the ether as a "spirit" implied that it could be *im*material. He then perceptively discussed the function and structure of such an ether. He pointed out that if the Newtonian ether, which caused gravity, was material, what then caused the "elasticity and repulsion" of its parts? We must either postulate another ether and another in an infinite regress, "or we must suppose these Qualities innate to them, and to have been impress'd on them immediately by the *first* and *supreme Cause.*" But the split between material and spiritual was not therefore abrupt. Rather, Cheyne postulated a Platonic hierarchy of "Intermediates between *pure, immaterial Spirit* and *gross Matter.*" He tentatively identified Newton's ether as an "intermediate material Substance," the "*Medium* of the Intelligent Principle," which "may make the Cement between the human Soul and Body," as well as cause numerous other functions. This resembles the function of Nicholas Robinson's "animal aether," but Cheyne's concept of the ether called for the direct and active involvement of God in contrast to the vague deistic role of Robinson's "self-moving" soul, which only implied God's action. While Cheyne may not have understood Newton's intent in reintroducing the ether any better than did Robinson, he was more aware of the central role played by God in Newton's system.[52]

Cheyne intended in formulating this theory to steer the argument away from the deistic aspects of both Robinsons' interpretations of Newton's ether. To Cheyne, atheism, not enthusiasm, presented the clear and present danger. The social disintegration he had witnessed in his younger days in the London of Queen Anne was closely tied, in his mind, to a devaluation of the spiritual, and the crisis had only grown nearer, three decades later. To support this, let us return to Cheyne's crisis of 1705, described in *The English Malady.* His illness at that time led to a profound depression. His high-living friends abandoned him to his fate, "leaving me," he wrote, "to pass the melancholy Moments with my own *Apprehensions* and *Remorse.*" This melancholy and remorse were religious in origin; and having retired to the country—probably to his native

Aberdeen—Cheyne found ample time to engage in religious medita-
tion.[53]

In his *Philosophical Principles of Natural Religion,* published in
1705 shortly before his illness, Cheyne imitated the latitudinarian
arguments of such low-church Whig apologists as the Boyle lecturers Richard Bentley and Samuel Clarke, who in their reconciliation of religion and natural philosophy attempted to counter the
rationalist arguments of the deists. In the *Philosophical Principles,*
Cheyne found that the contrivance of the world, and especially the
Newtonian principles thereof, happily supported religion, if a rather
pared-down religion whose precepts consisted solely of a supreme
being, free will, and an immortal soul.[54] But in his country reflections Cheyne realized that his proof of a Deity by natural philosophy came dangerously close to deism itself and its denial of
revelation. Such beliefs moreover were sorely inadequate for his
spiritual needs. We discern the glimmer of enthusiasm in his mind.
This glimmer undoubtedly was fueled by Cheyne's contact with a
friend of his youth, the Aberdeen clergyman George Garden.[55]

In his account, Cheyne identifies Garden only as "a worthy and
learned *Clergyman* of the *Church of England.*" Garden was indeed
an Anglican, but a most unconventional one. He had lost his official
positions in Aberdeen—including the rectorship of its cathedral—
after the Glorious Revolution, when he refused the oaths to William
and Mary. He was strongly influenced by the personal religion
preached earlier in the seventeenth century by his fellow Aberdo-
nians John Forbes and Henry Scougall, and in the late 1690s, he
discovered the works of the French mystic Antoinette Bourignon.
By the time Cheyne reencountered him in 1705–6, Garden had
translated several of Bourignon's works and published an *Apology
for Mme. Bourignon,* which led to his suspension from the ministry.
Like Forbes and Scougall, Bourignon emphasized a personal, mystical union with God. Bourignon believed herself to have been
divinely inspired.[56]

Cheyne's conversations with Garden confirmed his growing spiritual discomfort, a discomfort mirrored in his bodily unease. He
found that while he had observed the outward rules of Christianity,
"from *abstracted* Reasonings, as well as from the best *natural
Philosophy,*" he now began to wonder whether this was enough.[57]
Clearly, he was sinking into a religious melancholy, that "dark
lanthorn of the Spirit," closely related to religious enthusiasm.[58] In
an earlier work, his popular *Essay of Health and Long Life* (1724),
Cheyne had discussed "That kind of *Melancholy,* which is called

Religious, because 'tis conversant about matter of *Religion;* although, often the Persons so distempered have little *solid Piety.*" This harsh judgment was undoubtedly directed against himself, and he stated that the ailment—which manifested itself in physical symptoms—arose from "a *Disgust* or *Disrelish* of worldly *Amusements* and *Creature-comforts.*" In such cases, physical regimen was first to be employed against the physical symptoms. But if these were insufficient, only God could effect a cure.[59]

In the *English Malady,* Cheyne did not try to explain away his melancholic symptoms as a "clog in the brain." Rather, he attempted a genuine spiritual self-exploration, seeking "higher, more noble, and more enlightening Principles," and "more encouraging and enlivening *Motives* proposed, to form a more extensive and *Heroic* Virtue upon, than those arising from *natural* Religion only." Lastly, he sought "some clearer Accounts discoverable of that *State* I was then apparently going into [i.e., his depression], than could be obtained from the mere Light of *Nature* and *Philosophy.*"[60] On Garden's recommendation, Cheyne turned to reading theological works, and found his mind much settled; in fact, he concluded, such reflection and reading "greatly contributes to forward the Cure of such *nervous* Diseases."[61]

Cheyne was not alone in his discontent with the overly intellectual apologetics of the established church. Garden and his mystical circle had placed themselves well outside the Anglican mainstream; but in the late 1720s, a young Oxford don named John Wesley similarly revolted against the "almost Christians" of the time in favor of a more direct and emotional relationship with God. The members of his "Holy Club" in Oxford called themselves "Methodists."[62] Cheyne and Wesley traveled paths that often crossed. While still an undergraduate in 1724, Wesley had read Cheyne's well-known *Essay of Health and Long Life.* A lifelong valetudinarian, Wesley took Cheyne's exhortations to moderation in that work to heart, and the regime he adopted at this time became a central feature of Methodist life. Indeed the "method" that gave the sect its name included not only regularity in spiritual matters but also regularity and moderation in daily life, including early rising and abstemiousness in food and drink.[63] Wesley commenced his own work on medicine, the immensely popular *Primitive Physick,* first published in 1747, with Cheyne's aphorisms on health.[64]

Cheyne's account of his religious awakening, written in 1733, in fact bears many of the marks of the standard Methodist testimony as it was later established, including the description of a sinful youth, a dramatic experience leading to a recognition of mortality, and an extended ordeal of conversion.[65] It is possible that Cheyne knew of Wesley, as Wesley knew of him, at this time. In the summer of 1732 a young member of the Holy Club, William Morgan, had died of an unknown affliction. He had exhibited symptoms of religious melancholy, however; according to his father, young Morgan "used frequently to say that enthusiasm was his madness, [and] repeated often 'Oh religious madness!' that they had hindered him from throwing himself out at the window."[66] The case was widely publicized, and a response to criticisms of the Wesleys arising out of the case was published in 1733. This work, *The Oxford Methodists,* contained the first extended account of Methodist views and may have been written by William Law, author of the popular devotional work *A Serious Call* (1728). Introduced to mystical circles by Garden, Cheyne by this time exchanged mystical literature with a wide circle of correspondents, among them Law, to whom he recommended the works of Jacob Boehme.[67] Cheyne may thus have heard of the Morgan case and noted its parallels with his own, in concluding that the problem was not too much religion, as the Wesleys' critics maintained, but rather too little.

Cheyne knew John Wesley well by the late 1730s. Wesley returned from Georgia and his encounter with the Moravian sect in February 1738, and a few months later experienced his London "conversion," which led to the establishment of his Fetter Lane congregation. By the end of that year the congregation counted among its members Selina Hastings, Countess of Huntingdon. Lady Huntingdon had been Cheyne's patient since 1730, as was her sister-in-law, Lady Betty Hastings, who may have met Wesley as early as 1729.[68] Cheyne's conversations and correspondence with Lady Huntingdon covered, not surprisingly, both medical and spiritual matters. Both, like Wesley, found the two topics to be closely related. In 1741 Lady Huntingdon commented of Cheyne that "the people of Bath says [sic] I have made him a Methodist, but indeed I receive much light and comfort from his conversation."[69] In a 1747 letter, Wesley mentioned a conversation with Cheyne that appears to have taken place in 1738.[70] Although there is no evidence that Cheyne became a true follower of Wesley, he was certainly sympathetic to Methodist views, which coupled a strict physical regimen with a deeply emotional and personal spiritual life.

Many of his contemporaries considered Wesley to be a wild enthusiast. Methodist meetings, which included open-air sermons to thousands, were stormy and emotional affairs.[71] David Hume may have been thinking of the "raptures and transports" of Methodist meetings when he defined a religious enthusiast as a "fanatic madman."[72]

I wish to make two points in conclusion. First, far from rationalizing insanity—or at least religious melancholy—with a mechanical explanation, Cheyne instead sought spiritual solutions to what he perceived to be a spiritual problem. Second, he nonetheless sought to integrate his view of Newtonian science with this spiritual world. His last works, *An Essay on Regimen* (1740) and *The Natural Method of Cureing the Diseases of the Body and the Disorders of the Mind Depending on the Body* (1742)—works much admired by Wesley—reiterated these ideas.[73] Science and religion were not, to his mind, incompatible, but inseparable. Newton's ether, both as substance and as concept, was central to this integration as an intermediate substance between matter and spirit, the occult and the supernatural, bridging the gap; and as evidence of God's direct intervention in the world. Cheyne did not reject the spiritual for the mechanical; nor did he reject Newtonianism for Methodism: he believed both.[74] England may have outlawed witchcraft trials in 1736, but the life of the spirit was far from dead.

Acknowledgments

Research for this article, and for the longer project on George Cheyne of which it is part, was supported by National Science Foundation grant SES-86-19503 and by a Fletcher Jones fellowship at the Huntington Library, for which I am grateful. I also wish to thank Michael Osborne, Roy Porter, and Ron Sawyer for their comments on earlier drafts, and Robin Overmier, formerly curator of the Wangensteen Library, University of Minnesota, for access to materials.

Notes

1. On seventeenth-century insanity, see particularly Michel Foucault, *Madness and Civilization,* trans. Richard Howard (New York: Random House, 1965); H. C. Erik Midelfort, "Madness and Civilization in Early Modern Europe: A Reappraisal of Michel Foucault," in *After the Reformation,* ed. Barbara Malament (Philadelphia: University of Pennsylvania Press, 1980), pp. 247–65; George Rosen, *Madness in Society* (Chicago: University of Chicago Press, 1968); T. H. Jobe, "Medical Theories of Melancholia in the Seventeenth and Early Eighteenth Centuries," *Clio Medica* 11 (1976): 217–31; Michael MacDonald, *Mystical Bedlam* (Cambridge: Cambridge University Press, 1981); idem, "Religion, Social Change,

and Psychological Healing in England, 1600–1800," in *The Church and Healing*, ed. W. D. Sheils (Oxford: Blackwell, 1982), pp. 101–25; David Harley, "Mental Illness, Magical Medicine and the Devil in Northern England, 1650–1700," in *The Medical Revolution of the Seventeenth Century*, ed. Roger French and Andrew Wear (Cambridge: Cambridge University Press, 1989), pp. 114–44.

2. Thomas Willis, *Cerebri anatome* (1664); idem, *Pathologiae cerebri, et nervosi generis specimen* (1667); translated in *The Remaining Medical Works of . . . Dr. Thomas Willis*, trans. S[amuel] P[ordage] (London, 1681). Subsequent citations will be from this edition; it should perhaps be noted that Pordage, the translator, was the son of a prominent disciple of Jacob Boehme. See also Robert G. Frank, *Harvey and the Oxford Physiologists* (Berkeley: University of California Press, 1980), pp. 182–83, 222–23; Alfred Meyer and Raymond Hierons, "On Thomas Willis's Concepts of Neurophysiology," *Medical History* 9 (1965): 1–15, 142–55.

3. Thomas Willis, *The Anatomy of the Brain*, in *Remaining Medical Works*, pp. 95–197, 131–36; Edwin C. Clarke, "The Doctrine of the Hollow Nerve in the Seventeenth and Eighteenth Centuries," in *Science, Medicine, and Culture*, ed. Lloyd G. Stevenson and Robert Multhauf (Baltimore: Johns Hopkins University Press, 1968), pp. 123–41.

4. Thomas Willis, *Of Musculary Motion*, in *Remaining Medical Works*, pp. 34–49 (translation of *De motu musculari* [1670]); Frank, *Harvey and the Oxford Physiologists*, pp. 222–23; E. Bastholm, *History of Muscle Physiology* (Copenhagen: Munksgaard, 1950), pp. 202–9. Bastholm argues that Willis viewed the nerves as solid cords through which animal spirits passed "like a wind" (pp. 202–3), but Willis clearly defined them rather as spongy and porous; the physiological effects remain the same in any case. See Meyer and Hierons, "Willis's Concepts," p. 2.

5. Thomas Willis, *An Essay on the Pathology of the Brain and Nervous Stock: In Which Convulsive Diseases are Treated of*, in *Remaining Medical Works*, separate title page (London, 1681), pp. 1–12; Jobe, "Theories of Melancholia," pp. 221–24; Meyer and Hierons, "Willis's Concepts," pp. 11–13; Ilza Veith, *Hysteria: The History of a Disease* (Chicago: University of Chicago Press, 1965), pp. 132–37; Stanley W. Jackson, *Melancholia and Depression: From Hippocratic Times to Modern Times* (New Haven: Yale University Press, 1986).

6. Willis, *Pathology of the Brain*, pp. 48–49.

7. Keith Thomas, *Religion and the Decline of Magic* (1971; rpt. ed., Harmondsworth: Penguin, 1973), pp. 524–25, 686n, 690; George Rosen, "Enthusiasm: 'A Dark Lanthorn of the Spirit,' " *Bulletin of the History of Medicine* 42 (1968): 394–96; see also R. S. Westfall, *Science and Religion in Seventeenth-Century England* (New Haven: Yale University Press, 1958).

8. Rosen, "Enthusiasm," p. 397; H. R. Trevor-Roper, "The European Witch-craze of the Sixteenth and Seventeenth Centuries," in *The European Witch-craze of the Sixteenth and Seventeenth Centuries and Other Essays* (New York: Harper and Row, Harper Torchbooks, 1969), pp. 167–68; Wallace Notestein, *A History of Witchcraft in England* (1911; rpt. ed., New York: T. Y. Crowell, 1968), pp. 322–30; Phyllis J. Guskin, "The Context of Witchcraft: The Case of Jane Wenham (1712)," *Eighteenth-Century Studies* 15 (1981): 48–71. See also Joseph Klaits, *Servants of Satan: The Age of the Witch Hunts* (Bloomington: Indiana University Press, 1985), pp. 159–76.

9. Cotton Mather, *Wonders of the Invisible World* (Boston, 1693; rpt. ed., London: John Dunton, 1693). See also his *Memorable Providences, Relating to Witchcrafts and Possessions* (Boston: Printed by R. P., sold by Joseph Brunning, 1689). On the "Surey Demoniack," see Notestein, *Witchcraft*, pp. 315–19 (who

notes the connection with the Mathers); Harley, "Mental Illness, Magical Medicine," pp. 131–40.

10. Michael Heyd, "The Reaction to Enthusiasm in the Seventeenth Century: Towards an Integrative Approach," *Journal of Modern History* 53 (1981): 258–80, at p. 259.

11. See Christopher Hill, *The World Turned Upside Down* (1971; rpt. ed., Harmondsworth: Penguin, 1975). The Puritan Richard Baxter commented in 1651, "Doth he think they knew it by Enthusiasm or Revelation from Heaven?" (OED, s.v. "enthusiasm").

12. Heyd, "Reaction to Enthusiasm," pp. 259–60; see also George Williamson, "The Restoration Revolt against Enthusiasm," *Studies in Philology* 30 (1933): 571–603.

13. Quoted in Williamson, "Restoration Revolt," p. 575. Dryden wrote his poem in 1682.

14. Margaret C. Jacob, *The Newtonians and the English Revolution, 1689–1720* (Ithaca: Cornell University Press, 1976); idem, *The Radical Enlightenment* (London: George Allen and Unwin, 1981); Michael MacDonald, "Insanity and the Realities of History in Early Modern England," *Psychological Medicine* 11 (1981): 11–25; idem, *Mystical Bedlam*, pp. 1–12, 217–31; Roy Porter, "The Rage of Party: A Glorious Revolution in English Psychiatry?" *Medical History* 27 (1983): 35–50, at pp. 43–48; this view is modified in idem, *Mindforg'd Manacles: A History of Madness in England from the Restoration to the Regency* (Cambridge: Harvard University Press, 1987).

15. See J. C. D. Clark, *English Society 1688–1832* (Cambridge: Cambridge University Press, 1985); Joanna Innes, "Jonathan Clark, Social History, and England's 'Ancien Regime,'" *Past and Present* 115 (1987): 165–200; Roy Porter, "Medicine and Religion in Eighteenth-Century England: A Case of Conflict?" *Ideas and Production* 7 (1987): 4–17; John Redwood, *Reason, Ridicule and Religion: The Age of Enlightenment in England, 1660–1750* (Cambridge: Harvard University Press, 1976); Gordon Rupp, *Religion in England 1688–1791* (Oxford: Clarendon Press, 1986).

16. Biographical references for Freind include *Biographia Britannica* (London: J. Meadows et al., 1747–63), 3:2024–44; *Dictionary of National Biography (DNB)*, s.v. "Freind, John"; *Dictionary of Scientific Biography (DSB)*, 5:156–57; Anita Guerrini, "The Tory Newtonians: Gregory, Pitcairne and Their Circle," *Journal of British Studies* 25 (1986): 288–311. For the influence of Newtonian matter theory on physiological theory, see Anita Guerrini, "Newtonian Matter Theory, Chemistry, and Medicine, 1690–1713," (Ph.D. diss., Indiana University, 1983).

17. John Freind, "Epistola de spasmi rarioris historia," *Philosophical Transactions* 22 (1701): 799–804, at p. 799. My translation.

18. Ibid., pp. 803–4.

19. Ibid., p. 804.

20. Biographical references for Mead include: *DNB*, x.v. "Mead, Richard"; *Biographia Britannica*, 5:3077–86; Arnold Zuckerman, "Dr. Richard Mead," (Ph.D. diss., University of Illinois, 1965). For Pitcairne, see Anita Guerrini, "Archibald Pitcairne and Newtonian Medicine," *Medical History* 31 (1987): 70–83.

21. Richard Mead, *A Mechanical Account of Poisons in Several Essays* (London: J. R. for R. South, 1702), Preface, n.p.; pp. 13–14.

22. Ibid., pp. 64–66, 93–96.

23. Richard Mead, *Of the Power and Influence of the Sun and Moon on Humane Bodies* (London: R. Wellington, 1712), pp. 7–9, 13–16, 22–24.

24. Ibid., pp. 30–36. On Newton's formulation of his theory, see B. J. T. Dobbs,

The Foundations of Newton's Alchemy (Cambridge: Cambridge University Press, 1975), ch. 6; R. S. Westfall, "The Influence of Alchemy on Newton," in M. Hamen et al., eds., *Science, Pseudo-science, and Society* (Ottawa: Wilfred Laurier University Press, 1980), pp. 145–69. On the definition of *occult* in this period, see Keith Hutchison, "What Happened to Occult Qualities in the Scientific Revolution?" *Isis* 73 (1982): 233–53; in what follows I assume his interpretation of *occult* as "insensible but not unintelligible." See also Ron Millen, "The Manifestation of Occult Qualities in the Scientific Revolution," in *Religion, Science, and Worldview,* ed. M. J. Osler and P. L. Farber (Cambridge: Cambridge University Press, 1985), pp. 185–216.

25. Isaac Newton, *De natura acidorum,* in *The Correspondence of Isaac Newton,* ed. H. W. Turnbull et al. (Cambridge: Cambridge University Press, 1959–77), 3:205–14.

26. John Freind, *Chymical Lectures* (London: Philip Gwillim for Jonah Bowyer, 1712; translation of *Praelectiones chymicae,* 1709), Preface, n.p.; John Keill, "Epistola ad Cl. virum Gulielmum Cockburn, Medicinae Doctorem. In qua leges attractionis aliaque physices principia traduntur," *Philosophical Transactions* 26 (1708): 97–110. On the role of analogy in Newton's thought, see J. E. McGuire, "Atoms and the 'Analogy of Nature': Newton's Third Rule of Philosophizing," *Studies in the History and Philosophy of Science* 1 (1970): 3–58.

27. I. B. Cohen and A. Koyré, eds., *Isaac Newton's Philosophiae naturalis principia mathematica,* 2 vols. (Cambridge: Cambridge University Press, 1972), 2:764–65; R. S. Westfall, *Never at Rest* (Cambridge: Cambridge University Press, 1980), pp. 792–94; Newton's earlier concept of the aether is detailed in his "Hypothesis of Light" (1675), in Newton, *Correspondence,* 1:362–92. See also Marie Boas Hall and A. R. Hall, "Newton's Electric Spirit: Four Oddities," *Isis* 50 (1959): 473–76; Alexandre Koyré and I. B. Cohen, "Newton's 'Electric and Elastic Spirit,'" *Isis* 51 (1960): 337; Joan L. Hawes, "Newton and the 'Electrical Attraction Unexcited,'" *Annals of Science* 24 (1968): 121–30.

28. Isaac Newton, *Opticks* (rpt. of fourth ed., 1730, New York: Dover, 1952), Queries 17–24, pp. 347–54; Westfall, *Never at Rest,* pp. 792–94; Joan L. Hawes, "Newton's Revival of the Aether Hypothesis and the Explanation of Gravitational Attraction," *Notes and Records of the Royal Society* 23 (1968): 200–212.

29. Westfall, *Never at Rest,* p. 794.

30. P. M. Heimann and J. E. McGuire, "Newtonian Forces and Lockean Powers: Concepts of Matter in Eighteenth-Century Thought," *Historical Studies in the Physical Sciences* 3 (1971): 233–306, at pp. 242–43; Newton, *Opticks,* Query 21, p. 352.

31. See Arnold Thackray, "'Matter in a Nut-shell': Newton's *Opticks* and Eighteenth-Century Chemistry," *Ambix* 15 (1968): 29–53.

32. Westfall, *Never at Rest,* pp. 508–9, 645; Dobbs, *Foundations of Newton's Alchemy,* pp. 161–63. See also Guerrini, "Newtonian Matter Theory," ch. 6. On Newton's ontology, see J. E. McGuire, "Force, Active Principles, and Newton's Invisible Realm," *Ambix* 15 (1968): 154–208.

33. B. Mandeville, *A Treatise of the Hypochondriack and Hysterick Diseases,* 2d ed. (London, 1730), p. 182. On the medical marketplace, see Harold J. Cook, *The Decline of the Old Medical Regime in Stuart London* (Ithaca: Cornell University Press, 1986), ch. 1; Roy Porter, *Health for Sale* (Manchester: Manchester University Press, 1989), ch. 1. On Mead and Radcliffe, see William MacMichael, *The Gold-Headed Cane* (London: John Murray, 1827).

34. Thomas Hearne, *Remarks and Collections,* ed. C. E. Doble et al., vol. 7 (Oxford: Clarendon Press, 1906), pp. 271, 304.

35. William Cowper, *Myotomia Reformata . . . , to which is prefix'd An Introduction concerning Muscular Motion* (London: Robert Knaplock et al., 1724), p. lxxii. For Pemberton, see *DNB*, 15:725–26; *DSB*, 10:500–501; his essay is discussed in T. M. Brown, "Medicine in the Shadow of the *Principia*," *Journal of the History of Ideas*, 48 (1987): 629–48, at pp. 643–44.

36. Little is known of Robinson's origins. The biographical accounts in *DNB*, s.v. "Robinson, Nicholas"; and William Munk, ed., *Roll of the Royal College of Physicians of London* (London, 1878): 2:108, are sketchy. Later in life, Robinson was active in evangelical circles.

37. Nicholas Robinson, *A New Theory of Physick and Diseases, founded upon the Newtonian Philosophy* (London: for C. Rivington, J. Lacy, and J. Clarke, 1725), p. ix.

38. Ibid., p. 10.

39. *DNB*. Robinson moved from Wood Street in the City of London to Warwick Court off Holborn, adjacent to Gray's Inn.

40. Nicholas Robinson, *A New System of the Spleen, Vapours and Hypochondriack Melancholy* (London: for A. Bettesworth, W. Innys, and C. Rivington, 1729), pp. 2–4, 13–17.

41. Porter, *Mind-forg'd Manacles,* p. 193; Robinson, *Spleen,* pp. 18–23, 62.

42. Robinson, *Spleen,* pp. 22–23, 66–67.

43. Ibid., pp. 27–30.

44. Ibid., pp. 78–79; Newton, "Hypothesis of Light," p. 369.

45. Robinson, *Spleen,* pp. 35–42.

46. George Cheyne, *The English Malady: or, A Treatise of Nervous Diseases of All Kinds* (London: G. Strahan and T. Leake, 1733), pp. 4, 14.

47. Ibid., p. 325f.; see also Henry Viets, "George Cheyne (1673–1743)," *Bulletin of the History of Medicine* 23 (1949): 435–52; Lester S. King, "George Cheyne, Mirror of Eighteenth Century Medicine," *Bulletin of the History of Medicine* 48 (1974): 517–39; Geoffrey Bowles, "Physical, Human and Divine Attraction in the Life and Thought of George Cheyne," *Annals of Science* 31 (1974): 473–88; Anita Guerrini, "Isaac Newton, George Cheyne, and the *Principia Medicinae*," in *The Medical Revolution of the Seventeenth Century,* ed. Andrew War and Roger French (Cambridge: Cambridge University Press, 1989), pp. 222–45.

48. Cheyne, *English Malady,* p. 339. On Cheyne's medical practice, see C. F. Mullett, ed., *The Letters of Dr. George Cheyne to the Countess of Huntingdon* (San Marino: Huntington Library, 1940); idem, ed., *The Letters of Doctor George Cheyne to Samuel Richardson (1733–1743),* University of Missouri Studies, vol. 18, no. 1 (Columbia: University of Missouri Press, 1943); Anita Guerrini, "Medical Practice and the Birth of the Consumer Society," (Paper delivered at a Clark Library seminar, 3 February 1989).

49. Cheyne's patients included the Walpole family, the earls of Huntingdon, and several families of the Scottish aristocracy.

50. Cheyne, *English Malady,* p. 69; Newton, *Opticks,* p. 404.

51. Cheyne, *English Malady,* p. 75.

52. Ibid., pp. 85–89. See also Hélène Metzger, *Attraction universelle et religion naturelle chez quelques commentateurs anglais de Newton* (Paris: Hermann, 1938), part 3; Westfall, *Science and Religion,* ch. 8.

53. Cheyne, *English Malady,* p. 328. David Gregory reported in his diary that Cheyne intended to spend the summer of 1705 in Scotland: W. G. Hiscock, ed., *David Gregory, Isaac Newton and Their Circle* (Oxford, 1937), p. 23. My interpretation of these events differs from that of Bowles, "Physical, Human and Divine Attraction," and especially from that of G. S. Rousseau, "Mysticism and Mille-

narianism: 'Immortal Dr. Cheyne,'" in *Millenarianism and Messianism in English Literature and Thought, 1650–1800* (Leiden: Brill, 1988), pp. 81–126.

54. Cheyne, *English Malady*, pp. 330–31; idem, *The Philosophical Principles of Natural Religion* (London: George Strahan, 1705), esp. ch. 3, pp. 73–281. Gregory commented that Cheyne "has stoln a great deal of his book of Religion" from Bentley's Boyle lectures: Hiscock, *Gregory*, p. 25. On the relationship between deism and latitudinarianism, see Roger L. Emerson, "Latitudinarianism and the English Deists," in *Deism, Masonry, and the Enlightenment*, ed. J. A. Leo Lemay (Newark: University of Delaware Press, 1987), pp. 19–48.

55. For Garden, see *DNB*, s.v. "Garden, George"; G. D. Henderson, *Mystics of the North-east* (Aberdeen: Third Spalding Club, 1934), pp. 32–39. Cf. Rousseau, "Mysticism and Millenarianism," pp. 92–95.

56. Cheyne, *English Malady*, p. 331; Henderson, *Mystics of the North-east*, pp. 32–36. On Bourignon, see Salomon Reinach, *Cultes, Mythes et Religions*, vol. 1 (3d ed., Paris: Leroux, 1922), pp. 426–58; Marthe van der Does, *Antoinette Bourignon* (Amsterdam: Holland University Press, 1974). The distinctly feminine brand of Bourignon's mysticism fit in well with Cheyne's largely female clientele and his therapeutic emphases. See Anita Guerrini, "Women, Animals, and the Mechanical Philosophy," paper delivered at the History of Science Society meeting, October 1990.

57. Cheyne, *English Malady*, p. 330.

58. Rosen, "Enthusiasm," p. 412; MacDonald, "Insanity and the Realities of History"; Porter, *Mind-forg'd Manacles*, pp. 76–81. Porter's comment (p. 80) that Cheyne "hardly touch[es] on religious melancholy" in *The English Malady* is certainly incorrect.

59. Cheyne, *An Essay of Health and Long Life* (London: George Strahan, 1724), p. 157. See also Porter, *Mind-forg'd Manacles*, pp. 62–81.

60. Cheyne, *English Malady*, p. 331.

61. Ibid., pp. 332–34.

62. On early Methodism, see V. H. H. Green, *The Young Mr. Wesley: A Study of John Wesley and Oxford* (London: Edward Arnold, 1961).

63. Ibid., pp. 148, 164–67; John Telford, ed., *The Letters of John Wesley* (London: Epworth Press, 1931), 1:11 (Wesley to Susanna Wesley, 1 November 1724).

64. John Wesley, *Primitive Physick* (1747), rpt. as *Primitive Remedies* (Santa Barbara: Woodbridge, 1975), pp. 19–22. See G. S. Rousseau, "John Wesley's *Primitve Physick* (1747)," *Harvard Library Bulletin* 16 (1968): 242–56; Guerrini, "Medical Practice," p. 19.

65. See E. P. Thompson, *The Making of the English Working Class* (1963; rpt. ed., Harmondsworth: Penguin, 1968), pp. 402–3.

66. Quoted by Green, *Young Mr. Wesley*, p. 170.

67. Mullet, *Letters of Cheyne to Richardson*, pp. 11–112 and passim; see also Rousseau, "Mysticism and Millenarianism," p. 99 n. 59; on Law, see *DNB*, s.v. "Law, William"; Rupp, *Religion in England*, pp. 218–42.

68. *DNB*, s.v. "Hastings, Selina"; *The Life and Times of Selina, Countess of Huntingdon* (London, 1848), pp. 31–47; Cheyne, *Letters to the Countess of Huntingdon*, pp. v–vi; Green, *Young Mr. Welsey*, p. 121. Several letters of Lady Betty Hastings to her brother and sister-in-law, which mention Cheyne, are among the Hastings MSS, Huntington Library.

69. Quoted in Cheyne, *Letters to the Countess of Huntingdon*, p. vii.

70. *Letters of Wesley*, 2:285 (Wesley to Edmund Gibson, Bishop of London, 11 June 1747).

71. See Rosen, "Enthusiasm," pp. 400–401; Bernard Semmel, *The Methodist Revolution* (New York: Basic Books, 1973), pp. 5, 35.

72. Quoted by Semmel, *Methodist Revolution,* p. 15. On Methodism and madness, see MacDonald, "Insanity and the Realities of History"; idem, "Religion, Social Change, and Psychological Healing in England, 1600–1800"; Porter, *Mind-forg'd Manacles,* pp. 67–78.

73. George Cheyne, *An Essay on Regimen* (London: C. Rivington, 1740), dedicated to the Earl of Huntingdon; idem, *The Natural Method of Cureing the Diseases of the Body and the Disorders of the Mind Depending on the Body,* 2d ed. (London: G. Strahan and John and Paul Knapton, 1742), dedicated to the Earl of Chesterfield. Wesley mentioned the latter work in his 1747 letter to the Bishop of London (see n. 70 above).

74. For contrary views, see Porter, "A Rage of Party?" pp. 43–48; Jacob, *Radical Enlightenment,* pp. 96–97.

Newton and Lavoisier—From Chemistry as a Branch of Natural Philosophy to Chemistry as a Positive Science

ARTHUR DONOVAN

What is the historical relationship between the Scientific Revolution of the seventeenth century and the Chemical Revolution of the eighteenth century? Did the latter of these events build directly on the success of the former and essentially fulfill its program and promise? If one assumes that scientific knowledge grows progressively through time, it seems reasonable to suppose that these notable achievements represent successive applications of the scientific method of the study of natural phenomena, the latter building on the former while adding to it. But one should be cautious in interpreting these two revolutions as the working out of a common program and process. As the eminent historian Herbert Butterfield pointed out years ago, it is the historian's special responsibility to emphasize how events from different eras, or for that matter events in the past and the present, are *not* like one another.[1] And indeed there is considerable evidence of a radical disjunction between the Newtonian legacy, as it was understood in the eighteenth century, and the research program Lavoisier followed when transforming the science of chemistry. Thus there are good grounds for supposing that the relationship between the Newtonian and Lavoisian revolutions was not one of shared aims and methods, but rather one of critical reaction and new departure.

This indeed is the conclusion reached by Arnold Thackray in his book *Atoms and Powers—An Essay on Newtonian Matter-Theory and the Development of Chemistry*.[2] Thackray was primarily interested in British chemistry and therefore treated related developments in France as secondary issues, yet his conclusions are carefully documented and in the main correct. He describes two themes that were central to the Newtonian program for chemistry, one being the development of a concept of matter capable of providing an adequate theory of chemical elements, the other being the development of a quantified theory of the attractive force between chemical molecules. But as Thackray points out, neither of these themes played a significant part in Lavoisier's transformation of chemistry. With regard to theories of matter, Lavoisier simply "refused to make a frontal assault on those Newtonian categories that underlay the physicalist interpretation of chemistry," even though his new theories clearly constituted a challenge to "the whole Newtonian tradition."[3] A similar discontinuity occurred in the study of elective attractions. Thackray notes that "the late 1770's and the 1780's were the heroic days of Newtonian chemistry" and that "the aim of a quantified science of affinities was most keenly pursued in France," yet by the end of the 1780s it was clear that the experimental program that informed this effort had failed. Here too, Thackray reports, the conceptual problems and experimental investigations from which Lavoisier's new theories emerged owed little to Newtonianism: "Whatever his own momentary hopes in the early 1780's, Lavoisier did not proceed with the quantification of chemistry."[4]

If Lavoisier was not, in the filiation of scientific ideas, one of Newton's direct descendants, how should we characterize the research program he pursued? This is the question addressed in the subtitle of this essay. My claim, stated most generally, is that Newton's achievement should be seen as the culmination of the tradition of natural philosophy as it had developed up to the end of the seventeenth century, whereas Lavoisier's achievement stands at the beginning of the tradition of the positive sciences that became increasingly prominent in the latter half of the eighteenth century. To render this claim plausible, which is all that can be hoped for in a single essay, I will focus on three themes in Newton's science and three comparable themes in Lavoisier's science, my purpose being to highlight contrasts between important elements of their respective research programs. I realize, of course, that I provide no justification for treating the research programs pursued by these two men as representative of the science of their ages, but given their historical stature, I consider this a problem of secondary importance. My more significant claim is that by contrasting the research programs of Newton and Lavoisier, we can begin to understand how natural philosophy, the focal concern of the Scientific Revolution, was transformed into a set of autonomous positive sciences, a conception of science that remains central to our modern understanding of nature.[5]

Three Themes in Eighteenth-Century Newtonianism

THE PRIMACY OF MATTER

Debates over the theory of matter were of central importance in eighteenth-century Newtonianism. As Robert Schofield has pointed out, Newton's successors considered him an authority on many topics in natural philosophy, among them being "the nature of space and time, the uses of hypotheses, of mathematics, and of experiments, and even the role of God in the universe." Yet there is another, more fundamental theme in Newtonianism, "one that frequently comprehends the others and one which can reasonably be regarded as central both to Newton's work and to that of natural philosophy in general. That is, Newton's theory of matter and its action, which pervades both the *Principia* and the *Opticks* and, consciously or unconsciously, most of the scientific writing of the eighteenth century as well."[6]

Schofield gives two reasons for focusing on Newton's theory of matter. The first is that it links his account of Newtonian natural philosophy to a more comprehensive conception of the development of Western science: "The importance of establishing a theory of matter and its action has been evident to all natural philosophers since the days of the Greeks. . . . Indeed, one of the chief successes of the Scientific Revolution [of the seventeenth century] was that it found a new matter theory to replace the old one."[7] His other reason for focusing on Newton's theory of matter is that it enables one to explain many of the tensions within the Newtonian tradition as arising from a continuing debate between two rival conceptions of matter. Schofield calls the first of these the theory of dynamic corpuscularity, which explains natural phenomena in terms of the movement of atoms acted upon by forces of inertia, attraction, and repulsion. The other is the theory of the material aether, "a medium which expands through all space, filling it and pervading the pores of gross bodies by virtue of its great elasticity and because of its great subtlety."[8] By moving back and forth between these two theories of matter, Newton created a legacy of "conflicting alternatives for eighteenth-century natural philosophers looking for an authoritative, Newtonian, theory of matter." Schofield calls these theories the mechanistic and the materialist, and he organizes his account of eighteenth-century Newtonianism around their linked histories.

Schofield's account of eighteenth-century Newtonianism has been severely criticized from a variety of points of view, and it cannot be denied that he attempts to encompass more aspects of Newtonianism than can be accounted for in terms of the debate among rival theories of matter alone.[9] Yet if one concentrates on those areas of investigation that evolved into what came to be called the physical sciences, Schofield's emphasis on matter theory does not seem misplaced. And what is interesting for our purposes is that when Schofield looked at the Chemical Revolution from this vantage point, he saw it as part of a revolution against rather than within Newtonianism.

"Truly to assess Lavoisier's accomplishment against the background of eighteenth-century chemistry," Schofield writes, "one must begin a study of that chemistry at its beginning—with the failure of Robert Boyle's corpuscular philosophy."[10] This failure, evident by the end of the seventeenth century, arose from the absence of an independent means for determining the properties of the particles of matter that are fundamental to the corpuscular philosophy. The corpuscular strategy for solving the problem of transdiction was to posit a one-to-one relationship between observable chemical properties, such as acidity and alkalinity, and the shapes and motions of hypothetical material particles in the microscopic world, but the undeniably ad hoc character of such reasoning eventually undermined the research program of corpuscular chemistry. The program was revived, however, with the advent of Newtonianism. In this later form, the shapes and motions of the particles, which had been considered fundamental, were explained in terms of attractive and repulsive forces. But by the middle of the eighteenth century, Schofield notes, "dynamic corpuscularity appeared to be foundering on the same rock that sank kinematic corpuscularity—the inability to devise an independent measure of critical parameters."[11]

In those areas of natural philosophy that were eventually transformed into the modern science of physics, this collapse of dynamic corpuscularity led to the revival of aether theories. Schofield argues that within chemistry, however, the response involved a turn toward the antimechanistic and antireductionist approach of the Stahlian chemists.[12] Thus while Schofield still credits Lavoisier with leading a revolution in chemistry, it was a revolution against the Newtonian program rather than a fulfillment of it.

Far from freeing chemistry from Stahl, he [Lavoisier] ordered and rationalized Stahlian chemistry, and in doing so, changed the emphasis of future chemists' activities from a futile attempt at an overly sophisticated physical reductionism to the jig-saw puzzle problems of permutations and combinations of elements. He did not do for chemistry what Newton had done for mechanics, but what Linnaeus had done for botany.[13]

Schofield's conclusions concerning the radical divergence between Newtonian natural philosophy and Lavoisian chemistry, and on the role played by the Stahlian approach to chemistry, are in broad agreement with several recent studies on the origins of Lavoisier's new chemical theories.[14] It thus appears that the Newtonian tradition, with its focus on theories of matter, and the chemical tradition, at least those parts of it that by the middle of the eighteenth century had become militantly antimechanistic and antireductionist, evolved quite independently. Although reunited to a certain extent in the nineteenth century by Dalton's theory of chemical atomism, these two traditions were effectively uncoupled when Lavoisier was pursuing his program for the reform of chemistry.

THE LAWS OF NATURE

Discovering fundamental and comprehensive laws of nature was the primary goal of the Newtonians. Precisely how Newton and his heirs intended to explain the system of the world is still a matter of historical debate, and the differences among the various Newtonian factions deserve as much attention as do their areas of agreement. The point I wish to emphasize here is that Newton and his disciples were natural philosophers in the sense that they set out to construct systems of theories that were at least potentially capable of rendering intelligible nature as a whole.[15] In this they differed from most, but certainly not all, of the leading scientists of the nineteenth century, most of whom worked within discrete disciplines whose boundaries they by-and-large accepted. And in this regard the eighteenth-century Newtonians were also unlike Lavoisier who, like his nineteenth-century successors, set more modest goals for himself and never seriously addressed the general problem of formulating a comprehensive philosophy of nature.

The aims of the eighteenth-century Newtonians have been obscured by two subsequent philosophical interpretations of Newton's own achievement. The first of these proceeds from a tendentious reading of Newton's statements about science. The goal of this interpretation was to rescue Newton from what were thought to be his speculative excesses, not to mention those of his disciples; the method employed was to portray Newton as being solely interested in discovering descriptive laws of nature. The second obscuring interpretation evolved within a philosophical tradition that rigorously criticized Newtonian epistemological claims. The overall goal of this interpretation was to demonstrate that it is impossible, given the way the human senses and the human mind function, to ever know anything certain about nature itself. This skeptical response to Newtonianism was eventually brought into uneasy conjunction with the descriptive reading of his achievement and reinforced what became the predominant nineteenth-century interpretation of his program for scientific research. But, as recent scholarly studies of Newton and his disciples have emphasized, eighteenth-century natural philosophy was fundamentally different from the positive sciences of the nineteenth century.

Despite Newton's famous disclaimer that he feigned no hypotheses, modern students of his work have demonstrated that his aims were, as one would expect, those of a seventeenth-century natural philosopher. His underlying aims and guiding assumptions have been reconstructed through careful studies of his investigative practice and the ways he presented his findings. And, as Cohen pointed out many years ago, "there is a great contrast between what Newton said about method and what Newton did in science."[16] This recognition of divergence between precept and practice has led some to argue that Newton's investigations should be seen as consisting of two sorts. His mathematical science, most notably the *Principia*, was inspired by a desire to discover laws whose veracity is demonstrated mathematically, while eschewing the search for metaphysical entities or causes. Newton's experimental science, especially the *Opticks*, exemplifies positive science in the sense that it does not advance theories whose truth cannot be demonstrated by experiment. But this dichotomization, which ignores the deeper integrity of Newton's research program, has not survived careful scrutiny. As Robert Schofield has pointed out, "the *Opticks* is not an empirical work. In fundamental theory it does not differ from the *Principia*, and, moreover, it was not the experiments of the *Opticks* which inspired their study, but the queries, what Priestley was subsequently to call Newton's 'bold and eccentric thoughts.' "[17]

In emphasizing the underlying unity of Newton's investigations I do not mean to imply that he did not go far in establishing the truth of the system of the world he proposed. But, great as the Newtonian achievement is, even when measured by the criteria of nineteenth-century positive science, the legacy he left the eighteenth century must be understood in the context of natural philosophy. As Ephraim Chambers wrote in his 1728 *Cyclopedia*, the Newtonian philosophy is, among other things, a "Doctrine of the Universe," and Chambers listed Newtonianism, along with Cartesianism and Peripateticism, as one of the three sects in modern philosophy.[18] Thus, as Cohen has emphasized, Newton's allegedly positive account of science should be read as an admission of failure rather than as a statement of methodological policy.

> Newton's insistence that it is enough to be able to predict the celestial and terrestial motions and the tides of the sea was, in fact, less a battle-cry of the new science than a confession of failure. For what Newton was saying in essence is that his system should be accepted in spite of his failure to discern the cause or even to understand universal gravity, because its results accord so well with the data of observation and experiment.[19]

Newton's achievement fell short of the goals he considered central to natural philosophy. But by the middle decades of the eighteenth century the empiricist and idealist critique of seventeenth-century metaphysics had led to a radical revaluation of these very goals.[20] And, in one of those paradoxes that makes history so fascinating, this critical attack on Newton's ambitious philosophical program drew much of its power from the anomalous status of the Newtonian achievement within philosophy itself. It was David Hume who most famously turned Newton's appeal to experiment against the fundamental rationalism of Newton's own conception of natural philosophy. Hume announced his programmatic commitment to the primacy of experiment in 1739–40 when to his *Treatise on Human Nature* he added the subtitle "An Attempt to introduce the experimental Method of Reasoning into Moral Subjects." Hume also said he hoped to do for moral philosophy what Newton had done for natural philosophy.[21] And Hume elsewhere characterized Newton as an experimental philosopher when, in his *History of England*, he described the philosophical consequences of the Newtonian achievement:

> In Newton this island may boast of having produced the greatest and rarest genius that ever rose for the ornament and instruction of the species. [He was] cautious in admitting no principle but such as were founded on experiments. . . . While Newton seemed to draw off the veil

from some of the mysteries of nature, he shewed at the same time the imperfections of the mechanical philosophy; and thereby restored her ultimate secrets to that obscurity in which they ever did and ever will remain.[22]

Hume showed how the Newtonian achievement could be interpreted so as to render implausible if not utterly insupportable Newton's own philosophical aspirations, and this highly restrictive reading of the Newtonian heritage subsequently had a profound impact on other eighteenth-century Scottish philosophers. Colin Maclaurin, the foremost British Newtonian and mathematician of his generation and professor of mathematics in the University of Edinburgh from 1726 to his death in 1746, wrote as follows in his *Account of Sir Isaac Newton's Philosophical Discoveries,* a work he completed shortly before his death:

> It is not the business of philosophy, in our present situation in the universe, to attempt to take in at once, in one view, the whole scheme of nature, but to extend, with great care and circumspection, our knowledge, by just steps, from sensible things, as far as our observations or reasonings from them will carry us, in our enquires concerning either the greater motions and operations of nature, or her more subtile and hidden works. In this way Sir Isaac Newton proceeded in his discoveries.[23]

The next generation of Scottish professors, and especially Thomas Reid, the leader of the Scottish school of Common Sense philosophy, continued to grapple with the Humean interpretation of Newtonianism. According to Larry Laudan, "Reid was the first major British philosopher to take Newton's opinions on induction, causality, and hypotheses seriously."[24] Thus the ground was well laid in the eighteenth century for the methodological interpretation of Newtonianism, an interpretation that burgeoned in the nineteenth century, as science and its philosophical foundations came under increasing scrutiny. But within the broad stream of eighteenth-century Newtonianism this restrictive and skeptical reading of Newton's achievement was but one rivulet, although an especially clear one, in a roiling flood of competing interpretations. Indeed, most of Newton's many eighteenth-century disciples understood his achievement within the continuing and philosophically more ambitious tradition of natural philosophy, a tradition that derived much of its significance from its intimate involvement with the related tradition of natural theology.

Lavoisier appears to have been untouched by this largely British debate over the philosophical meaning of the Newtonian achievement. Evidently he acquired what I have been calling a positive view of science without either passing through a Pascalian dark night of the soul or struggling with a more comprehensive philosophical program for science.[25] So far as I know, there is no evidence of any theological concern on Lavoisier's part, and when he encountered problems in epistemology and ontology, his responses were largely opportunistic. This is not to say that Lavoisier did not see his study of chemistry as part of a larger personal and political undertaking. It appears, however, that the context of meaning in which the larger purposes of his investigations were defined had little to do with seventeenth-century attempts to construct philosophical systems of the world. Thus, in terms of his scientific aims, as well as in his selection of problems and concepts, Lavoisier seems to have functioned entirely outside the eighteenth-century Newtonian tradition.

NATURAL PHILOSOPHY AND POLITICAL ORDER

The transformation of natural philosophy into positive science can also be followed in debates about the social and political implications of natural knowledge. There is considerable dispute as to whether Newton's theological and political views significantly shaped the content of his science. Be that as it may, there is considerable evidence that in Great Britain primarily conservative political lessons were drawn from Newton's system of natural philosophy. As Margaret Jacob has written,

> From its inception during the 1690s Newtonianism was tied to a social ideology, which can best be understood in its historical context, that is, by reconstruction of the social and political reality within which the Newtonians perceived the meaning and purpose of the new science. At its earliest popularization and explication, Newton's science was enlisted, with the consent of the master himself, in the attempt to justify and to explain in Christian terms the post–1688–89 order.[26]

That Newton's natural philosophy played a prominent part in political and social discourse will not come as a surprise to students of early modern European culture. The linkages among natural philosophy, natural theology, and natural law were very close in the seventeenth and early eighteenth centuries and one cannot fully understand the impact of the Newtonian achievement without exploring its implications in these related branches of philosophy.

Furthermore, as several scholars have recently pointed out, the alliance between Newtonian natural philosophy and conservative political doctrines led to the articulation of a wide range of explicitly anti-Newtonian approaches to the understanding of nature.[27]

Among the many fascinating turns in this complex story, and one that is especially germane to the concerns of this essay, was the convergence of chemistry and anti-Newtonianism around the middle of the eighteenth century. These two facets of natural philosophy were brought into close conjunction by a widely held conviction that the material substratum of nature is active rather than passive, as Newton had believed. The supposition that matter itself is active fired the imagination of the Scottish medical student James Hutton while he was studying in Paris and Holland, and it thereafter informed his sustained attempt to explain a variety of natural systems without violating the skeptical epistemological beliefs of his philosophical cronies in Edinburgh.

Writing as a philosopher of mind, Hutton composed a tedious theory of knowledge that in some ways anticipated Kant's analysis of the categories of thought. Writing as a natural philosopher, he propounded cyclical theories of the earth, of rain, and of agriculture, theories he attempted to make intelligible by appealing to the laws of latent heat discovered by his great friend Joseph Black. Although not unmindful of the need to account for natural phenomena as they actually appear in the world, Hutton was inspired by a speculative urge that led him to scorn the experimental verification that Lavoisier and his foremost Scottish disciple James Hall considered fundamental to scientific chemistry. Hutton is a fascinating historical figure, not so much because he was an anti-Newtonian who believed that matter is active and natural systems are self-sustaining, but because he struggled to extract theories from these beliefs that would appear persuasive to colleagues who did not share his faith in man's ability to gain certain knowledge about nature. In the larger context of Newtonianism, Hutton's aims seem less anomalous than they do in the more restricted context of Scottish philosophy. Like Newton, Hutton was a natural philosopher, but unlike Newton the providentialist, Hutton was a radical naturalist. He believed that by beginning with the active matter of the chemists rather than the passive matter of the Newtonians, one could penetrate to the inmost secrets of nature and acquire true understanding of the many systems by which the habitable world is maintained.[28]

The career of Newtonianism in France in some ways replicated the British experience, but the differences are important also.[29]

Whereas in Great Britain Newtonianism was enlisted in the defense of the theological and political orthodoxy established in 1688–89, in France this reading of Newtonianism appeared politically subversive in the first half of the eighteenth century. Voltaire clearly appreciated this. In the famous letters he wrote while a political exile in England, he transformed the Newtonian achievement and the English celebration of it into a comprehensive indictment of all that was outmoded and insupportable in France.[30] In doing so he was simultaneously addressing issues in natural philosophy, natural theology, and natural law, and the growing acceptance of Newtonian doctrines within natural philosophy made his rhetorical invocation of the Newtonian achievement especially effective.

But the more extreme ideas spawned in the English Revolution were also finding advocates in France, and by the middle of the century many of the more radical proponents of Enlightenment were beginning to move beyond the political limitations of orthodox Newtonianism. Thus during the 1750s and 1760s a more radical account of matter and nature began to surface in France as well as in Britain, an account that in its social and theological implications was clearly anti-Newtonian. These more extreme naturalists have been characterized as follows:

> When it suited their purposes the radical advocates of Enlightenment hailed science, whether that of Descartes or Newton, as the antidote to Christianity and supernatural knowledge. They did so, however, in complete defiance of the new scientific tradition that began with Boyle, Wilkins, and Barrow and culminated in the Newtonians. With the clear exception of 'sGravesande, the circle represented by Toland in England and by Marchand, Picart, Sallengre, and Saint-Hyacinthe on the Continent rejected any model of the universe that depended for its operation on the separation of matter and motion and therefore upon the active participation of God in natural and human affairs.[31]

In France, as in Great Britain, anti-Newtonian naturalism drew heavily on theories of active matter, especially as formulated by certain chemists and biologists.[32] As Lester Crocker has observed, "the scientific method practiced in eighteenth-century France was, to a considerable extent, non-Newtonian or even anti-Newtonian," and Jean-Claude Guédon has examined the anti-Newtonian strategy that guided Diderot's deployment of chemical doctrines in his materialistic philosophy.[33] A striking critique of the tyranny of physics can also be found in the long article on chemistry that the prominent Montpellier physician G. F. Venel wrote for the *Encyclopédie*.[34] Venel's quarrel was not with Newtonianism as such, but rather with the larger enterprise of physics of which it was a part. Building on the Stahlian reaction to corpuscular theories of matter, Venel insisted that chemists alone "can penetrate to the interior of certain bodies, while physics can only describe their surface and external figure." He called for a "revolution that will bring chemistry to the level that it deserves, which is at least equal to that of mathematical physics," and he expected this revolution to be led by "a new Paracelsus."

While Venel wrote, as he said, "solely to stimulate interest in chemistry,"[35] his trenchant indictment of the hegemony of physics within natural philosophy was also read as a warrant for a more general heterodoxy, one that a certain number of critics of the old regime who were motivated by a potent combination of personal resentment and democratic fervor found attractive.[36] Thus in the 1760s the relative advantages of pursuing alternative programs for the study of natural phenomena, especially those championed by chemists and biologists, were being hotly debated in both Great Britain and France.

With Lavoisier it appears that the connection between politics and natural philosophy has been severed and that we suddenly enter the world of modern science. It seems that during the final decades of the old regime at least some French investigators, Lavoisier being the most prominent among them, simply turned their backs on the multiple entanglements of natural philosophy, natural theology, and natural law and concentrated instead on the dialectic of experiment and theory.[37] The murky and multilayered debate over Newtonianism suddenly seems to yield to a post-Newtonian conception of science, one in which the aims of the investigators and the implications of the knowledge sought were vastly reduced and carefully delimited. This shift to more narrowly defined disciplinary investigations of nature occurred with striking abruptness in chemistry, at least in France, and it was central to many of the differences that separated French and British science until at least the 1830s. Explaining how and why this shift took place when and where it did is, I suggest, a problem of central importance in the history of modern science.

Three Themes in the Chemical Revolution

THE METHOD OF EXPERIMENT

While both Newton and Lavoisier made use of experimental evidence, there are significant differences in the roles such evidence played in the ways they practiced science. Experimental investigation was central in all stages of Lavoisier's science. Joseph Priestley excelled in the experimental discovery of new phenomena and Newton had made brilliant use of a crucial experiment when choosing between two competing theories of color. Lavoisier's use of experiment was distinguished by his insistence that the interaction of reason and experiment must be close and constant. Lavoisier was especially skillful in designing, performing, and interpreting analytic and synthetic experiments, and he accepted that, at least in principle, every inference and prediction of the oxygen theory of combustion ought to be confirmed by experimental tests. Indeed, it would not be an overstatement to say that in Lavoisier's science experiment played the same role that mathematics played in Newton's science, for it provided the surest evidence that his central theoretical claims were true.[38]

How did Lavoisier come to be so convinced of the importance of experiment? Chemists, of course, could point with pride to a long tradition of experimentation, but Lavoisier's insistence on precision, testing at every step, and verification made experiments far more important in theoretical chemistry than had previously been the case. It therefore appears there were other factors that helped shape Lavoisier's views on experimentation. Lavoisier scholars have long noted that he considered himself as much an experimental physicist as a chemist. In the 1760s Lavoisier attended the Abbé Jean-Antoine Nollet's public lectures on experimental physics, and, in 1772, when formulating the research plan that led to his new chemical theories, Lavoisier set himself the task of applying to the study of air the investigative methods of experimental physics. These were the methods that the Abbé Nollet had employed with such success and to such acclaim in Paris during the middle decades of the century.[39]

Since eighteenth-century experimental physics is widely considered part of the Newtonian tradition, Lavoisier's adoption of its methods may appear to have been thoroughly Newtonian in origin and character. But linking these two traditions in this way obscures many of the most interesting areas of divergence in eighteenth-century natural philosophy. The modern experimental tradition in physical science goes back at least to the middle of the seventeenth century, and, so far as the appeal to experiment is concerned, Newton was himself working within this flourishing tradition. And, while many of the best-known experimental physics of the eighteenth century considered themselves Newtonians, their very devotion to experiment reveals that by-and-large they were adding a veneer of Newtonian interpretation to their descriptions of experiments rather than seriously examining the fundamental assumptions of Newton's natural philosophy.[40] The method of experiment was considered important because it provided an empirical warrant for belief that was convincing to those who were ignorant of or unpersuaded by mathematical demonstration. In practice most eighteenth-century experimentalists were lecturers, demonstrators, and popularizers, and appeals to experimental evidence served their purposes well. It was a method Lavoisier also found both convincing and useful when he set out to transform chemistry into a proper science.

LANGUAGE AND SCIENCE

If the method of experimental physics was the means Lavoisier chose to transform chemistry, his goal was to make chemistry a theoretical science. Earlier chemists, such as Lemery, Boerhaave, and Macquer, had been well aware of the importance of theoretical reasoning in the investigation of nature. But when one examines the ways in which theoretical problems are treated in their published treatises, it becomes clear that their interest in theory was largely heuristic and episodic. This is hardly surprising, for between the collapse of the Boylean program of corpuscular chemistry and the triumph of the Chemical Revolution, the primary reason for studying chemistry had been to acquire an extensive knowledge of chemical substances and of techniques for altering their properties. The standard format for the organization and presentation of chemistry was the lecture series, with Macquer's use of the dictionary form being an interesting but not radically innovative departure from the norm.[41] To give their lectures a sense of order, chemists organized their courses according to the traditional categories of natural history and supplemented this format with a discussion of the "instruments" of chemistry. Theories were interspersed eclectically within this structure to provide plausible explanations of specific phenomena, with disagreements over theoretical interpretations occasionally being acknowledged as well. But, since instruction in chemistry was being offered to enlighten the curious and render accessible knowledge that would be useful in medicine, agriculture, commerce, and industry, the main focus at all times was on the presentation of extensive knowledge rather than on the intensive examination of theoretical problems.

Lavoisier's interest in chemistry was quite different. Like Joseph Priestley, he brought to chemistry the intellectual concerns of a philosopher, not the career concerns of a practical chemist. From the very beginning Lavoisier set out to understand a relatively restricted set of chemical phenomena. As he wrote in the famous research memorandum with which he opened a new series of laboratory notebooks in 1773, his goal in undertaking a course of experiments on the fixation and release of air was

> to link our knowledge of the air that goes into combustion or that is liberated from substances, with other acquired knowledge, and to form a theory. The results of the other authors whom I have named, considered from this point of view, appeared to me like separate pieces of a great chain; these authors have joined only some links of the chain.[42]

Lavoisier was fortunate in having the freedom to pursue whatever interested him within chemistry, and it was theory, not pedagogy or the solution of practical problems, that attracted him to the subject in which he distinguished himself.

But while Lavoisier wanted to attain theoretical understanding of the subjects he studied, he did not harbor the vaunting ambitions that motivated more traditional natural philosophers, and in this he differed from Priestley. As was pointed out above, Lavoisier acquired his understanding of the proper aims and methods of science in a milieu in which several of the central assumptions of seventeenth-century natural philosophy had been largely discredited. Unlike Diderot and Venel, he did not need to rail against the hegemony of Newtonianism and mathematical physics, and he was free to take up chemical problems without fearing that a comprehensive new physical theory might suddenly render his efforts irrelevant. More constructively, Lavoisier realized that while there is indeed a natural world to be investigated and rendered intelligible, scientific knowledge is a creation of the human mind. The only means by which knowledge of nature can be acquired are by the use of experiment and reason, and, to attain an understanding of the general laws of nature, one must begin by articulating restricted theories that explain carefully delimited sets of natural phenomena. By the middle of the nineteenth century, when the transformation of natural philosophy into a set of autonomous experimental sciences had been largely completed, Lavoisier's view of chemistry had become a commonplace of science. But in the eighteenth century it constituted a revolutionary abandonment of the traditional aims of natural philosophy.

While Lavoisier understood that the construction of theories requires a creative act on the part of the individual scientist,[43] he also believed that the structure of science reveals the general principles that govern the way the mind works. He further believed that to grasp the inner logic of science, we need to focus our attention on the nature of language. Of course Lavoisier was not being original in this, and as is well known, he forthrightly declared himself a disciple of Condillac's account of language and logic.[44] Condillac provided Lavoisier with a philosophical rationalization for a set of methodological principles he had long employed in his research. Among the foremost of these were the importance of experiment in gaining factual knowledge about the world, the need to ground all abstract reasoning in empirical fact, the need to formulate analytic languages that are specific to different empirical domains, and the assurance that the proper use of reason and experiment will insure the compatibility and ultimate unity of science. These principles include a large part of what came to be known as the positivist program for science. We must of course be careful not to commit the anachronism of calling Lavoisier a Comtean positivist. Yet there are good reasons for thinking that the way in which he reconstructed the science of chemistry provided an important paradigm for the theory of science that came to be called positivism.[45]

SCIENCE AND ADMINISTRATIVE REFORM

When attempting to grasp the personality of a great scientist, it is especially important that his activities not be interpreted in terms of a concept of science appropriate to a different age. Newton scholars continue to debate the significance of his writings on theology and alchemy. They rightly oppose claims that the importance of these subjects was in proportion to the time Newton devoted to them, yet they also realize they can no longer simply dismiss these investigations as obviously irrelevant to his achievement as a scientist. Lavoisier scholars face a similar problem. While Lavoisier was a masterful chemist, he was also a remarkably active public figure in Paris during the final decades of the old regime. His reforming efforts in the Academy of Sciences and his service on innumerable committees and commissions are well documented, but their relation to his achievement as a chemist remains problematic.[46] What, we would like to know, was the relationship between his particular program for the reform of chemistry and his activities as a reforming administrator?

To answer this question we need to focus on the shift in political sensibilities that occurred as the great figures of the High Enlightenment, such as Voltaire, Diderot, D'Alembert, and Rousseau, passed from the scene and were replaced by the last generation of prerevolutionary reformers, men such as Condorcet and Lavoisier.[47] The earlier phase of the Enlightenment, which culminated in the publication of literary and philosophical masterpieces, was followed by a later phase that stressed the advancement of the sciences and administrative reform. The younger men no longer needed to insist that French society should be refashioned along more natural lines. Their first responsibility was to carry the study of nature beyond the metaphysical naturalism of their precedecessors, for they had to demonstrate that it was indeed possible to acquire the kind of natural knowledge needed to carry out specific programs of reform. Their second and closely related responsibility was to demonstrate through their achievement as scientists and through their effectiveness as reformers that they were the "natural" leaders among all those who were clamoring for reform.[48]

The two most prominent reform programs of the late Enlightenment were Turgot's campaign to consolidate the royal administration and the economic program of the Physiocrats. Lavoisier was deeply involved in both these activities. While he clearly derived great satisfaction from his chemical studies, it appears they also supported his efforts at administrative reform by exemplifying the power of the methods he employed and by demonstrating that those who best knew how to acquire reliable knowledge of nature were best able to serve the nation. The central issue was the authority of science, not its immediate material utility. When the old social and political order of France collapsed in the turmoil of revolution, the alternatives to Lavoisier's type of reform program were displayed for all to see. To understand the connections between Lavoisier's chemistry and his other activities, we therefore need to spell out the meaning of his achievement within this larger prerevolutionary political context. And to do that, we need to remember that Lavoisier devoted a great deal of his life to administrative reform, that he was a *philosophe* for whom the reform of chemistry was but one of the many reform campaigns to which he committed himself.

Notes

Preparation of this paper was supported in part by grants from the National Science Foundation's Program for the History and Philosophy of Science and the National Endowment for the Humanities' Research Division, and by a Fellowship from the University of Edinburgh's Institute for Advanced Studies in the Humanities.

1. Herbert Butterfield, *The Whig Interpretation of History* (New York: Norton,

1965 [first published in 1931], p. 10. Students of the Chemical Revolution have long lamented that Butterfield failed to follow his own advice in his later book, *The Origins of Modern Science 1300–1800,* revised ed. (New York: Collier, 1962), ch. 11, "The Postponed Scientific Revolution in Chemistry."

2. Arnold Thackray, *Atoms and Powers—An Essay on Newtonian Matter-Theory and the Development of Chemistry* (Cambridge: Harvard University Press, 1970).

3. Ibid., pp. 197–98.

4. Ibid., pp. 214–17; cf. also Thackray's summary statement on p. 4:

the brilliant experimental work and conceptual reformulations associated with his [Lavoisier's] name were not directly aimed at either denying or replacing the Newtonian categories of thought, views of matter, or research priorities so widespread among his contemporaries. Hence his work is less than central to a study of Newtonian matter-theory and the development of chemistry. The task Lavoisier set himself was to reorganize the superstructure of the science rather than to deny its very Newtonian frame. That the impact of his art was far different is . . . one of history's ironies.

See also I. Bernard Cohen, *The Newtonian Revolution* (Cambridge: Harvard University Press, 1980), pp. 9–10: "Newton had a number of brilliant insights into the structure of matter and the process of chemical reaction, but the true revolution in chemistry did not come into being until the work of Lavoisier, which was not directly Newtonian."

5. For a slightly longer but still programmatic discussion of this transformation, see my "Newton, Lavoisier and Modern Science," in P. B. Scheurer and G. Debrock, eds., *Newton's Scientific and Philosophical Legacy* (Dordrecht: Kluwer, 1988), pp. 219–25.

6. Robert E. Schofield, *Mechanism and Materialism—British Natural Philosophy in the Age of Reason* (Princeton: Princeton University Press, 1969), p. 4.

7. Ibid., p. 5.

8. Ibid., p. 14.

9. See Simon Schaffer, "Natural Philosophy," in G. S. Rousseau & Roy Porter, eds., *The Ferment of Knowledge—Studies in the Historiography of Eighteenth-Century Science* (Cambridge: Cambridge University Press, 1980), pp. 55–91; G. N. Cantor and M. J. S. Hodge, "Introduction: major themes in the development of ether theories from the ancients to 1900," in G. N. Cantor and M. J. S. Hodge, eds., *Conceptions of Ether* (Cambridge: Cambridge University Press, 1981), pp. 1–60, on pp. 35–37.

10. Robert E. Schofield, "The Counter-Reformation in Eighteenth-Century Science—Last Phase," in Duane H. D. Roller, ed., *Perspectives in the History of Science and Technology* (Norman: Oklahoma University Press, 1971), pp. 39–54, on p. 40. See also the commentaries by Robert J. Morris, Jr., and Robert Siegfried, pp. 55–66.

11. Ibid., p. 45.

12. Ibid., p. 46.

13. Ibid., p. 50.

14. See Hélène Metzger, *Newton, Stahl, Boerhaave et la doctrine chimique* (Paris: Blanchard, 1930); Rhoda Rappaport, "Rouelle and Stahl—The Phlogistic Revolution in France," *Chymia* 7 (1961): 73–102; Martin Fichman, "French Stahlism and Chemical Studies of Air, 1750–1770," *Ambix* 18 (1971): 94–122; and J. B. Gough, "Lavoisier and the Fulfillment of the Stahlian Revolution in Chemistry," in Arthur Donovan, ed., *The Chemical Revolution: Essays in Reinterpreta-*

tion, vol. 4 of *Osiris*, second series (Philadelphia: History of Science Society, 1988), pp. 15–33. See also Frederick Gregory, "Romantic Kantianism and the End of the Newtonian Dream in Chemistry," *Archives internationales d'historie des sciences* 34 (1984): 108–23.

15. See my "Buffon, Lavoisier and the Transformation of French Chemistry," in J. Gayon, ed., *Buffon 88: Actes du colloque international*, (Paris: Vrin, in press).

16. I. Bernard Cohen, *Franklin and Newton* (Philadelphia: American Philosophical Society, 1956), p. 17.

17. Schofield, *Mechanism and Materialism*, p. 3.

18. Ephraim Chambers, *Cyclopedia: or An Universal Dictionary of Arts and Sciences*, 2 vols. (London, 1728), 2:628 ("Newtonian Philosophy"); 2:42 ("Sect").

19. Cohen, *Newtonian Revolution*, p. 131.

20. For a discussion of the impact on Newtonianism of this "intellectual revolution which was philosophical in character," see P. M. Heimann and J. E. McGuire, "Newtonian Forces and Lockean Powers: Concepts of Matter in Eighteenth-Century Thought," *Historical Studies in the Physical Sciences* 3 (1971): 233–306.

21. David Hume, *A Treatise of Human Nature*, ed. L. A. Selby-Bigge (Oxford, 1896), p. xx.

22. David Hume, *History of England* (London, 1780), 8:326.

23. Colin Maclaurin, *An Account of Sir Isaac Newton's Philosophical Discoveries* (London, 1748), p. 19.

24. Larry Laudan, "Thomas Reid and the Newtonian Turn of British Methodological Thought," in Robert E. Butts and John W. Davis, eds., *The Methodological Heritage of Newton* (Toronto: Toronto University Press, 1970), pp. 103–31, on p. 106.

25. The separation of religious issues, and especially theology, from philosophy, and the attack on the spirit of system, as opposed to the systematic spirit, were two of the main foci of the Enlightenment. In the *Encyclopédie*, with which Lavoisier must have been intimately familiar, the attack on religion is carried on primarily by ridicule and irony, the chief rhetorical device employed being cross-referencing among articles. In place of the spirit of system, the guiding spirit of traditional natural philosophy, the *Encyclopédie* propagandized for the encyclopedic method of constructing an open circle of knowledge, an image that reminds one of Lavoisier's search for theoretical chains with which to create a science of chemistry. See Diderot's article "Encyclopédie" in the *Encyclopédie*. See also Robert Darnton, "Philosophers Trim the Tree of Knowledge: The Epistemological Strategy of the *Encyclopédie*," in his *The Great Cat Massacre* (New York: Basic Books, 1984), pp. 190–213; and Wilda C. Anderson, *Between the Library and the Laboratory—The Language of Chemistry in Eighteenth-Century France* (Baltimore: Johns Hopkins University Press, 1984).

26. Margaret C. Jacob, "Newtonianism and the Origins of the Enlightenment: A Reassessment," *Eighteenth-Century Studies* 11 (1977–78): 1–25, on p. 2; see also her *The Newtonians and the English Revolution, 1689–1720* (Ithaca: Cornell University Press, 1976).

27. For some examples of anti-Newtonianism in British natural philosophy, see A. J. Kuhn, "Glory or Gravity: Hutchinson versus Newton," *Journal of the History of Ideas* 22 (1961): 302–22; C. B Wilde, "Hutchinsonianism, Natural Philosophy and Religious Controversy in Eighteenth-Century Britain," *History of Science* 18 (1980): 1–24; and A. J. Kuhn, "Nature Spiritualized: Aspects of Anti-Newtonianism," *English Literary History* 41 (1974): 400–412.

28. For a general discussion of Hutton's natural philosophy, see my introduc-

tion to *James Hutton's Medical Dissertation* (Philadelphia: American Philosophical Society, 1980), *Transactions* vol. 70, part 6; see also my "Scottish Responses to the New Chemistry of Lavoisier," in *Studies in Eighteenth-Century Culture*, vol. 9, ed. R. Runt (Madison: University of Wisconsin Press, 1979), pp. 237–49, and Peter Jones, "An Outline of the Philosophy of James Hutton (1726–1797)," in *Philosophers of the Scottish Enlightenment*, ed. Vincent Hope (Edinburgh: Edinburgh University Press, 1984), pp. 182–210.

29. On Newtonianism in France, see Pierre Brunet, *L'Introduction des théories de Newton en France au XVIIIᵉ siècle—avant 1738* (Paris: Blanchard, 1931); A. R. Hall, "Newton in France: A New View," *History of Science* 13 (1975): 233–50; Henry Guerlac, "Some Areas for Further Newtonian Studies," *History of Science* 17 (1979): 75–101; and Henry Guerlac, *Newton on the Continent* (Ithaca: Cornell University Press, 1981).

30. Voltaire, *Lettres philosophiques,* first French edition, 1734; trans. Leonard Tancock as *Letters on England* (Harmondsworth: Penguin, 1980).

31. Jacob, "Newtonianism and the Origins of the Enlightenment," p. 22.

32. See Jacques Roger, *Les sciences de la vie dans la pensée française du XVIIIᵉ siècle* (Paris, 1963); Shirley A. Roe, "Voltaire versus Needham: Atheism, Materialism, and the Generation of Life," *Journal of the History of Ideas* 46 (1985): 65–87; John W. Yolton, *Thinking Matter: Materialism in 18th-Century Britain* (Minneapolis: University of Minnesota Press, 1983).

33. Lester G. Crocker, "Recent Interpretations of the French Enlightenment," *Cahiers d'histoire mondiale* 8 (1964): 426–56, on p. 431; Jean-Claude Guédon, "Chimie et matérialisme—La Stratégie anti-Newtonian de Diderot," *Dix-Huitième siècle* (1979): 185–200.

34. Gabriel-Françoise Venel, "Chymie," in Denis Diderot and Jean d'Alembert, eds., *Encyclopédie, ou Dictionnaire raisoné des sciences, des arts et des métiers, par une société de gens de lettres* (Paris, 1753), 3 : 408–47, on pp. 408–10.

35. Ibid., p. 437.

36. On "Jacobin science," see Marshall Clagett, ed., *Critical Problems in the History of Science* (Madison: University of Wisconsin Press, 1969), ch. 9; Robert Darnton, *Mesmerism and the End of the Enlightenment in France* (Cambridge: Harvard University Press, 1968); and Charles C. Gillispie, *Science and Polity at the End of the Old Regime* (Princeton: Princeton University Press, 1980), ch. 4.

37. Thomas Kuhn and John Heilbron have suggested that a narrowing of the range of subjects addressed is one sign of the emergence of a mature scientific discipline; see Evan Melhado, "Chemistry, Physics, and the Chemical Revolution," *Isis* 76 (1985): 195–211, on p. 196, n. 2.

38. For detailed studies of Lavoisier's use of experiments in the construction of novel theories, see Marcellin Berthelot, *La révolution chimique—Lavoisier* (Paris, 1890; 2d ed. 1902); Henry Guerlac, *Lavoisier—The Crucial Year* (Ithaca: Cornell University Press, 1961); Frederic L. Holmes, *Lavoisier and the Chemistry of Life* (Madison: University of Wisconsin Press, 1985); C. E. Perrin, "Continuity and Divergence of Research Traditions: Lavoisier and the Chemical Revolution," in Donovan, ed., *The Chemical Revolution,* pp. 53–81; and C. E. Perrin, "The Chemical Revolution: Shifts in Guiding Assumptions," in Arthur Donovan, Larry and Rachel Laudan, eds., *Scrutinizing Science—Empirical Studies of Scientific Change* (Dordrecht: Kluwer, 1988), pp. 105–24.

39. See my "Lavoisier and the Origins of Modern Chemistry," in Donovan, ed., *The Chemical Revolution,* pp. 214–31. On experimental physics in the eighteenth century, see Jean Torlais, "La physique expérimentale," in René Taton, *Enseignement et diffusion des sciences en France au XVIIIᵉ siècle* (Paris: Hermann, 1964),

pp. 619–45; J. L. Heilbron, *Electricity in the 17th and 18th Centuries: A Study in Early Modern Physics* (Berkeley: University of California Press, 1979); R. W. Home, "Out of a Newtonian Straitjacket: Alternative Approaches to Eighteenth-Century Physical Science," in R. F. Brissenden and J. C. Eade, eds., *Studies in the Eighteenth Century* (Canberra: Australian National University Press, 1979), 4:235–49; R. W. Home, "The Notion of Experimental Physics in Early Eighteenth-Century France," in Joseph C. Pitt, ed., *Change and Progress in Modern Science* (Dordrecht: D. Reidel, 1985), pp. 107–31; and Thomas L. Hankins, *Science and the Enlightenment* (Cambridge: Cambridge University Press, 1985), ch. 3.

40. See Pierre Brunet, *Physiciens hollandais et la méthode expérimentale en France au XVIIIᵉ siècle,* (Paris: Blanchard, 1926), p. 22, where he defines a scientific school, such as the Cartesian and the Newtonian, as "a group attached to a given system."

41. J. R. Christie and J. V. Golinski, "The Spreading of the Word: New Directions in the Historiography of Chemistry 1600–1800," *History of Science* 20 (1982): 235–66.

42. This memorandum was first published in Berthelot, *Révolution chimique,* pp. 46–49; reprinted in Guerlac, *Lavoisier—The Crucial Year,* pp. 228–30; English translation in Andrew Norman Maldrum, *The Eighteenth Century Revolution in Science—The First Phase* (Calcutta: Longmans, Green, n.d. [1930]), pp. 8–10. On the dating of this memorandum, see Henry Guerlac, "The Chemical Revolution: A Word from Monsieur Fourcroy," *Ambix* 23 (1976): 1–4, n. 2. The passage quoted is taken from Meldrum, p. 9.

43. In 1792 Lavoisier characterized his contributions to the Chemical Revolution as follows: "This theory is not, as I have heard it said, the theory of French chemists, it is *mine,* and it is a property that I claim from my contemporaries and from posterity." A. Lavoisier, *Oeuvres de Lavoisier,* 6 vols., vols. 1–4 ed. J. B. Dumas, vols. 5–6 ed. Edouard Grimaux (Paris: Imprimerie Impériale, 1862–93; reprint ed., New York: Johnson Reprint, 1965), 2:104.

44. See the preface to Antoine-Laurent Lavoisier, *Traité élémentaire de Chimie, présenté dans un ordre nouveau et d'après les découvertes modernes* (Paris, 1789); see also William Randall Albury, "The Logic of Condillac and the Structure of French Chemical and Biological Theory, 1780–1801" (Ph.D. dissertation, Johns Hopkins University, 1972), and Anderson, *Between the Library and the Laboratory.*

45. August Comte, *Cours de philosophie positive,* 1st ed., 6 vols. (Paris, 1830–42). The first half of vol. 3, published in 1838, is devoted to chemistry. Cf. Maurice Crosland, "Comte and Berthollet: A Philosopher's View of Chemistry," in *Actes du XIIᵉ Congrès International d'histoire des sciences, Paris, 1968,* vol. 6, *Histoire de la chimie depuis le XVIIIᵉ siècle* (Paris: Albert Blanchard, 1971), pp. 23–27.

46. See Gillispie, *Science and Polity,* pp. 58–73 and passim. On p. 64 Gillispie quotes a biographical fragment in which Lavoisier's wife reports that her husband worked at chemistry from 6:00 to 8:00 in the morning and from 7:00 to 10:00 at night, with one whole day a week being set aside for experiments. The rest of his time was devoted to other affairs. On Lavoisier's attempts to reform the Academy of Sciences, see Roger Hahn, *The Anatomy of a Scientific Institution* (Berkeley: University of California Press, 1971); for overviews of Lavoisier's life see Douglas McKie, *Antoine Lavoisier* (New York: Collier, 1962), and my *Lavoisier: Scientific and Political Revolution in Eighteenth-Century France* (London: Blackwell, in press).

47. See Gillispie, *Science and Polity,* p. 196; see also Keith Michael Baker,

Condorcet (Chicago: University of Chicago Press, 1975). For a preliminary discussion of this topic, see my "Lavoisier's Politics: The Scientist as Administrator," *Bulletin for the History of Chemistry* 5 (1989): 10–14.

48. See Baker, *Condorcet,* ch. 5; Peter Gay, *The Enlightenment, An Interpretation,* vol. 2 (New York: Knopf, 1969), chs. 7–10 and Finale.

Isaac Newton and Adam Smith: Intellectual Links between Natural Science and Economics

NORRISS S. HETHERINGTON

Introduction

Similarities between the works of Isaac Newton and Adam Smith are obvious and have not escaped notice. The Newtonian example, the discovery of general laws in natural philosophy, excited searches for general laws in other realms, including economics and an understanding of the state. In this branch of philosophy, Smith was the Newton.

Smith's efforts to discover general laws of economics paralleled Newton's efforts in physics. The likeness is not fortuitous. Smith was directly inspired by Newton's success. Before reading Newton, Smith was skeptical of hypotheses; he perceived them as but arbitrary creations of history. After reading Newton, Smith believed in the reality of invisible chains binding together disjointed objects; he now sought to discover the real connecting principles of economics.

Smith's work in economics was inspired by Newton's work in natural philosophy and also was shaped in places by Newton's strategy. The inspiration is apparent in Smith's juvenile literary fragment on the history of astronomy and in his early writings on economics, particularly in frequent and explicit references to the "real chains which Nature makes use of to bind together her several operations." General philosophical considerations are less openly paraded in Smith's mature work. Structural similarities, though, between Newton's *Mathematical Principles of Natural Philosophy* and Smith's *Inquiry into the Nature and Causes of the Wealth of Nations* help reveal the extent of Smith's intellectual debt to Newton.

Smith's Intellectual Milieu

The traditional list of influences on Smith includes mercantilistic theory, against which Smith reacted, and the school of Physiocrats in France, a reaction against mercantilism with which Smith sympathized. There is a problematic link between Smith and Bernard de Mandeville, who defended the utility of vices.[1]

More certain intellectual links are to be found between Smith and Francis Hutcheson, who supported the usefulness of wealth and argued in defense of the legitimacy of the natural appetites of man, and between Smith and David Hume, who argued for free trade. Hutcheson and Hume were members of a Scottish school of scientific or philosophical history using induction to frame universal principles from observations of a vast number of particular cases.[2]

The framing of universal principles by induction evokes the image of Newton in the Age of Reason. At Edinburgh, where Smith taught during the 1751–52 university year, intellectual life "was nourished in great measure by the writings of Bacon and Newton."[3] John Millar, who attended Smith's lectures on moral philosophy and was later Smith's colleague at the University of Glasgow, came to regard Montesquieu as the Lord Bacon of the understanding of the state and Smith as the Newton.[4]

Smith was interested in natural philosophy. A fellow student of Smith's at Glasgow, a Dr. Maclain of the Hague, later remembered that Smith's favorite pursuits at the university were mathematics and natural philosophy.[5] And Dugald Stewart, Smith's biographer and son of a distinguished mathematician,[6] remembered discussions between his father and Smith concerning a geometrical problem of considerable difficulty.[7] It may be that a fellow student in a natural philosophy course could have exaggerated Smith's interest in that subject, as one biographer has argued,[8] or that Stewart, raised in an environment placing a high importance on science, might have unduly emphasized Smith's interest in mathematics and natural science. Other evidence, however, also testifies to Smith's interest in science.

Smith associated with Benjamin Franklin, who had an international reputation as a writer and as an inventor,[9] and Smith's literary executors were Joseph Black, known for his discovery of carbon dioxide and the formulation of the concepts of specific and latent heat, and the geologist James Hutton.[10] Moreover, Smith possessed many books on the physical and biological sciences, including volumes of the *Philosophical Transactions* of the Royal

Society of London through 1780 and Newton's *Principia, Arithmetica, Method of Fluxions,* and *Opticks.*[11]

Smith, a posthumous son, as was Newton, but a son who enjoyed a close and uninterrupted early life with his mother, went to the University of Glasgow at the age of fourteen and to Balliol College, Oxford, at the age of seventeen, as a Snell exhibitioner. There Smith stayed for six years, from 1740 to 1746. At Oxford there was not opportunity to receive instruction in science, for the university was then, and for the remainder of the century, "sunk deep in intellectual apathy, a muddy reservoir of sloth, ignorance, and luxury from which men sink as by a law of gravitation into the still lower level of civil and ecclesiastical sinecures."[12] Balliol's library was well stocked in ancient Greek and Latin classics, though, and Smith read them.[13] Later, Smith would list among the claims of the University of Glasgow upon his gratitude that it had sent him to Oxford.[14] Smith's introduction to natural philosophy and Newton, though, owes more to his Scottish years, before and after the Oxford interlude.

Smith's History of Astronomy and the Image of Newton

Leaving Oxford in 1746, Smith spent two years at his mother's home in Kirkcaldy. This may be when Smith wrote his history of astronomy,[15] though the dating of this work, only published posthumously in 1795, five years after Smith's death, is not without controversy.

Smith discussed in his history of astronomy the predicted return of Halley's comet in 1758. In a footnote to the manuscript, Smith added:

> It must be observed, that the whole of this Essay was written previous to the date here mentioned; and that the return of the comet happened agreeably to the prediction.[16]

Ordinarily, such a statement would settle the question of chronology and position the history of astronomy unambiguously prior to the *Wealth of Nations,* which was published in 1776 and existed in embryo in Smith's lectures on revenue of approximately 1763.[17] Confusing matters, though, is a letter from Smith to Hume in 1773 asking, should Smith die very suddenly on his trip to London, carrying his manuscript of the *Wealth of Nations,* that Hume might destroy without examination Smith's remaining papers except "a fragment of a great work which contains a history of astronomical systems that were successively in fashion down to the time of Descartes." Whether that might be published, Smith left to Hume's judgment, in the event of Smith's sudden demise.[18] This letter suggests a later date for the section on Newton, who followed Descartes. It may have been, however, that Smith had completed the concluding section on Newton but was dissatisfied with it and thus described the essay to Hume as only going down to the time of Descartes. Some notes and memoranda left behind by Smith, when he did finally die, indicated to his editors, Black and Hutton, that Smith regarded the section on Newton of his history of astronomy as imperfect and in need of additions.[19]

Evidence internal to the manuscript also argues for an early completion date, though the argument becomes circular in nature. In his history of astronomy, Smith cites explicitly the real chains of Nature binding together her operations. In his early writings on economics, Smith explicitly displays a continuing interest in general principles and laws. Growing in maturity, Smith suppressed the explicit appearance of such passages in *The Wealth of Nations*. No longer did he find it necessary to write out repeatedly his basic philosophical assumptions. Also, epistemological convention may have dictated that Smith emphasize fact rather than hypothesis.

Smith, though, always thought in a broad context, even in his history of astronomy. He viewed and titled it as an essay on "The Principles which lead and direct Philosophical Enquiries; illustrated by the History of Astronomy." He began with an examination of the sentiments of wonder and surprise, excited by what is new and unexpected:

> It is evident that the mind takes pleasure in observing the resemblances that are discoverable betwixt different objects. . . .
>
> When one accustomed object appears after another, which it does not usually follow, it first excites, by its unexpectedness, the sentiment properly called Surprise. . . .
>
> When two objects, however unlike, have often been observed to follow each other, and have constantly presented themselves to the senses in that order, they come to be so connected together in fancy, that the idea of the one seems, of its own accord, to call up and introduce that of the other. If the objects are still observed to succeed each other as before, this connection, or, as it has been called, this association of their ideas, becomes stricter and stricter, and the habit of imagination to pass from the conception of the one to that of the other grows more rivetted and confirmed.[20]

Hypotheses may give some coherence to the appearance of nature, but Smith doubted the reality of such associations.

Hypotheses were arbitrary, with a history of one replacing another and then being supplanted in turn:

> The same orders of succession, which to one set of men seem quite according to the natural course of things, and such as require no intermediate events to join them, shall to another appear altogether incoherent and disjointed, unless some such events be supposed: and this for no other reason, but because such orders of succession are familiar to the one, and strange to the other. . . .
>
> Philosophy is the science of the connecting principles of nature. Nature, after the largest experience that common observation can acquire, seems to abound with events which appear solitary and incoherent with all that go before them, which therefore disturb the easy movement of the imagination. . . . Philosophy, by representing the invisible chains which bind together all these disjoined objects, endeavors to introduce order into this chaos of jarring and discordant appearances, to allay this tumult of the imagination. . . . Let us examine, therefore, all the different systems of nature, which . . . have successively been adopted by the learned and ingenious; and, without regarding their absurdity or probability, their agreement or inconsistency with truth and reality, let us consider them only in that particular point of view which belongs to our subject; and content ourselves with inquiring how far each of them was fitted to sooth the imagination, and to render the theatre of nature a more coherent, and therefore a more magnificent spectacle, than otherwise it would have appeared to be.[21]

Although not explicitly denying the reality of the invisible chains that bind together disjointed objects, Smith at the beginning of his survey of western astronomical theories perceived such theories as merely fictions to sooth the imagination.

His attitude at the end of the essay was reversed. In the process of writing the essay and in the contemplation of Newton's work, Smith apparently became persuaded that there are real connecting principles to be discovered. He concluded that:

> The superior genius and sagacity of Sir Isaac Newton, therefore, made the most happy, and, we may now say, the greatest and most admirable improvement that was ever made in philosophy, when he discovered, that he could join together the movements of the Planets by so familiar a Principle of connection, which completely removed all the difficulties the imagination had hitherto felt in attending to them. . . . Having thus shown that gravity might be the connecting principle which joined together the movements of the Planets, he endeavored next to prove that it really was so. . . .

Such is the system of Sir Isaac Newton, a system whose parts are all more strictly connected together, than those of any other philosophical hypothesis. Allow his principle, the universality of gravity, and that it decreases as the squares of the distance increase, and all the appearances, which he joins together by it, necessarily follow. Neither is their connection merely a general and loose connection, as that of most other systems in which either these appearances, or some such like appearances, might indifferently have been expected. It is every where the most precise and particular that can be imagined, and ascertains the time, the place, the quantity, the duration of each individual phaenomenon, to be exactly such as, by observation, they have been determined to be. . . . His principles, it must be acknowledged, have a degree of firmness and solidity that we should in vain look for in any other system. The most sceptical cannot avoid feeling this. They not only connect together most perfectly all the phaenomena of the Heavens, which had been observed before his time, but those also which the persevering industry and more perfect instruments of later Astronomers have made known to us; have been either easily and immediately explained by the application of his principles, or have been explained in consequence of more laborious and accurate calculations from these principles, than had been instituted before. And even we, while we have been endeavoring to present all philosophical systems as mere inventions of the imagination, to connect together the otherwise disjointed and discordant phaenomena of nature, have insensibly been drawn in, to make use of language expressing the connecting principles of this one, as if they were the real chains which Nature makes use of to bind together her several operations. Can we wonder then, that it should have gained the general and complete approbation of mankind, and that it should now be considered, not as an attempt to connect in the imagination the phaenomena of the Heavens, but as the greatest discovery that ever was made by man, the discovery of an immense chain of the most important and sublime truths, all closely connected together by one capital fact, of the reality of which we have daily experience.[22]

Smith's interest in the principles or laws of nature, whetted by a new belief in their reality, was apparently expressed in his university lectures at Glasgow between 1762 and 1764 on justice, police, revenue, and arms. Some qualification is necessary when presenting Smith's beliefs, because most of Smith's papers were destroyed by his executors. Most of what is known of his lectures comes from a 1766 copy of notes made earlier by one of Smith's students. (Booksellers' shops during the eighteenth century commonly stocked manuscript copies of popular professors' lectures.) Students' lecture notes are of uncertain accuracy. It further appears that the copyist was not the original note-taker and introduced additional errors.[23]

Questionable as it is, the copy of notes taken at Smith's lectures is all that remains of his lectures. The manuscript does show traces of interest in the laws of nature. It has Smith defining jurisprudence as "that science which inquires into the general principles which ought to be the foundations of the laws of all nations" and, regarding the division of labor, posing the question of "what gives occasion to the division of labor, or from what principles in our nature it can be accounted for."[24]

Newtonian Echoes in *The Wealth of Nations*

More certain is Smith's search for general laws in his *Inquiry into the Nature and Causes of the Wealth of Nations,* even if general philosophical considerations, so apparent in his earlier essay on the history of astronomy, are not explicitly stated in the *Wealth of Nations.* One can search in vain for phrases such as "the real chains which nature makes use of to bind together her several operations." Their absence, though, may be attributed to Smith's growing maturity and to epistemological convention rather than to any waning of interest by Smith in general laws of nature and the Newtonian mechanical world view.

Smith's interest in general laws of nature might be attributed either to general Newtonian aspects of Smith's overall intellectual milieu or specifically to a reading and understanding of Newton's science. Striking similarities between Smith's and Newton's work suggest, albeit without definitively establishing beyond all doubt, a close connection.

Smith searched for general laws or principles. In the *Wealth of Nations,* he investigates "the principles which regulate the exchangeable value of commodities"[25] and he notes that "the profits of stock constitute a component part altogether different from the wages of labour and regulated by quite different principles."[26]

Newton in his *Mathematical Principles of Natural Philosophy* presents a mechanistic view of the world. From the phenomena of motions he investigates the forces of nature, and then from these forces he demonstrates other phenomena that follow. He would derive, if he could, all the phenomena of Nature from mechanical principles.[27] Smith, too, in passages in the *Wealth of Nations,* presents a mechanical system, of push and shove, of action and reaction, of regulation by principles or natural laws. Smith writes that:

> The proportion between capital and revenue, therefore, seems everywhere to regulate the proportion between industry and idleness. Wherever capital predominates, industry prevails; wherever revenue, idleness. Every increase or diminution of capital, therefore, naturally tends to increase or diminish the real quantity or industry, the number of productive hands, and consequently the exchangeable value of the annual produce of the land and labour of the country, and the real wealth and revenue of all its inhabitants.[28]

and:

> There is in every society or neighbourhood an ordinary or average rate both of wages and profit in every different employment of labour and stock. This rate is naturally regulated. . . . There is likewise . . . an ordinary or average rate of rent, which is regulated too.[29]

and:

> The natural price, therefore, is, as it were, the central price, to which the prices of all commodities are continually gravitating. Different accidents may sometimes keep them suspended a good deal above it, and sometimes force them down even somewhat below it. But whatever may be the obstacles which hinder them from settling in this centre of repose and continuance, they are constantly tending towards it.
>
> The whole quantity of industry annually employed in order to bring any commodity to market, naturally suits itself in this manner to the effectual demand. It naturally aims at bringing always that precise quantity thither which may be sufficient to supply, and no more than supply, that demand.[30]

The reference to gravity was no accident. Smith had Newton in mind. Two pages later Smith repeats the phrase. And, lest his readers fail to see the connection to Newton, Smith coyly calls attention to his wording:

> But though the market price of every particular commodity is in this manner continually gravitating, if one may say so, towards the natural price.[31]

Smith's use of the term *gravitating* could be merely metaphorical, though it also seems intended as descriptive of an economic law corresponding to a real chain of nature. Either usage is evidence of an intellectual debt to Newton, albeit not necessarily more than a very minor debt if the usage is solely metaphorical.

In addition to the common theme of regulation by principles of nature in Newton's *Principia* and Smith's *Wealth of Nations,* there are also important similarities of structure, or strategy, linking the two presentations. The method of moving from phenomena by induction to the framing of principles, and then the deduction of phenomena from the principles, is found in both Newton's and Smith's books.

After describing four rules of reasoning in Book 3 of the *Principia,* the *System of the World,* Newton lists phenomena. These include the observation that Jupiter's satellites sweep out areas (as determined by lines drawn from the satellites to Jupiter) proportional to the times of the motion (equal areas in equal times) and that the times of revolution about Jupiter are as the 3/2 th power of the distances of the satellites from Jupiter; that the same motions exist for Saturn's satellites; that the planets move similarly about the sun; and that the moon sweeps out an area proportional to the time of the motion.[32] (With only one satellite for the earth, the proportional relation between period and diameter could not be established, as it was in the other cases.)

As Newton begins Book 3 of the *Principia* with a list of phenomena, so Smith begins the *Wealth of Nations.* To more easily understand the effects of the division of labor, Smith considers the manner in which labor operates in some particular manufactures. The often-cited description of the manufacture of pins follows. Smith had seen ten men manufacture pins at a rate of 48,000 per day. One man drew the wire; a second man straightened the wire; a third man cut the wire; a fourth man pointed the wire; a fifth man ground the wire to receive the head, etc.[33]

After listing phenomena, Newton demonstrates that the principle of gravity is in agreement with the phenomena.[34] Newton presented his work in a manner to imply that the principle or law of gravity had been obtained by induction from the phenomena. The possibility, if not the detailed mathematical demonstration, of such a law was suggested by men before Newton;[35] in this case the logic of justification may differ from the logic of discovery.

Smith, having presented the phenomenon of the division of labor in the pin factory, gives the general principle. He sees the division of labor as the necessary result of a human propensity to exchange one thing for another.[36]

Once Newton has his general principle, the law of gravity, he then demonstrates by deduction that all the observed phenomena follow from it. Not only the observations of satellites, from which the law of gravity was supposedly induced, but other phenomena, including the orbits of comets, anomalies in the motion of the moon, the precession of the equinoxes, the tides, and even the motion of pendulum clocks, are all shown to be encompassed within the universal law of gravity.[37]

Having discovered a natural law or human propensity, of the division of labor, Smith next shows that the law encompasses additional phenomena. The propensity to exchange items and the resulting division of labor is present in tribes of hunters and shepherds as well as in the pin factory.[38]

Unlike Newton's force of gravity, the phenomenon of the division of labor is not universal. Smith recognized that the difference, the failure of the phenomenon to assert itself universally, requires explanation. He states that the extent of the division of labor must always be limited by the extent of the market. If the market is small, there is not opportunity to exchange all the surplus part of one's labor; thus encouragement to dedicate oneself entirely to a single job is lacking.[39] Seemingly more universal was the number of useful and productive laborers. According to Smith, the number, "it will hereafter appear, is everywhere in proportion to the quantity of capital stock which is employed in setting them to work, and to the particular way in which it is so employed."[40]

There are obvious similarities in the work of Newton and Smith. Both present phenomena, next obtain general laws, ostensibly by induction, and then deduce the original and further phenomena from the general laws. There is also an important similarity in Newton's and Smith's answers to the question of the ultimate nature of their discovered general principles.

Newton was unable to present a mechanical model to explain the working of gravity. He was forced to argue that it was enough for him to show that his law of gravity accounted for the observed phenomena:

> hitherto I have not been able to discover the cause of those properties of gravity from phenomena, and I frame no hypotheses. . . . And to us it is enough that gravity does really exist, and act according to the laws which we have explained, and abundantly serves to account for all the motions of the celestial bodies, and of our sea.[41]

Smith was unable to determine whether the propensity to exchange one thing for another was an original principle of human nature or a necessary consequence of the faculties of reason and speech. Rather than frame an uncertain hypothesis, Smith adopted

Newton's strategy and adapted it to his purpose. As had Newton, Smith perceived universal or original principles in no need of further account. Whether the division of labor was such an original principle or a consequence of a more fundamental principle, Smith did not know. He acknowledged his inability to answer the original question and then reframed the question. He asserted that it was enough to know that the principle was universal, or common to all men:

> Whether this propensity be one of those original principles in human nature, of which no further account can be given; or whether as seems more probable, it be the necessary consequence of the faculties of reason and speech, it belongs not to our present subject to inquire. It is common to all men, and to be found in no other race of animals, which seem to know neither this nor any other species of contracts.[42]

Similarities between Newton's *Principia* and Smith's *Wealth of Nations* are not universal. Smith did not advance all the way to the modern definition of economics as a science discovering universal laws. Instead, when defining political economy, Smith considered it a science proposing to enable people to provide revenue or subsistence for themselves and to supply the state or commonwealth with a revenue sufficient for the public services.[43] Economics for Smith still included an allowance for human intervention. It was not yet a completely abstract science, a physics of weightless strings and frictionless pulleys. In its numerous examples and anecdotes, Smith's *Wealth of Nations* perhaps looked more ahead to Darwin than back to Newton. Still, Smith's intellectual indebtedness to Newton is tremendous.

Conclusion

The influence of Newton's *Principia* on Smith and the *Wealth of Nations,* while occasionally noted, has not often received its due emphasis. This failure is partly attributable to circumstances. Smith's history of astronomy, in which his appreciation of Newton is most readily apparent, was published in 1795, five years after Smith's death and two years after the first presentation, before the Royal Society of Edinburgh, of Stewart's biographical memoir of Smith.[44] The memoir appeared in the *Transactions* of that society in 1794 and was reprinted in 1795 and in 1811. Though at the end of later editions a note was added on Smith's essay on the history of astronomy, many subsequent biographies of Smith gave little attention to the scientific aspect of his life and work. John Rae in his 1895 *Life of Adam Smith* mentions Smith's history of astronomy only indirectly, quoting the letter from Smith to Hume and commenting that "Black and Hutton were his literary executors, and published in 1795 the literary fragments which had been spared from the flames."[45]

Later biographers picked up Smith's appreciation of Newton. As early as 1904, Francis Hirst in his *Adam Smith* reported Smith's

> enthusiastic description of Sir Isaac Newton's discovery as the greatest ever made by man. He had acquired "the most universal empire that was ever established in philosophy," and was the only natural philosopher whose system, instead of being a mere invention of the imagination to connect otherwise discordant phenomena, appeared to contain in itself "the real chains which nature makes use of to bind together her several operations."[46]

And in his 1969 *Adam Smith*, E. G. West noted that Smith had given Newton "a special place of honor. Newton, claimed the young Smith, was the only natural philosopher whose system contained in itself "the real chains which nature makes use of to bind together her several operations.'"[47] Biographers of Smith have moved his history of astronomy from a posthumous publication to its proper position in Smith's early life, and have begun to comment on Smith's appreciation of Newton and natural laws.

Economists have gone beyond the readily apparent fact of Smith's appreciation of Newton and have begun to look for influences of Newton in Smith's writings. In 1940, Henry Bitterman observed that "Adam Smith's methodology was essentially empirical, deriving its inspiration from Newton and Hume."[48] More recent investigations by economists emphasize the importance of Smith's essay on the history of astronomy for an understanding of Smith's *Wealth of Nations,* though without detailed textual comparisons to Newton's *Principia.*[49]

A third group of scholars, historians of science, has taken as its starting point Newton rather than Smith and has sought to determine the extent of Newton's influence. That Smith came under the shadow of the eighteenth-century image of Newton seems likely, a priori, and the suggestion has been developed. John Greene in his study of Darwin remarked that:

> the idea of creating a social science by applying the methods of natural science to the study of man and society is nearly as old as modern science itself. Adam Smith took Newton's conception of nature as a law-

bound system of matter in motion as his model when he represented society as a collection of individuals pursuing their self-interest in an economic order governed by the laws of supply and demand.[50]

And Gerd Buchdahl in his *Image of Newton and Locke in the Age of Reason* wrote that:

Smith's imagination had been kindled early by the vistas opened up by the Newtonian Revolution. It is perhaps no accident that he should have begun his career as a writer with an historical account of the world's great astronomical and physical systems, concluding with an enthusiastic description of Newton's discoveries.[51]

Subsequent studies have taken Smith's enthusiasm for Newton as a starting point and sought its expression in Smith's *Wealth of Nations*.[52]

It is probably only the youthfulness of the discipline of the history of science that explains why its viewpoint is yet to be incorporated fully into the literature on Adam Smith. The theft of Smith by gypsies when he was a baby, albeit more dramatic than Smith's interest in science, will be given a lesser place in his life when connections between science and other intellectual endeavors are studied.

Smith's cultural and intellectual milieu was dominated by the image of Newton, and Smith was subject to that influence. Furthermore, Smith enjoyed a first-hand understanding of Newton's philosophy, and expressed a great appreciation of Newton's achievement in the essay on the history of astronomy. Newton's influence on Smith's economic thinking is further apparent in a long-overdue compaison of Smith's *Wealth of Nations* with Newton's *Principia*.

Notes

1. Alexander Gray, *Adam Smith* (London: G. Philip, 1948), pp. 8–11.

2. Andrew Skinner, "Natural History in the Age of Adam Smith," *Political Studies* 15 (1967): 32–48; W. L. Taylor, *Frances Hutcheson and David Hume as Precursors of Adam Smith* (Durham, N.C.: Duke University Press, 1965).

3. John Veitch, "A Memoir of Dugald Stewart," in Sir William Hamilton, ed., *The Collected Works of Dugald Stewart*, vol. 10 (Edinburgh: T. Constable and Co., 1858), p. xi.

4. John Millar, *An historical view of the English government, from the settlement of the Saxons in Britain to the revolution in 1688: to which are subjoined, some dissertations connected with the history of the government, from the revolution to the present time,* 4th ed. (London: J. Mawman, 1818). See also Andrew Skinner, "Economics and History—The Scottish Enlightenment," *Scottish Jour-*

nal of Political Economy 12 (1965): 1–22.

5. Dugald Stewart, "Account of the Life and Writings of Adam Smith, LL.D.," in Hamilton, ed., *Collected Works,* p. 7.

6. Veitch, "Memoir of Stewart," pp. v–xii.

7. Stewart, "Account of the Life and Writings of Adam Smith, LL.D.," p. 7.

8. Veitch, "Memoir of Stewart," pp. v–xii.

9. C. R. Fay, *Adam Smith and the Scotland of His Day* (Cambridge: Cambridge University Press, 1956), pp. 125–27.

10. Ibid.

11. For a complete checklist of Adam Smith's library, see Hirochi Mizuta, *Adam Smith's Library: A Supplement to Bonar's Catalogue with a Checklist of the Whole Library* (London: Cambridge University Press, 1967). Mizuta's volume includes the information contained in James Bonar, *A Catalogue of the Library of Adam Smith* (London: Macmillan and Co., 1894; 2d. ed. 1932), and also a bibliography of journal articles on the topic of Smith's library.

12. Francis W. Hirst, *Adam Smith* (New York: Macmillan, 1904), p. 11. On the contrasting situation in Scotland, see J. B. Morrell, "The University of Edinburgh in the Late Eighteenth Century: Its Scientific Eminence and Academic Structure," *Isis* 62 (1971): 158–71. Smith himself commented on the dismal state of education in England and its causes, in his *An Inquiry into the Nature and Causes of the Wealth of Nations,* Book 5, Part 3, Article 2, "Of the Expense of the Institutions for the Education of Youth."

13. E. G. West, *Adam Smith* (New Rochelle, N.Y.: Arlington House, 1969), p. 42.

14. James E. Thorold Rogers, "Editor's Preface," in Adam Smith, *An Inquiry into the Nature and Causes of the Wealth of Nations* (Oxford: Clarendon Press, 1869), 1 : ix.

15. Hirst, *Adam Smith,* pp. 16–17.

16. Adam Smith, "The Principles which lead and direct Philosophical Enquiries; illustrated by the History of Astronomy," in Adam Smith, *Essays on Philosophical Subjects* (Edinburgh: 1795); reprinted in J. R. Lindgren, ed., *The Early Writings of Adam Smith* (New York: A. M. Kelley, 1967), pp. 30–108. The footnote is on p. 106 in the 1967 reprint.

17. Adam Smith, *Lectures on Justice, Police, Revenue and Arms,* ed. Edwin Cannan (Oxford: Clarendon Press, 1896).

18. Letter from Adam Smith to David Hume, 16 April 1773; quoted in John Rae, *The Life of Adam Smith* (London: Macmillan & Co., 1895; reprint ed., New York: A. M. Kelley, 1965), pp. 262–63.

19. See the introduction to Smith, *Essays on Philosophical Subjects.*

20. Smith, ". . . History of Astronomy," pp. 10–15; pp. 36 and 39 in the 1967 reprint.

21. Ibid., pp. 18–21; pp. 43–46 in the 1967 reprint.

22. Ibid., pp. 83, 91–93; pp. 100 and 107–8 in the 1967 reprint.

23. Edwin Cannan, ed., introduction to Adam Smith, *Lectures on Justice, Police, Revenue, and Arms,* vol. 1 (Oxford: Clarendon Press, 1896; reprint ed., New York: A. M. Kelley, 1964), pp. xi–xx.

24. Ibid., p. 168.

25. Smith, *Wealth of Nations,* 2 : 29.

26. Ibid., p. 51.

27. Isaac Newton, *Mathematical Principles of Natural Philosophy,* trans. Andrew Motte, rev. Florian Cajori (Berkeley: University of California Press, 1966), 1 : xvii–xviii.

28. Smith, *Wealth of Nations*, 1:340.
29. Ibid., p. 57.
30. Ibid., p. 60.
31. Ibid., p. 62.
32. Newton, *Principia*, 2:401–5.
33. Smith, *Wealth of Nations*, 1:6.
34. Newton, *Principia*, 2:406–14.
35. Alexandre Koyré, *Newtonian Studies* (London: Chapman & Hall, 1965), pp. 180–84.
36. Smith, *Wealth of Nations*, 1:14.
37. Newton, *Principia*, 2:460–63.
38. Smith, *Wealth of Nations*, 1:18.
39. Ibid.
40. Ibid.
41. Newton, *Principia*, 2:547.
42. Smith, *Wealth of Nations*, 1:14.
43. Ibid., 2:1.
44. Stewart, "Account of the Life and Writings of Adam Smith, LL.D.," pp. 97–98.
45. Rae, *Life of Adam Smith*, p. 436.
46. Hirst, *Adam Smith*, p. 17.
47. West, *Adam Smith*, pp. 44–45.
48. Henry J. Bitterman, "Adam Smith's Empiricism and the Law of Nature," *Journal of Political Economy* 48 (1940): 487–520 and 703–34.
49. Herbert F. Thomson, "Adam Smith's Philosophy of Science," *Quarterly Journal of Economics* 79 (1965): 212–33; Andrew S. Skinner, "Adam Smith: Philosophy and Science," *Scottish Journal of Political Economy* 129 (1972): 307–19; Stephen T. Worland, "Mechanistic Analogy in Smith's Theory of Policy," in William R. Morrow and Robert E. Stebbins, eds., *Adam Smith and the Wealth of Nations, 1776–1976* (Richmond: Eastern Kentucky University, 1978), pp. 94–112. There exists an argument that Smith had an even greater enthusiasm for Descartes than for Newton: Vernard Foley, "Smith and the Greeks: A Reply to Professor McNulty's Comments," *History of Political Economy* 7 (1975): 379–89. For another example of an intellectual link between science and economics, between François Quesnay's careers in economics and in medicine, see V. Foley, "An Origin of the Tableau Economique," *History of Political Economy* 5 (1973): 121–50.
50. John C. Greene, *Darwin and the Modern World View* (Baton Rouge: Louisiana State University Press, 1961), pp. 88; reprinted (New York: 1963), p. 80.
51. Gerd Buchdahl, *The Image of Newton and Locke in the Age of Reason* (London: Sheed & Ward, 1961), p. 26. See also Serge Moscovici, "A propos de quelques travaux d' Adam Smith sur l'histoire et de la philosophie des sciences," *Revue d'histoire des sciences* 9 (1956): 1–20.
52. Norriss S. Hetherington, "Isaac Newton, Adam Smith, and the Concept of Natural Laws in Economics," in Morrow and Stebbins, eds., *Adam Smith and the Wealth of Nations, 1776–1976*, pp. 415–30; "Isaac Newton's Influence on Adam Smith's Natural Laws in Economics," *Journal of the History of Ideas* 44 (1983): 497–505.

A Modern Look at Newton's Final Queries

FRANK WILCZEK

I love to go back and read the masters. Not for the details, but to get a fresh perspective on the world. The world-view of physics can largely be summarized in a few powerful ideas. These ideas may come to seem stale through familiarity, but by watching how great minds wrestled with them, we can appreciate them afresh.

I will now illustrate these remarks by reference to three of the major themes permeating modern physics:

> uniformity of parts: that the universe is made of the same building-blocks, everywhere
>
> transformations: that the basic changes in the properties of matter—not only forces—characterize the fundamental interactions in nature
>
> inevitability: that the content of the world—not only its behavior—is dictated by the laws of physics.

Newton's work provides some striking, surprising perspectives on each of these themes. It is especially interesting to read the Queries from the *Opticks* in this regard, since in the Queries Newton relaxes from his usual austere precision and lets us see the unfinished, provisional perimeter of his picture of the world.

Uniformity of parts is the idea that the same building blocks occur throughout the universe, and in particular that we can learn the laws that govern the cosmos by studying material in earthly laboratories. Newton's law of *universal* gravitation, which he applied alike to terrestrial and celestial, is of course suggestive of this view, but it is far from conclusive since it deals only with one limited aspect of matter. And so I find Query 11 especially impressive:

> Do not great Bodies conserve their heat the longest, their parts heating one another, and may not great dense and fix'd Bodies, when heated

beyond a certain degree, emit Light so copiously, as by the Emission and Re-action of its Light, and the Reflexions and Refractions of its Rays within its Pores to grow still hotter, till it comes to a certain period of heat, such as is that of the Sun? And are not the Sun and fix'd Stars great Earths vehemently hot, whose heat is conserved by the greatness of the Bodies, and the mutual Action and Reaction between them, and the Light which they emit, and whose parts are kept from fuming away . . . by the vast weight and density of the Atmospheres incumbent upon them . . .?[1]

For this query clearly expresses the essence of the great idea that the same physical laws, acting on the same materials, will produce two such different things as terrestrial matter and stars—and even sketches how this might happen.

Actually Newton's views on the uniformity of parts are not always so straightforward, as we shall see shortly. But now I want to move on to discuss what seems to me to be the most extraordinary and complex of his struggles with an unborn theme, that is, with the theme of transformations. Newton had a vision of the organic unity of light and matter, as appears in Query 30:

> Are not gross Bodies and Light convertible into one another, and may not Bodies receive much of their Activity from the Particles of Light which enter their Composition? . . .
>
> The changing of Bodies into Light, and Light into Bodies, is very conformable to the Course of Nature, which seems delighted with Transmutations.[2]

in Queries 5–9, where the interconversion of light with motion and heat is discussed, and also in Query 14, where even physiology is brought into the synthesis:

> May not the harmony and discord of Colours arise from the proportions of the Vibrations propagated through the Fibres of the optick Nerves into the Brain, as the harmony and discord of Sounds arise from the proportions of the Vibrations of the Air?[3]

and in many other passages.

Newton's intuition for unity, I believe, was ultimately responsible for his adopting a particle theory of light, even though he had evidence of its periodicity in hand—indeed, Young argued for his wave theory largely from Newton's observations of interference colors—and Huygens's contemporary wave theory as a model.

Newton's vision of the organic unity of light and matter was largely abandoned, for good pragmatic reasons, by most physicists for the next 150 to 200 years. Optics on the one hand and the theory of particles and forces on the other developed separately. The ideal of the physics of particles and forces was given its classical expression by Helmholtz, in his master work of 1847, wherein the principle of conservation of energy was first clearly enunciated:

> In science we call bodies with unchangeable forces (indestructible qualities) chemical elements. If we think of the universe as consisting of elements with inalterable qualities, the only possible changes in such a system are spatial ones, that is, movements. . . . Thus we see that the problem of the physical sciences is to trace natural phenomena back to inalterable forces of attraction and repulsion, the intensity of the forces depending upon distance. The solution of this problem would mean the complete comprehensibility of nature.[4]

Here *forces* alone are recognized; process of transformation must somehow be explained away as rearrangements. Now since the generation and absorption of light are common observations, Helmholtz's view required that light be treated as a thing apart— that physics be divided into matter-physics, with no transformations, and light-physics, with transformations. As optics and electromagnetic theory developed further, a second profound dichotomy between matter and light emerged: particle versus wave, or discrete versus continuous. Einstein, who like Newton grasped for unity, opened his 1905 paper on the photoelectric effect by alluding to these dichotomies:

> There exists an essential formal difference between the theoretical pictures physicists have drawn of gases and other ponderable bodies and Maxwell's theory of electromagnetic processes in so-called empty space.[5]

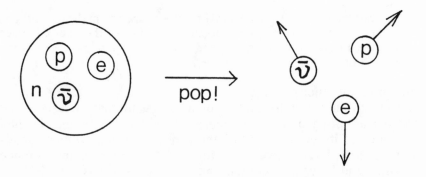

Figure 1

As is well known, the particle-wave dichotomy was resolved by the development of quantum mechanics. Less well known, perhaps, is the difficulty physicists had, even after this development, in accepting transforming principles among the basic laws of physics. The theory of radioactive ß-decay was seriously impeded by the unconscious assumption that a decay like $n \longrightarrow pe\bar{v}$ should be interpreted as the break-up of an entity into its preexisting components. This led to futile attempts to locate electrons inside atomic nuclei, and other wrong ideas.

Fermi made an utterly simple, but great, contribution when in 1934 he postulated that *transformations* (specifically $n \longrightarrow pe\bar{v}$) as well as *forces* must be included among the basic laws of physics.[6] This theme has since come to pervade our understanding of all the basic laws of physics, especially the strong and weak interactions.

The psychological difficulty physicists had with accepting that transforming principles underlay ß-decays is particularly ironic, because the example of the photon was before them.

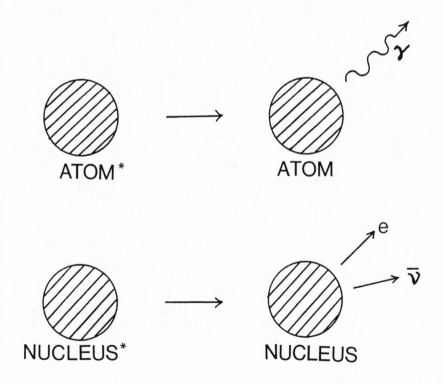

Figure 2

Evidently, even in the 1930s light was regarded as a thing apart. It took a genius like Fermi, pushed by experiment, to adopt it as his model for matter—closing the circle on Newton's attempts to do the converse 250 years before.

So in the end, after many years and many travails, Newton's vision of the organic unity of light and matter has been realized. It was necessary to learn much more about light and matter separately, before the synthesis could be achieved. And perhaps it was his perception of the gap between his vision and the confusing facts at hand, that led him to state his ideas in the form of queries—and not always consistently, at that.

Is there a methodological moral to be drawn here? Only that the search for unity may be either fruitful, or premature and misleading, at different times. General principles at this level are not reliable aids in solving the concrete problems of developing science.

Finally I want to touch on Newton's relationship to a truly modern theme, the theme of inevitability. In physics there has traditionally been a definite division between the basic laws and the initial conditions. In other words, if you tell what exists and how it is arranged, I will tell you what will happen in the future.

Now a great emergent theme in modern physics has been to push back the line between laws and initial conditions. In other words, we attempt to show that the contents and structure of the present universe—not just its future development—follow inevitably from the basic laws. (In practice we cannot yet get by with no initial conditions at all—but amazingly simple and definite ones seem adequate, as I will discuss.) For example, the relative abundance of different chemical elements, the overall density of matter in the universe, and its distribution into galaxies and uniformity on large scales are all now thought to result inevitably from the workings of microphysical laws—observable in earthly laboratories—in an evolving universe.

(By the way, of course, inevitability is inconceivable without transforming principles: without transforming principles, the material content of the world is forever frozen in.)

Newton's views were quite different. He was very much concerned with the historical-cosmological implications of his theories, as shown in this important extract from Query 31 (the last one):

And thus Nature will be very conformable to her self and very simple, performing all the great Motions of Heavenly Bodies by the Attraction of Gravity which intercedes those Bodies, and almost all the small ones of their Particles by some other attractive and repelling Powers which intercede the Particles. . . .

Some other Principle was necessary for putting Bodies into Motion; and now they are in Motion, some other Principle is necessary for conserving the Motion,[7]

and he gives an example of how motion is got or lost:

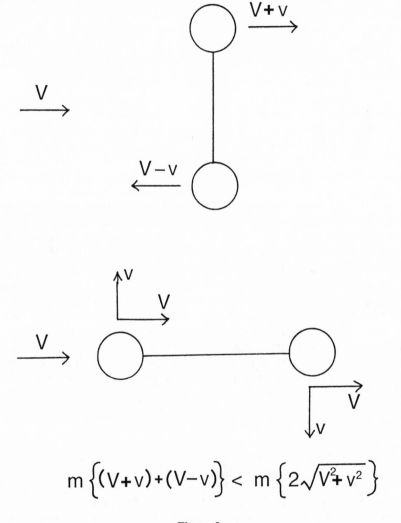

$$m\left\{(V+v)+(V-v)\right\} < m\left\{2\sqrt{V^2+v^2}\right\}$$

Figure 3

A dumbbell revolves around its center of mass and at the same time its center of mass moves with constant velocity in the plane of rotation. If the quantity of motion is defined to be the sum of the absolute value of the momenta of the end masses, then it does indeed change. (But of course the more meaningful measure, namely energy, does not change.)

This example is unfortunate, but in the extended discussion of "loss and motion" that follows it becomes clear that the concept he is groping toward is "loss of free energy" and his concern is with what in the nineteenth century would be known as the "heat death of the Universe." At any rate, he continues:

> Seeing therefore the variety of Motion we find in the World is always decreasing, there is a necessity of conserving and recruiting it by active Principles. . . .
>   . . . all material Things seem to have been composed of the hard and solid Particles above-mention'd, variously associated in the first Creation by the Counsel of an intelligent Agent. For it became him who created them to set them in order. And if he did so, it's unphilosophical to seek for any other Origin of the World, or to pretend that it might arise out of a Chaos by the mere Laws of Nature; though once form'd, it may continue by those Laws for many Ages.[8]

Clearly we have made great strides in cosmology since Newton's time, but in some sense his ultimate problem still remains. Why is the universe not in the state of maximum entropy, or lowest free energy? That would be a universe populated only by black holes and a little radiation, very different from ours.

In today's cosmology, our working assumption is that at some very early time the universe was in a state of perfect equilibrium for all *nongravitational* interactions, but the hot matter was distributed as uniformly as possible. This latter assumption means, however, that the early universe was far out of equilibrium for gravitation, which seeks to form clumps. So in effect, our modern version of Newton's "Intelligent Agent" has been reduced to just one task— that is, to turn off gravity, briefly.

Before leaving this theme, I cannot resist sharing one last remarkable quotation from Query 31

> And since Space is divisible *in infinitum*, and Matter is not necessarily in all places, it may also be allow'd that God is able to create Particles of Matter of several Sizes and Figures, and in several Proportions to Space, and perhaps of different Densities and Forces, and thereby to vary the Laws of Nature, and make Worlds of several sorts in several Parts of the Universe. At least, I see nothing of Contradiction in all this.[9]

As I mentioned earlier, Newton's commitment to uniformity of parts was not itself entirely uniform. Newton allowed God the right to tinker—but assumed He only did so far away!

Now in trying to understand what seems to be a strange or bizarre idea, it can be helpful to try to make a model—that is, an arrangement or interpretation of familiar things that embodies or illustrates the idea. In this spirit, let me offer a model of Newton's cosmology. Let us suppose that computer technology and biotechnology continue to progress rapidly. Eventually, then, our descendants will learn how to seed whole planets with self-reproducing machines that turn whole planets into gigantic computers of incalculable power. Such computers would offer impressive resources for simulations. It is not beyond belief that they could simulate entire universes—or, rather, what would appear to be universes, to their inhabitants! And, of course, in this case both the laws of nature and the initial conditions would be up to the manufacturers, within the limits of consistency. And if this is possible, is it not likely that some more ancient civilization has achieved it already and is simulating *us?* Still speculative, but more physically based ideas in inflationary cosmology offer other prospects that different fundamental physical laws hold in distant regions of space-time.[10] And so possibly Newton was right after all.

At least, I see nothing of contradiction in all this.

Notes

1. Sir Isaac Newton, *Opticks or A Treatise of the Reflections, Refractions, Inflections, & Colours of Light* (New York: Dover Publications, Inc., 1952), pp. 343–44.

2. Ibid., p. 374.

3. Ibid., p. 346.

4. Hermann von Helmholtz, "The Conservation of Force: A Physical Memoir," in *Selected Writings of Hermann von Helmholtz,* ed. Russell Kahl (Middletown, Conn.: Wesleyan University Press, 1971), pp. 5–6.

5. Albert Einstein, "On a Heuristic Point of View about the Creation and Conversion of Light," in *The Old Quantum Theory,* ed. D. Ter Harr (Oxford: Pergamon Press Ltd., 1967), p. 91.

6. Enrico Fermi, "Versuch einer Theorie der ß-Strahlen," in *The Collected Papers of Enrico Fermi,* ed. E. Segrè et al. (Chicago and London: University of Chicago Press, 1962), 1:575–90.

7. Newton, p. 397.

8. Ibid., pp. 399–402.

9. Ibid., pp. 403–4.

10. See A. Linde, *Inflation and Quantum Cosmology* (New York: Academic Press, 1990).

Overview: Newton's Place in History

DUDLEY SHAPERE

We have been treated to a feast of insights into the contributions of a very great man. We have heard Newton's own work examined closely: his methods, his mathematics, his alchemical studies, his chemical experiments, his rhetoric and style. We have heard about a variety of relations between Newton, or Newtonianism, and economics, politics, theology, art, and even insanity. But when all is said and done, it was as a scientist that Newton was important for later ages. And so my "overview" of Newton's place in history will focus on an examination of his scientific achievement and specifically on three aspects of that achievement: first, some of the contrasts between his work and his way of approaching it, and those of his predecessors; second, some issues that were left unsettled by his work and some new features that entered science in the first two centuries following it; and third, some fundamental ways in which Newton's (or "Newtonian") science relates to the physics of the present day. In discussing these three topics, I want to take the following famous passage, from Query 31 of the *Opticks,* as my central theme.

> To tell us that every species of things is endowed with an occult specific quality by which it acts and produces manifest effects is to tell us nothing, but to derive two or three general principles of motion from phenomena, and afterward to tell us how the properties and actions of all corporeal things follow from those manifest principles, would be a very great step in philosophy, though the causes of those principles were not yet discovered; and therefore I scruple not to propose the principles of motion above mentioned, they being of very general extent, and their causes to be found out.[1]

To tell us how the properties and actions of a variety of things follow from two or three general principles of motion: that was indeed a very great step. The late medieval doctrine of substantial forms, an outgrowth of Aristotle's notion of essences or natures of things, had explained entities and their behavior in terms of hidden powers that were irreducible to anything more fundamental. By the seventeenth century it had become common to criticize this view, not simply as giving incorrect explanations, but as not providing any explanation at all: "To tell us that every species of things is endowed with an occult specific quality by which it acts and produces manifest effects is to tell us nothing." To assert that man is rational because his specific difference from other types of things is his possession of a rational soul is no more an explanation, the seventeenth-century critics of scholasticism held, than is the claim in Molière's spoof of such purported explanation, that opium puts people to sleep because it has a dormative power; both are explanations of the "because-it's-the-nature-of-the-beast" sort, and that, they maintained, is fundamentally nonexplanatory.

To many of his contemporaries, Newton's explanations in terms of gravity seemed to be of the same type: things gravitate because they have a power of gravitation. They missed the Newtonian point: while the scholastics had used a different occult quality, a different substantial form, for the explanation of each supposedly irreducible type of phenomenon, Newton's gravity, together with his "two or three general principles of motion," explained phenomena of many different types—indeed, "the properties and actions of all corporeal things": not only "all the motions of the celestial bodies," but also "of our sea,"[2] and the motions of bodies on earth, whether free-falling or hurled. It is true that Descartes had insisted that all corporeal phenomena must be explained in terms of one fundamental material substance in motion; but it was Newton who showed us how to give such unifying explanations.

Newton's contemporaries had yet another criticism of his concept of gravitation: it appeared to violate the ancient dictum, accepted by the Cartesians and Leibnizians, that all causal influence must be by contact—that there could be no action at a distance. Again they missed the point: it was, as Newton noted, a manifest fact that bodies gravitate toward one another according to the law he had written down. It did not matter that gravitation was itself mysterious, that a causal explanation of it was lacking; that fact, which Newton acknowledged, could not detract from the "very great step in philosophy" that the concept made possible, of showing "how the properties and actions of all corporeal things follow from those manifest principles," even though the cause of the gravity itself were not known. We can hope that Bernard Cohen has—once and for all—laid to rest the view that Newton's "hypotheses non fingo" was an injunction to excise such questions as the

explanation of gravity from legitimate science.[3] The question, Newton thought, was a legitimate one, but one he himself could not
answer, though he tried; and given his failures in that regard, it was
a question whose answering would have to be left to a later age.

Westfall has emphasized that Newton was the culmination of the
seventeenth-century scientific revolution.[4] There is important truth
in this: the debate as to whether matter should be considered
atomistic or continuous, a space-filling plenum, was effectively
ended, at least as far as the mathematical physics of the following
century was concerned, by the success of Newton's physics. Contrary to Descartes, matter and space would have to be distinguished
sharply and fundamentally: matter is not identical with extended-
ness, being present wherever there is space and vice versa; on the
contrary, matter is something different from space; it is something
that occupies space, and by no means all of it. The way mathematics entered into physics was shown by example if not entirely
explicated in principle. And as both Westfall and Cohen have noted,
though Newton did not invent the experimental method, his *Opticks* gave his successors a model of how to go about it.

Still, if there were respects in which Newton's work was a culmination, it also left much undone. I have mentioned the matter of
gravity; but also unresolved was a larger question, a legacy of the
ancient Greek problem of the relative roles of the active and the
passive in any understanding of nature. The seventeenth-century
version of this problem ran roughly like this: are passive, inert
matter, space, and time sufficient to explain the workings of nature,
or is it necessary also to introduce active agencies, active forces,
into nature? For the Cartesians, once the material universe had
been created and endowed with motion and the laws governing it,
the very character of that matter (or, for Descartes himself at least,
matter/space) in motion would do all the rest. Newton himself
maintained that in addition to passive, inert matter, active forces
were necessary. His archrival Leibniz went still further, holding
that purely passive matter, space, and time have no fundamental
role to play, they being only manifestations, "well-founded phenomena," of a fundamentally active substance, to be explained
ultimately in terms of the only truly intelligible, truly explanatory
concept, that of active force. Arthur Donovan, following Schofield
and others, has reminded us of how that issue continued to divide
science in the eighteenth century,[5] and Leibniz would be vindicated
in the nineteenth, when Hegelians, covert Hegelians like Faraday
(as Arthur Molella called him) and others—all really descendants of

Leibniz—would rebel against notions like that of passive matter, replacing them with immaterial or energetic ones. Yet other openings were left by the mechanical philosophy of nature, and particularly by its Newtonian variant, and not all these open issues were matters of substantive claims about nature. Some of the most important had to do with the proper methods and aims of science. Westfall is right, for example, in noting that "it remained for the scientists of the seventeenth century to realize fully that what is in the senses need not be confined to the phenomena that nature presents," that we can probe nature to learn its secrets.[6] But it was well beyond the seventeenth century before they learned how to legitimate, in a scientific sense, the introduction of concepts that referred to entities and processes behind the scenes of perceivable phenomena. Again, it would be a long time before the methodology of idealization[7] would be conceived in its modern form—that is, before what were considered to be idealized treatments could be considered such for scientific reasons, rather than on the basis of such philosophically-inherited dicta as "All real bodies are extended." And whatever Newton's own reasons may have been for presenting the *Principia* in the form of an axiomatic treatise on the model of Euclid's *Elements*—if that was indeed Newton's model[8]— we must not forget, in talking of Newton as having aided in the rise of the experimental method, that men no less than d'Alembert, Euler, and Kant, among others, attempted to show that the axioms of the *Principia,* the laws of motion, or at least certain aspects or variations of them, could be deduced a priori, no appeal to experience or experiment being necessary. This was only one extreme response, conditioned perhaps by a Cartesian heritage, to the ambiguities left by Newton's work regarding the relative roles of theory and experiment in science. Those ambiguities, and the consequent variety of Newtonian and mechanistic themes, were aggravated by the fact that there were other traditions concerned with inquiry into nature; particularly important was that of what we call chemistry and its predecessors, a tradition that already in Boyle's time opposed speculation about atoms. It was an attitude that continued to run deep, even if it was not universal, among chemists and others well into the twentieth century. Given the large openings left by Newton himself and by the mechanical philosophy of nature in general, and given the other traditions developing at least partly in independence of that tradition, it should not surprise us that, as Arthur Donovan has emphasized, there was much in eighteenth-century thought, both substantive and methodological, that was

different and in some important respects antithetical to that of
Newton. A variety of ways were available in which the science of
Newton and his contemporaries might be extended further.

I want to focus on one particular difference that Donovan has
noted between seventeenth- and eighteenth-century science, particularly, as Donovan presents it, in the way Newton and Lavoisier
approached science. Donovan points out, quite rightly, that much of
seventeenth-century science was concerned "to construct systems
of theories that were at least potentially capable of rendering nature
as a whole intelligible." As Bernard Cohen puts it, "Newton's
ultimate goal was to construct a complete philosophy of nature
along the lines that had proved so successful in the *Principia*."[9] In
contrast, Donovan sees Lavoisier and many of his nineteenth-
century successors as having had "little to do with seventeenth-
century attempts to construct philosophical systems of the world,"
preferring "a more narrowly conceived investigation of nature,"
one of working "within discrete disciplines whose boundaries they
by-and-large accepted."[10] One can perhaps attribute more nar-
rowness to such men than was there, but certainly there is a great
deal to Donovan's point: scientists and science did tend, more and
more as the eighteenth and nineteenth centuries wore on, to focus
their attention on specific subject-matters, investigating specific
types of phenomena, in isolation from other domains, and developing specific theories for their specific domain, using methods and
concepts forged for use in that domain.[11] Scientists became specialists rather than universalists, adopting a piecemeal rather than a
holistic approach to the investigation of nature. There is another
aspect of the paper by Donovan that I do think it important to
reconsider. As Donovan sees things, the concerns of Lavoisier and
his successors of the piecemeal approach "had little to do with the
seventeenth-century attempts to construct systems of the world";
"the aims of the investigators and the range of implications associated with the knowledge sought were vastly reduced and carefully
delimited."[12] And with that narrowing, in Donovan's view, "the
tradition of natural philosophy is over and we enter the world of
modern science."[13]

I think we need to take a closer look at this question: to what
extent does the science of today continue to observe a fragmented,
piecemeal approach and to eschew the search for a system that is
"potentially capable of rendering nature as a whole intelligible,"
that is even "trying to explain everything"? I must agree with the
historians who believe that Isaac Newton had such an aim: he
wished "that we could derive the rest of the phenomena of Nature

by the same kind of reasoning" that he had employed in *Principia,* and he claimed to have been induced to that wish by many reasons. Hence I will return to the question of whether modern science is a piecemeal or a "natural philosophy" type of enterprise in the context of discussing my third major point, the relations between Newtonian physics and the physics of the present day.

In the past several decades, physicists have gradually come to the understanding that, besides gravity, there are three other fundamental forces in nature: the electromagnetic, responsible for most of the phenomena of the everyday world, not only Newton's cohesion of bodies, and chemistry in general, but also light; the strong force, governing the properties and behavior of atomic nuclei and much of those of elementary particles; and the weak force, responsible for radioactivity, but far more importantly involved in the conversion of the simple primordial elements, hydrogen and helium, into the carbon, nitrogen, oxygen, and other more complex chemical elements that go to make up solid objects, including ourselves. The great achievement of physics since the 1930s, and particularly since the end of the 1960s, has been the taking of large steps toward constructing a unified theory embracing those three forces. We now have a highly successful unification—not quite complete, but close—of the electromagnetic and weak forces into a theory of the "electroweak" interaction. Modeled on that theory, a new and workable theory of the strong force, in the development of which Professor Wilczek played a central role, has been developed; and that theory has been brought into close association with those of the electroweak interaction in what has come to be known as the "standard model" of elementary processes. Application of that model to Big Bang cosmology has resulted in a vastly deepened understanding of the earliest processes in the history of the universe, processes in which, as the temperature of the early universe cooled with its expansion, the strong force broke from its unity with the electroweak, and at a still later epoch—all still within the first second of the universe's history—the electroweak unity also was broken, giving us three of the four distinct forces we know today. It is one of the ironies of history that the fourth fundamental force, gravitation, about which Isaac Newton taught us so much, has for a long time been left out of this increasingly unified picture of nature. In that respect, at least, the legacy of Newton has been peripheral to the main line of development of physics since the 1920s; indeed, it is fair to say that today we have a better understanding of the other three forces than we do of gravitation, despite the fact that we were presented with a quite good theory of it three hundred years ago, long before we had any comparable understanding of electricity and magnetism, and longer still before we even suspected the existence of the strong and weak forces. There are other respects, of course, in which his heritage remains central: we must not forget, for example, that he taught us much (though not everything) about how the investigation of nature should be done. Even that legacy which has been on the fringes of modern physics may not remain so for very much longer; for there are increasing signs that gravitation may yet be brought together with the other forces into a completely unified theory of nature. We can expect such a theory, if it can indeed be worked out, to carry the history of the universe back to a still earlier fraction of that first second, to a time when all four fundamental forces would have been unified, the gravitational force having broken from the unity of the remaining three in only one ten-millionth of the time that elapsed before the strong force in its turn broke off from the electroweak. Such a unified theory—again, if we get it—can be expected to throw profound light on the concepts Newton did so much with. For in the modern theory of gravitation, general relativity, a theory whose later development has been close to the heart of much of Professor Chandrasekhar's work, the gravitational force is reduced to a variation in the metric of space-time. We can thus expect a unification of that theory with the quantum field theory of elementary processes to give us a much-deepened understanding of space and time themselves. And if the unification takes the direction that many scientists today expect, the fundamental particles of matter may not be points, as Newton's were in approximation (light-atoms excluded), but rather extended strings.

But these are not the only ways in which we can expect the fundamental concepts of Newton's physics to be illuminated by the hoped-for unification. The fundamental concepts of mechanics, as elementary textbooks of physics tell us, are those of space, time, matter, and force. So it was also in the seventeenth century, though that age was marked by debates as to the proper interpretation of those four concepts, whether all four were really necessary in a "mechanical philosophy of nature," and which of the four were truly fundamental as explanatory concepts. The four were brought to a high degree of sophistication by Newton, who argued for the absoluteness of space and time, an atomistic conception of matter, and the necessity of active forces in nature, of which gravity was one example and the forces, whatever they are, responsible for cohesion of bodies another. Though they have had to be interpreted, sometimes in drastic ways, the four have continued to play important roles in physics. But it is not simply their interpretation that has been revised; the process of unification has erased much of the distinction between them as well. Special relativity fused the concepts of space and time into one, space-time, the separation into space *and* time being no more than a product of viewing nature from the perspective of a particular reference-frame. Quantum field theories view matter and force as very much on the same footing: what we have classically thought of as matter consists of particles called fermions (but "particles" in the quantum-theoretic, not the Newtonian-Laplacean, sense); while forces are carried by other sorts of particles, bosons. In the well-developed theories of contemporary physics, fermions and bosons are fundamentally distinct: fermions have half-integer spin or integral multiples thereof, bosons have integer spins; and that is what remains of the distinction between matter and force in those theories. Supersymmetry, if it turns out, as many expect, to be an ingredient in a superunified theory, would make even that difference superficial. In such a theory, fermions and bosons would be interconvertible, so that there would no longer be the absolute bifurcation of matter and force that the seventeenth century saw as a distinction between the passive and the active, and the nineteenth as one between the material and the immaterial. Finally, a fully unified theory of the fundamental forces may remove even the ultimate distinction between force-matter and four-dimensional space-time. Our best—by a long shot, now—theory of four-dimensional space-time, general relativity, is, as I noted, simultaneously a theory of gravitation, so that the symmetry breaking that separates the gravitational from the other forces also breaks four-dimensional space-time from a primordial unity with the fermions and non-gravitational bosons that we know today, in our cold universe, as matter and the remaining forces.

Bernard Cohen has related that Newton never said what sorts of things those "general principles" were from which the properties and actions of all corporeal things follow.[14] He could not have: in his day, too little was understood about nature to allow anything more than "hypotheses" about how nature is to be explained on the most fundamental level, hypotheses stemming largely from arguments about what explanation *must* be like, rather than from investigation of nature itself. Today, as a result of such investigation, the issue has been clarified: the fundamental principles are symmetry principles, in the sense made precise in group theory. More specifically, the groups of concern are Lie groups, and more specifically still, in many important cases, gauge groups. It is through such gauge symmetries, and the breaking of them, that, through a long and complex chain of developments in which Isaac Newton played a seminal role, contemporary physics has arrived at the hope—already realized to a remarkable if as yet incomplete extent—of gaining a unified theory of nature. In addition to the points already raised, the employment of gauge symmetries adds in one further, profoundly significant way to the increasingly interlocked relationship between the concept of force and other concepts. In Newton's physics, the general concept of force is introduced in the second law of motion. However, the specific forces that exist in nature are not given by the theory; they are independent of it and must be discovered, as Newton hoped they could be. (A parallel situation obtains in quantum mechanics.) But in contemporary quantum field theories, the characteristics of a (local) gauge field *require* the existence and specific characteristics of corresponding forces.

Regardless of the ultimate outcome, what has been accomplished so far in recent physics has already been a very great step in philosophy. Now, in the light of these developments, we can return to the question I raised earlier: does today's science remain piecemeal, or is it more comparable to the natural philosophy of Newton? My answer should by now be clear: it is a vast oversimplification, indeed a misconception of the scientific process, to view the introduction of a piecemeal approach to the investigation of nature as a sign of modernity in scientific method, whether our hero in the introduction of such modernity is Newton or Lavoisier or someone else. For as things have turned out, modern scientists, while the sheer vastness of the material requires them to specialize in limited areas, are far from having a piecemeal attitude toward their work and its goals: rather, they fit quite closely Donovan's description of the Newtonians as "natural philosophers," setting out "to construct systems of theories that are at least potentially capable of rendering nature as a whole intelligible." Like Cohen's Newton, they seek a complete philosophy of nature. And the subject-matter of contemporary physics itself is certainly not fragmented, as it was in the intervening period of piecemeal methodology. If the adoption of a natural-philosophy approach was premature in the seventeenth century—if a narrower piecemeal approach replaced it to a large extent in the late eighteenth and nineteenth centuries—that was not because a piecemeal approach is the kind that scientists ought always, forevermore, to take. Rather, it appears to have been a step, a necessary one perhaps, that is leading to its transcendence of itself: that is, toward the achievement of a goal that Newton himself would have embraced, the goal of achieving a comprehensive theory capable of rendering nature as a whole intelligible.

One final moral ought to be drawn. I hope my discussion has illustrated by example an important point: that the distinction between a scientific and a nonscientific question, between what is internal to science and what is external to it, is not fixed and firm, but evolves.[15] What were once questions excluded from science can, and in many cases have, become scientific. The same may be said about methods: at one time a piecemeal approach may be the one that leads to results, at another time we may be justified in asking larger questions. Science is a developing enterprise, not one whose boundaries and methods are handed down from above, engraved on stone. Modern physicists, induced by many reasons, are now seeking a theory that many of them refer to unabashedly as a "theory of everything," and such it will indeed be if we get it, in the following sense: that it will provide a framework for understanding the history of the universe, wherein the evolution of galaxies, stars, chemical elements, and life will have a setting within which they must and can be understood. I do not know what an "ultimate why" question that would be the exclusive property of philosophers would be like. Perhaps if we get the superunified theory I have been discussing, it may give us an understanding of gravity, even if it seems inappropriate, in a quantum universe, to speak of it as giving a "cause" of gravity. And certainly the origins and nature of space and time, which we may also understand through the foreseeable theories, are pretty ultimate issues. If that is not enough, consider the advocates of an inflationary scenario for the early universe. If they have their way, perhaps even such a paradigmatically philosophical dictum as *ex nihilo nihil fit*, from nothing nothing can come, may be brought to scientific scrutiny. Their proposal is that perhaps everything *did* come from nothing. For, they suggest, everything might add up to exactly nothing, if the total energy of the universe, positive and negative, is zero. Though speculative, this proposal cannot be rejected as *merely* philosophical, as *clearly* unscientific; for the inflationary theory answers many scientific problems, and the proposal, in line with that theory, is consistent with contemporary quantum theory: the universe would, on that scenario, be a quantum fluctuation, one of those things, as Edward Tryon has remarked, that just happens every once in a while.[16] Again, if we get the superunified theory scientists are seeking, there may be no deeper questions to ask: for the constraints on the kind of theory it would be grow tighter with every journal issue, so that many physicists have begun to suspect that the ultimate answer, why this theory, this universe, rather than any other, will turn out to be that this is the only theory, the only universe, we could have. At least as far as questions about nature are concerned, the boundary between what is internal to science and what is external to it is not fixed and firm.

I cannot help but believe that if Isaac Newton could return to see what has happened in the subject he did so much to create, even that dour man would be overjoyed at the developments I have described, developments that seem to be nearing the fulfillment of his own aim, of telling us how the properties and actions of all corporeal things (corporeal in a much-broadened sense, of course) follow from a few general principles. And we too, as we come to the end of this magnificent conference, must be delighted by the rich variety of thought we have been given.

Notes

1. Isaac Newton, *Opticks* (New York: Dover, 1952), pp. 401–2.
2. Isaac Newton, *Mathematical Principles of Natural Philosophy*, trans. Andrew Motte, rev. Florian Cajori (1729; Berkeley: University of California Press, 1966), p. 547 (General Scholium).
3. Cohen, this volume.
4. Westfall, this volume.
5. Donovan, this volume.
6. An extended discussion of how modern science goes beyond the senses in its study of nature can be found Dudley Shapere, "The Concept of Observation in Science and Philosophy," *Philosophy of Science* 49 (1982): 485–525.
7. Harper, this volume. Harper discusses Newton's use of idealizations as if there were no problem as to *whether* the concepts under discussion are "idealizations," or how it is determined whether or not they are such, or how the methodology and use—and indeed the very concept—of idealization developed over the history of science. For discussion of these problems, see Dudley Shapere, *Reason and the Search for Knowledge* (Dordrecht: Reidel, 1984), chapter 13. For an updating of the discussion, see the Introduction to that volume, pp. xix, xli–xliii.
8. There appears to be a fundamental disagreement on this question in the papers of Mahoney and Westfall, this volume.
9. Cohen, this volume.
10. Donovan, this volume.
11. By the second half of the nineteenth century, of course, the isolation of many domains, for instance those of electricity, magnetism, and light, was beginning to disappear in the increasing trend toward unification, which I will discuss shortly.

The concept of a "domain" is discussed in Shapere, *Reason and the Search for Knowledge,* chapter 13, and is updated extensively in the Introduction to that volume.
12. Donovan, this volume.

13. Donovan, this volume.

14. Cohen, this volume.

15. This view is developed in Dudley Shapere, "External and Internal Factors in the Development of Science," *Science and Technology Studies* 4 (1986): 1–9 (with commentaries and reply, 10–23).

16. Edward Tryon, "Is This Universe a Vacuum Fluctuation?" *Nature* 246 (14 December 1973): 396–97.

Contributors

STEPHEN G. BRUSH, past president of the History of Science Society, is a professor in the Department of History and the Institute for Physical Science and Technology at the University of Maryland. Professor Brush received his Ph.D. from Oxford University, and his field is the History of Modern Physical Science. His recent publications include *The Kind of Motion We Call Heat: A History of the Kinetic Theory of Gases in the Nineteenth Century* (1976) and *Statistical Physics and the Atomic Theory of Matter, from Boyle and Newton to Landau and Onsager.*

I. BERNARD COHEN, Victor S. Thomas Professor (emeritus) of the History of Science, Harvard University, has been the editor of *Isis* and president of the History of Science Society and of the International Union of the History and Philosophy of Science. His writings include *Introduction to Newton's 'Principia'* (Harvard University Press and Cambridge University Press, 1971), *The Newtonian Revolution* (Cambridge University Press, 1980), *Revolution in Science* (Harvard University Press, 1985), and *The Birth of a New Physics* (W. W. Norton, 1960, revised ed. 1985), which has been translated into eleven languages. He also prepared (in collaboration with Alexandre Koyré and Anne M. Whitman) the recent edition of Newton's *Principia* with variant readings (Harvard University Press and Cambridge University Press, 1972). A foreign member of the British Academy, he has just been elected to the Accademia Nazionale dei Lincei (Rome), the national academy of Italy, which traces its roots back to Galileo.

BETTY JO TEETER DOBBS received her Ph.D. from the University of North Carolina at Chapel Hill in 1974 and is now professor of history at Northwestern University. Her publications include *The Foundations of Newton's Alchemy, or "The Hunting of the Greene Lyon," The Janus Faces of Genius: The Role of Alchemy in Newton's Thought,* and numerous articles.

ARTHUR DONOVAN is head of the Department of Humanities at the

U.S. Merchant Marine Academy, Kings Point, and adjunct professor of history, State University of New York, Stony Brook. His principal publications focus on chemistry and geology in the Enlightenment, especially in Scotland. He is currently editing a selection of Lavoisier's chemical memoirs and writing a biography of Lavoisier.

ANITA GUERRINI received her Ph.D. from Indiana University in 1983. She taught the history of biology and medicine at the University of Minnesota before moving to the University of California, Santa Barbara. She has published several articles on Newtonianism and medicine, and recently completed a book manuscript, *Theory and Practice in Eighteenth-Century Medicine: George Cheyne and Some Contemporaries.*

WILLIAM L. HARPER, professor of philosophy, University of Western Ontario, received his Ph.D. from the University of Rochester in 1974. He is author of articles on Kant, Newton, rational belief change, decision theory, and the philosophy of science, and co-editor of *Causation, Chance and Credence* (with Brian Skyrms), *Ifs* (with Ralf Meerbote), and six other edited volumes. He is on the board of the journal *Philosophy of Science,* of the Western Ontario Series in Philosophy of Science, and the Cambridge University Press series in the decision sciences. He has recently been elected to the governing board of the Philosophy of Science Association.

NORRISS S. HETHERINGTON is research associate at the Office for the History of Science and Technology, University of California. His book *Science and Objectivity* was published in 1988. He is a member of the American Association for the Advancement of Science Project on Liberal Education and the Sciences.

MICHAEL S. MAHONEY is professor of history and history of science at Princeton University, where he divides his teaching and research between the history of the mathematical sciences and the history of technology. He is the author of several studies on the development of algebra and analysis during the seventeenth century, as well as on ancient and medieval mathematics in general. Editor of the History Series of the ACM Press, he is currently engaged in a history of computer software during the 1950s and 1960s and in several related projects.

SIMON SCHAFFER, professor in the Department of History and

Philosophy of Science at the University of Cambridge, is a scholar in the field of seventeenth- and eighteenth-century physical science, particularly astronomy and physics. His many articles and essays include "Newton's Undergraduate Notebook," "Newton at the Crossroads," "Newton's Comets and the Transformation of Astrology," and "Natural Philosophy."

ADELE F. SEEFF is executive director of the Center for Renaissance and Baroque Studies, a research center within the College of Arts and Humanities at the University of Maryland at College Park. She is project director for two multiyear projects—one statewide and one extending through four states—sponsored by the National Endowment for the Humanities. In addition to articles on higher education, she has written two award-winning television scripts. She is currently a scriptwriter for a television-assisted course, Introduction to Literature, sponsored by the Corporation of Public Broadcasting/Annenberg Corporation. She is also working on a book-length project on *Henry IV, I and II*.

DUDLEY SHAPERE received his Ph.D. from Harvard University and is currently Z. Smith Reynolds Professor of the Philosophy and History of Science at Wake Forest University. His most recent book is *Reason and the Search for Knowledge,* published by Reidel.

PETER SPARGO read engineering at the University of the Witwatersrand, Johannesburg, later undertaking postgraduate work in the history of science and in education at the University of Cambridge. He is currently director of the Science Education Unit in the University of Cape Town. He has published equally in science education and in the history of science.

PAUL THEERMAN received his Ph.D. in history from the University of Chicago in 1980, and he has since worked at the Joseph Henry Papers, Smithsonian Institution. His dissertation was on the subject of James Clerk Maxwell. His main research interests are in scientific biography and scientific popularization in the nineteenth century, especially in Britain and the United States. Among his publications is "Unaccustomed Role: The Scientist as Historical Biographer—Two 19th-Century Portrayals of Newton."

RICHARD S. WESTFALL, professor of the history of science at Indiana University, has been primarily concerned with the scientific revolution of the sixteenth and seventeenth centuries. In addition to a biography of Newton *(Never at Rest),* he is the author of a general history of science during the seventeenth century *(The Construction of Science)* and a history of the science of dynamics during the seventeenth century *(Force in Newton's Physics).*

FRANK WILCZEK is professor in the School of Natural Sciences at the Institute for Advanced Study. Among his contributions of physics are the discovery of asymptotic freedom, the introduction of the concept of axions, and the introduction of the concept of fractional statistics. He has received numerous honors and awards, including a John A. and Catherine D. MacArthur Foundation fellowship and the J. J. Sakurai prize of the American Physical Society.

Index